Developments in Economic Geology, 3

WORLD MINERAL SUPPLIES
Assessment and perspective

Further titles in this series

1. *I.L. ELLIOTT and W.K. FLETCHER*
GEOCHEMICAL EXPLORATION 1974

2. *P.M.D. BRADSHAW*
CONCEPTUAL MODELS IN EXPLORATION GEOCHEMISTRY
The Canadian Cordillera and Canadian Shield

4. *R.T. SHUEY*
SEMICONDUCTING ORE MINERALS

5. *J.S. SUMNER*
PRINCIPLES OF INDUCED POLARIZATION FOR GEOPHYSICAL
EXPLORATION

Developments in Economic Geology, 3

WORLD MINERAL SUPPLIES

Assessment and perspective

Edited by

G.J.S. GOVETT

Professor, University of New Brunswick, Fredericton, N.B., Canada

and

M. H. GOVETT

Consultant, Fredericton, N.B., Canada

ELSEVIER SCIENTIFIC PUBLISHING COMPANY
Amsterdam — Oxford — New York   1976

ELSEVIER SCIENTIFIC PUBLISHING COMPANY
335 Jan van Galenstraat
P.O. Box 211, Amsterdam, The Netherlands

AMERICAN ELSEVIER PUBLISHING COMPANY, INC.
52 Vanderbilt Avenue
New York, New York 10017

ISBN: 0-444-41366-9

Printed in The Netherlands

PREFACE

It was perhaps appropriate that the first draft of this preface was written the same day that the dramatic film version of the Club of Rome's "Limits to Growth" was shown at our University. The audience — students and professors from the Arts, Science, and Engineering Faculties — was virtually unanimous in its reaction of shock and concern over mineral resource shortages and ecological disaster; not surprisingly, the solutions to the problem of exponential growth and environmental damage suggested in a forum after the showing of the film reflected the particular discipline of the speaker. To the scientist and engineer the unquestioning acceptance of the thesis of crisis and doom was unexpected, and many of the suggestions offered by the social scientist seemed utopian; to the non-scientist the remarks of the geologists and engineers seemed unnecessarily pedestrian.

Much of the outpouring of literature on the energy and mineral supply "crisis" which followed the "Limits to Growth" and the Ecologist's "Blueprint for Survival" is the result of further investigations by economists and social scientists; the impact of geologists, mining engineers, and mineral technologists has been limited primarily to the technical literature. The tendency to view the problem of resource adequacy from the politicians' and economists' point of view — while understandable — results in studies which list a number of options for society based on technological and geological assumptions which are not necessarily either correct or meaningful or which have been taken out of the original scientific context.

In Canada a recently announced study of the national impact of the mineral industry (to be undertaken by the Centre for Resource Studies at Queen's University) does not include geologists; the Ford Foundation's study on the energy situation in the U.S. (at least on the basis of a preliminary report, "Exploring Energy Choices", published in 1974) does include physcists and engineers, but the emphasis is largely economic and social. This tendency to assume that science and technology will provide the answers whatever scenario is adopted by national or international planners is dangerous, and there is little indication that it will change in the immediate future.

This book — which had its origins in a series of lectures on the "History of Science" planned by the editors in 1970 and in the enormous response to our article "Mineral Resource Supplies and the Limits to Economic Growth" published in *Earth-Science Reviews* in 1972 — is an attempt to present an assessment of some of the issues in the growing debate about whether there is an impending global shortage of mineral resources and whether technology and exploration can meet the needs of a rapidly growing rate of mineral consumption in the world. In choosing to bring together contributors who are specialists in the various fields of mineral resources we hope to avoid the glossy, popular approach which has increasingly surrounded the debate and to provide scientific and technical appraisals of the various fields which will determine the future of world mineral supplies.

We agree with the conclusions of Peter Flawn that those who, "... proceeding from analyses of seawater or average granite, triumphantly calculate the enormous quantities of various elements contained in a cubic mile of seawater or granite" are taking "... a wholly frivolous and misleading approach which contributes only a false sense of well-being and cannot withstand a thoughtful probe of the problem" (P.T. Flawn, *Mineral Resources*, Rand McNally, New York, 1966, p. 356). In trying to assess the present and future of mineral demand and supply we have compiled chapters written by people who we believe are competent to comment on the various aspects of the problem. These people were selected for their special knowledge, and not because they support the thesis we presented in *Earth-Science Reviews;* indeed, we had no prior knowledge of their personal views, and editorial instructions to each contributor only specified that they write a quiet, sober appraisal of the problem from their own point of view.

Therefore it is perhaps surprising that there is general unanimity in the book. We believe that this is a hopeful sign; if economists, a geochemist, a geophysicist, a geographer, economic and mining geologists, a mining engineer, and a mineral technologist from such diverse professional bodies as the U.S. Geological Survey, the United Nations, the Canadian Department of Energy, Mines and Resources, and various universities in North America and the U.K. can agree on the general problem and its solution, then it should be possible for the world to solve at least the technical problems of mineral resource supplies.

The most hopeful parts of the book are those dealing with technology. This reflects our own intuitive feeling that, while the present is dominated by the geologist in his search for new ore deposits and the conservationist in his attempts to reduce consumption and better plan the use of known resources, the future will be shaped by the technologist and his success in devising new means of mining and processing both conventional and non-conventional sources of minerals.

On balance the book is neither optimistic nor pessimistic. It is clear that the world faces a serious problem *if* current rates of economic growth, patterns of mineral consumption, and levels of exploration and spending on research continue. Resources are abundant, but their availability depends on exploration, conservation, and the development of new technology — and these in turn depend on man's attitudes towards the planet on which he lives and on his willingness to spend the money and talent needed for research and development. Whether new programmes will be undertaken — as a matter of urgency — is debatable. While scientists are in general agreement that a vast new effort should be made to find new ore deposits and to develop new methods of mining and processing minerals, governments — the main source of funds for research and development — are slow to recognize their vital role in funding research and in designing policies which will encourage and stimulate exploration and development. Even in a country as dependent on minerals as Canada there is an unwillingness to spend money on research into new exploration methods and to legislate measures to stimulate private investment in exploration. In fact, in many provinces of Canada, as well as on the national level, the current approach to the mining industry is one which is leading to retrenchment rather than growth; at the research level scientists are finding it increasingly difficult to undertake programmes to develop new techniques and methods. Hopefully this book will stimulate a new approach to research and development and will stimulate governments to recognize the immense importance of science and technology in man's desire for a better way of life.

The early planning and some of the actual writing of the book was done prior to the "energy crisis". The difficulties of writing about energy, and indeed about mineral demand and supply in general, during a year when the price of Middle Eastern oil increased four-fold, when a number of countries found themselves in serious balance of payments trouble, and when many of the developed countries entered a period of deepening recession were severe. The temptation to continually up-date and revise material was often irresistable; even as the publisher's deadline approached a number of chapters were reappraised and new material was incorporated.

Inevitably the enormity of the task of keeping up with world-wide mineral resource developments dictated a cut-off point; this was arbitrarily set at February 1975. Whether a number of contributors would have revised their chapters is not known. The editors — having the advantage of the last word — feel that the material presented here is a fair statement of the problems and potentials of mineral resource demand and supply. The chapters on energy might have been written from a somewhat different viewpoint — with more emphasis on the effects of a period of high-cost energy fuel — if Mr. Thomas had been able to revise them himself. As editors we tried to incorporate new material that became available after his untimely death in October, 1974. However, the problems of keeping abreast with the rapidly changing petroleum situation are such that almost any forecast about the future has an aspect of prophecy; with all due respect to the futurologists, it seems virtually impossible to settle on any one "scenario". Mr. Thomas's chapters contain the significant facts, and we are sure that he would have concurred in our view that the era of cheap energy is over for the immediate future; we are also sure he would have agreed that it would be foolhardy to predict the price of a barrel of oil five or ten years hence — and equally foolish to try to forecast what effect price changes will have on demand and supply, particularly on the future development of nuclear energy, the Canadian tar sands and other unconventional sources of petroleum. The geologist can make some reasonably good guesses about where petroleum might be found; the economist would be daring indeed to predict the economic and political changes which could make it economic to develop high-cost offshore and non-conventional supplies on a world-wide basis.

In editing and in writing part of this book we have become indebted to many people — firstly, of course, to our contributors who we thank for their good-natured reaction to our editorial comments and changes and, especially, for meeting our deadline. We would also like to thank Mr. M.J. Jones of the Institution of Mining and Metallurgy (London) for his many helpful suggestions. Special thanks are due to the reference and documents sections of the Harriet Irving Library of the University of New Brunswick, and particularly to Mrs. J. Colson, for hours of patient work to supply us with material; to Mr. G.E. Thomas for painstakingly transcribing his brother's draft into a useable manuscript; to Dr. A.A.A. Hussein, United Nations Research Associate at the University of New Brunswick in 1974—75 from the Geological Survey of Egypt, for some technical editorial assistance; and to Mr. J. Ross and Mr. R. McCulloch of the University of New Brunswick Department of Geology for preparing all the diagrams except for those in Chapters 5 and 7. Finally, we would like to express our sincere appreciation to Drs. F.W.B. van Eysinga of Elsevier Scientific Publishing Company for his constant support, encouragement — and non-interference — throughout the two years this book has been in preparation.

G.J.S. GOVETT and M.H. GOVETT

Fredericton, New Brunswick, Canada
March 1975

LIST OF CONTRIBUTORS

David B. Brooks. B.Sc. (geology) Massachusetts Institute of Technology; M.Sc. (geology) California Institute of Technology; Ph.D. (economics) University of Colorado. Dr. Brooks was Chief of the Division of Mineral Economics of the Bureau of Mines, U.S. Department of the Interior from 1967 to 1970 when he joined the Department of Energy, Mines and Resources (Canada) as Chief, Division of Mineral Economics Research. Since 1974 he has been Director, Office of Energy Conservation. Earlier appointments include five years with Resources for the Future Inc. and professor of economics at Berea College (Kentucky, U.S.A.).

Henry E. Cohen. B.Sc., M.Sc., Ph.D. (University of London). Professor Cohen joined the faculty at Imperial College of Science and Technology (London) in 1956, and was appointed Reader in Mineral Technology in the Department of Mining and Mineral Technology in 1966. In 1974 he was appointed Professor of Mineral Technology.

L.S. Collett. B.Sc. (physics and chemistry) McMaster University; M.A. (geophysics) University of Toronto. He joined the Geological Survey of Canada in 1953; since 1961 he has been Head, Electrical Methods Section. In 1974 he was appointed coordinator of environmental and engineering geophysics and in 1975 became responsible for the Seismic and Electrical Methods Section of the Resource Geophysics and Geochemistry Division of the G.S.C. During 1969—1974 he was Principal Investigator with NASA for Apollo Missions. Prior to joining the G.S.C. he was employed by Newmont Exploration Ltd.

Donald W. Gentry. B.Sc. (mining engineering), M.S. (mining engineering), University of Nevada; Ph.D. (mining engineering), University of Arizona. Dr. Gentry joined the Colorado School of Mines in 1972 and is now Associate Professor of Mining Engineering. He had previously been employed by Anaconda Co. He consults for the mining industry in many parts of the U.S. and Canada.

G.J.S. Govett. B.Sc. (geology and chemistry) University of Wales; D.I.C. (geochemistry) Imperial College; Ph.D. (geochemistry) University of London; D.Sc., University of Wales. Professor Govett joined the Research Council of Alberta in 1958 and the University of New Brunswick in a permanent capacity in 1968 where he has been Professor of Geology since 1970. He has served as consultant for the United Nations, OECD, and CIDA and for the mining industry in various parts of the world including Canada, the Philippines, Cyprus, Jordan, Ethiopia, Guyana, and Greece.

M.H. Govett. B.A. (economics), M.A. (economics) University of Washington. Mrs. Govett is an economic consultant and has taught economics at the English School (Cyprus) and the University of New Brunswick. She has also worked as an economist for the Government of Libya and the RAND Corp. and has published widely in the field of mineral economics.

Daniel A. Harkin. B.Sc. (geology), Ph.D. (geology) University of Glasgow. Dr. Harkin joined the United Nations in 1964 and since 1967 has been Technical Advisor to the Department of Economic and Social Affairs; while with the United Nations he has been concerned with mineral resource programmes in more than two dozen countries. Prior to joining the United Nations Dr. Harkin served with the British Overseas Civil Service in Tanganyika (Tanzania) where he was Deputy Director of the Geological Survey in 1957— 1964.

Evan Just. B.S. (geology) Northwestern University; M.S. (geology) University of Wisconsin; Ph.D. (Hon.) Montana School of Mines. Professor Just retired from the position as Head of the Department of Mineral Engineering at Stanford University in 1969 and is now Professor Emeritus and actively teaching. He has been editor of *Engineering and Mining Journal* and Vice-President of Cyprus Mines Corp. In his earlier career Professor Just was engaged in mineral exploration in the U.S., Brazil and the U.S.S.R., taught at Lehigh University and the New Mexico School of Mines, and has been production engineer for Carter Oil Co. and executive secretary for the Tri-State Zinc and Lead Ore Producers Association.

A.L. McAllister. B.Sc. (geology) University of New Brunswick; M.Sc., Ph.D. (geology) McGill University; Fellow of the Royal Society of Canada. Professor McAllister has been on the faculty of the University of New Brunswick since 1952; he is now Professor of Geology; he was Head of Department for 11 years until 1971. He is a member of the University Board of Governors and the Senate. Professor McAllister has served on many national and professional committees, is an active consultant to the mining industry, and is on the board of directors of five Canadian mining companies.

V.E. McKelvey. B.A. (geology) Syracuse University; M.S. (geology) and Ph.D. (geology) University of Wisconsin. Dr. McKelvey joined the U.S. Geological Survey in 1941 and has been Director since 1971. His entire career has been concerned with economic and geological aspects of mineral resources. He has been U.S. representative to the OECD Energy Committee, the Law of the Sea Committee, and has been chairman of the Project Independence Blueprint Integrating Oil Task Force. Dr. McKelvey has received the Department of the Interior Distinguished Service Award, the National Civil Service League Award, and the Rockerfeller Public Service Award.

T.M. Thomas (deceased Oct. 1974). B.Sc. (geography) and M.Sc. (meteorology) University of Wales. Mr. Thomas was Principal Scientific Officer for mineral planning at the Welsh Office, Cardiff. He was a prolific writer who gave his attentions to an extraordinarily wide range of subjects from mineral deposits of the world, meteorology, geomorphology, oil and gas exploration, to the effects of underground nuclear explosions.

F.M. Vokes. B.Sc. (mining engineering) and M.Sc. (geology) Leeds University; Ph.D., Oslo University. Dr. Vokes is Research Scientist with the Mineral Deposits Section of the Geological Survey of Canada. Between 1966 and 1974 he was Professor of Mining Geology at the Technical University of Norway (Trondheim). Earlier in his career he worked in Zambia, with the G.S.C. in Canada, for the Geological Survey of Norway, and as a consultant for the United Nations.

F.F.H. Wang. B.S. (geology) National SW University of China; Ph.D. (geology), University of Washington. Dr. Wang's work has been largely concentrated in the field of marine geology; he is Research Geologist with the U.S. Geological Survey and presently serves as Technical Advisor and Consultant to the Department of the Interior, the Department of State, the Marine Science Council, and as U.S. delegate to the United Nations on matters relating to the geologic framework and mineral resources of the continental margins and ocean basins. Before joining the U.S.G.S. he worked with the Gulf Oil Corp. and the International Mineral and Chemical Corp.

CONTENTS

PART I. AN ASSESSMENT OF WORLD MINERAL SUPPLIES

Developments in mineral exploration and exploitation

GLOSSARY

Wherever possible all measurements have been converted to metric units. In some instances where the original source did not specify whether long or short tons were used this was not possible. Furthermore, most of the units of natural gas are given in cubic feet (cu.ft) because this is still the most common unit used.

ABBREVIATIONS

g	gramme
kg	kilogramme
tonne	metric ton

μm	micrometre (= 10^{-6} metres = micron)
cm	centimetre
m	metre
km	kilometre

kW	kilowatt (= 10^3 watt)
kWh	kilowatt-hours
MW	megawatt (= 10^6 watt)
Btu	British thermal unit (= 0.252 kilocalorie)

cps	cycles per second
Hz	hertz
kHz	kilohertz

tph	tons per hour

ppm	parts per million (= per cent $\times$ 10^{-4})
ppb	parts per billion (= per cent $\times$ 10^{-7})

%	parts per hundred (= per cent)
‰	parts per thousand (= per mille)

UNITS OF MEASUREMENT AND CONVERSION FACTORS

1 inch (in) = 2.54 centimetres (cm); 1 cm = 0.394 in
1 foot (ft) = 0.305 metre (m); 1 m = 3.281 ft
1 statute mile = 1.609 kilometres (km); 1 km = 0.621 mile

1 square mile (sq. mi) = 2.5898 square kilometres (km^2); 1 km^2 = 0.3856 sq. mi
1 acre = 4046.9 square metres (m^2); 1 hectare = 10,000 m^2
1 cubic foot = 0.028 cubic metre (m^3); 1 m^3 = 35.714 cu.ft

1 pound (lb) = 0.454 kilogramme (kg); 1 kg = 2.205 lb
1 long ton = 1.016 tonnes; 1 tonne = 0.984 long ton
1 short ton = 0.907 tonne; 1 tonne = 1.103 short tons

1 imperial gallon = 4.546 litres (l); 1 l = 1.76 pints
1 U.S. gallon = 3.785 litres (l); 1 l = 2.113 pints

1 kilojoule (kJ) = 0.239 kilocalorie (kcal); 1 kcal = 4.183 kJ

1 barrel of oil = 42 U.S. gallons = 158.987 litres
1 barrel of oil $\simeq$ 5.11 tonnes coal equivalent

billion = 10^9 (1,000,000,000)
trillion = 10^{12} (1,000,000,000,000)

CHEMICAL SYMBOLS AND ELEMENTS

Ac	actinium	He	helium	Pt	platinum
Ag	silver	Hf	hafnium	Ra	radium
Al	aluminium	Hg	mercury	Rb	rubidium
Ar	argon	Ho	holmium	Re	rhenium
As	arsenic	I	iodine	Rh	rhodium
At	astatine	In	indium	Rn	radon
Au	gold	Ir	iridium	Ru	ruthenium
B	boron	K	potassium	S	sulphur
Ba	barium	Kr	krypton	Sb	antimony
Be	beryllium	La	lanthanum	Sc	scandium
Bi	bismuth	Li	lithium	Se	selenium
Br	bromine	Lu	lutetium	Si	silicon
C	carbon	Mg	magnesium	Sm	samarium
Ca	calcium	Mn	manganese	Sn	tin
Cd	cadmium	Mo	molybdenum	Sr	strontium
Ce	cerium	N	nitrogen	Ta	tantalum
Cl	chlorine	Na	sodium	Tb	terbium
Co	cobalt	Nb	niobium	Tc	technetium
Cr	chromium	Nd	neodymium	Te	tellurium
Cs	cesium	Ne	neon	Th	thorium
Cu	copper	Ni	nickel	Ti	titanium
Dy	dysprosium	O	oxygen	Tl	thallium
Er	erbium	Os	osmium	Tm	thulium
Eu	europium	P	phosphorus	U	uranium
F	fluorine	Pa	protactinium	V	vanadium
Fe	iron	Pb	lead	W	tungsten
Fr	francium	Pd	palladium	Xe	xenon
Ga	gallium	Pm	promethium	Y	yttrium
Gd	gadolinium	Po	polonium	Yb	ytterbium
Ge	germanium	Pr	praseodymium	Zn	zinc
H	hydrogen			Zr	zirconium

INTRODUCTION

G.J.S. GOVETT and M.H. GOVETT

As the title implies, this volume is divided into two parts. The first part is a brief assessment of world mineral supplies, and deals primarily with the geologic and economic factors which determine the demand for and the availability and distribution of mineral resources and reserves. The second part of the book is a more extensive and detailed view of the future of mineral resource supplies.

Since the book is written primarily by scientists, and particularly by geologists and engineers, the emphasis is on the technical problems of assuring adequate mineral supplies for the future. This emphasis has been dictated by the firm belief of the editors, one a geochemist and one an economist, that we do not need another book which reviews the gloomy projections of an overpopulated, greedy mineral-consuming world or another account of the economic and political problems of the less developed countries in conflict with the more affluent West. What we do need is a sober appraisal of the problems of, and possible solutions to, the present mineral resource situation. The problem of increasing mineral supplies to meet rapidly growing world consumption, although vitally affected by economic and political decisions, is ultimately technical in nature and requires a scientific solution.

Part I is, in some sense, "stage-setting". It is an attempt to arrive at a more rigorous definition of reserves and resources — how they are affected by economic, geologic, and technological factors, and how they are distributed throughout the world. Chapter 1 surveys the complexities and confusion that colour much of the discussion about the size and adequacy of world mineral supplies. The importance of distinguishing between "resources" — the total amount of an element in the earth's crust down to some grade higher than crustal abundance but lower than present economic grade — and "reserves" — minerals which can be extracted under present economic and technical conditions — is too often ignored in debates about the adequacy of mineral supplies; the distinction must be clearly made if future planning is to be soundly based.

In Chapter 2 Professor McAllister examines in more detail the creation of reserves from resources as a function of price, cost, ore grade, and technology, emphasizing the complexity of determining where the dividing line is at any given time. The point is made that calculations and predictions about the resources for most minerals are little more than guesswork because of the paucity of geological information; while predictions may be relatively realistic for stratabound deposits, there is an acute problem in making predictions for other types of deposits. Historically it is possible to show how changes in technology have converted resources into reserves, but this cannot be used in a predictive sense because there is no way of predicting the course of technological development. Moreover, as a number of other contributors have noted, development of marine resources may nullify all other predictions.

Dr. Vokes (Chapter 3) concentrates on the physical and chemical factors that determine the absolute abundance and the actual availability and distribution of resources. He points out that, while the chemical abundance in the earth's crust seems to be directly related to the availability of resources for man's use, this does not imply that the economically important minerals are available in amounts and in places where they can be easily extracted or processed. Technological factors are critical in determining the availability of both scarce and abundant elements; events in the geologic past, particularly the formation of "metallogenic provinces", dictate whether elements are concentrated in sufficient quantities, in amenable combinations, and in geographic areas which are accessible to allow mining under current economic conditions.

Chapter 4 surveys the actual geographic distribution among countries of both supply and demand, pointing out the remarkable degree of geographic concentration which characterizes mineral consumption and production. Mrs. Govett makes the point (which is again made by Dr. Harkin in Chapter 10) that the mineral "haves" and "have-nots" are not necessarily the developed countries and the less developed countries, respectively; the conflict between mineral producers and consumers today is not a simple split between the West and the "Third World". International trade and new forms of associations are vital to the continued prosperity of both the consumers and the producers.

While the first four chapters deal with energy inter alia, Chapter 5 is specifically addressed to the subject. The changing patterns of consumption (particularly the shift from coal to petroleum since World War II), the enormous increase in world demand for energy, and geopolitics have combined to produce the current energy "crisis". There seems little doubt that the era of cheap energy, which allowed the rapid industrial development of the first three-quarters of this century, is over. The current high price of petroleum may fall somewhat if recession and depression reduces demand and the politics of Middle Eastern oil change, but high-cost energy will mean that new consumption patterns and new sources of energy will have to be developed, both in the interests of national "self-sufficiency" and to meet projected long-term increases in energy demand. The existence of high-cost energy may be a positive factor to the degree that it accelerates the development of new energy sources and converts presently non-conventional or subeconomic resources into economic reserves.

The second part of the book — a general perspective on the future of world mineral supplies — is divided into two sections. The first section examines some of the problems of demand and supply, while the second section is essentially an account and assessment of future trends in mineral exploration, exploitation, and processing. No attempt is made to answer the question "Is there an imminent shortage of supplies of mineral raw materials?". This is, in our view, a rather fruitless debate, depending as it does on the base which is chosen for projections of demand, the rate of growth which is assumed, and the degree of optimism or pessimism about the economic and technological factors which determine supply. It is possible to be utterly pessimistic (as are Park, 1968; Meadows et al., 1972) or generally optimistic (Landsberg, 1964; Boyd, 1973) depending upon the definitions and the data used. As discussed in Chapter 1, the general lack of internationally comparable data and confusion of terms makes it exceedingly difficult to assess quantities of recoverable minerals, even in the context of today's economic and technical position; the problems of predicting future supplies are even greater (Chapter 2; Brobst and Pratt, 1973). We can go no further than we did in 1972 and conclude that "There is, of course, no absolute shortage of any one element — there is a crisis only within the present system of exploration and exploitation" (Govett and Govett, 1972, p. 288).

This conclusion does *not* imply that mineral resource appraisals are unnecessary; quite the contrary. This whole volume is an attempt to look at some of the technical problems which directly or indirectly affect the availability of mineral resources and to provide information for those who are presently trying to arrive at reserve and resource estimates in order to frame national or international policy toward resource development.

Both Professor Just (Chapter 6) and Dr. Brooks (Chapter 9) make the point that, in the economic sense, there can be no "problem" with the adequacy of mineral supplies; in the extreme case, if cost were of no importance, whole rock could be mined for its component parts (e.g., Brown, 1954; Skinner, 1969). If shortages arise, price would increase, but, provided that the consumer can pay, no element supply can be exhausted; the grade of ore processed would fall, processing costs would rise, but depletion as such could not occur. In this sense, there is little profit in discussing how many tons of copper or iron ore are estimated to be available in known or undiscovered deposits. This conclusion does not mean that we should consider mining whole rock, nor does it argue against the useful concept of "geologic availability" (Brobst and Pratt, 1973); it does, however, emphasize the role of "economic and technologic availability" in determining both short-term and long-term mineral supplies (cf. McKelvey, 1972). The most important point at issue is technological development — of finding, mining, and processing ores — and the capacity to pay.

Notwithstanding his projections which show that, at least until the year 2000, world affluence will grow somewhat more rapidly than the consumption of minerals, Professor Just takes a gloomy view of the effects of severe overpopulation (another three billion people by the end of the century now seems virtually certain) and the ability to find and pay for the exploration and exploitation of new mineral deposits to support increased consumption. Dr. Brooks is more optimistic; he expects increased exploration costs to be offset by the opening up of new *types* of deposits, and concludes that technology may be more important than the discovery of more deposits. In the long-term this conclusion will undoubtedly be valid; the future belongs more to developments in mineral processing than to improvements in exploration geology. However, in the short- to medium-term the geologist must solve the increasingly difficult problem of finding large quantities of ores in more or less conventional types of deposits.

It is difficult to convey to someone who is not a geologist the magnitude of the problem of finding a mineral deposit, even in a favourable region. To find near-surface deposits beneath the cover of glacial debris and the endless coniferous forests of Canada or beneath twenty metres of lateritic soil in tropical Africa is an awe-inspiring task. To expect to be able to find a deposit in such circumstances, and especially to find one located hundreds of metres below bedrock surfaces, implies an extraordinary faith in exploration methods. The two approaches to modern exploration which are most likely to be successful are geophysical and geochemical exploration techniques. Although they are operationally quite different — geophysical methods measure physical differences and geochemical methods measure geochemical differences in the earth's crust — they both have the capacity for detecting mineral deposits which are not exposed at the surface. Similarly, both techniques may be used on a reconnaissance scale (airborne geophysics and remote-sensing, and geochemical drainage surveys) for identifying broad areas of interest, and on a local detailed scale for locating individual deposits. Geophysics and geochemistry are commonly used together in mineral surveys, one being used to follow-up or confirm the results of the other.

Exploration geophysics is the older technique and Mr. Collett, in Chapter 12, concentrates on describing some of the newer instrumentation capable of greater depth penetration and on discussing the potential of various remote-sensing airborne and satellite techniques for mineral exploration. Geochemistry, which really only began to be considered seriously as an exploration technique in the western countries in the 1950s, entered a period of enormous development in the 1970s. The present capabilities of the method and the possibilities for its expanded use in all phases of exploration, particularly for deeply buried deposits, is examined by Professor Govett in Chapter 11. There is confidence that both geophysical and geochemical methods, together with modern geological theory and practice, are now available or are being developed which will allow the detection of ore deposits at depths of hundreds of metres; in both cases there is a pressing need for research into new and better methods.

The developing countries are generally the least well-prospected parts of the world; some offer promise in the search for new ore deposits, but private exploration effort in much of Latin America, Asia, and Africa is declining for political and economic reasons. Dr. Harkin (Chapter 10) discusses the significance of the problem and reviews the role of technical cooperation, particularly through the United Nations, in financing and providing mineral exploration in the developing countries. The apparent disagreement between Dr. Harkin and Mrs. Govett (Chapter 4) — he concludes that new sources of minerals will come primarily from the less developed countries, while she argues that Australia, Canada, South Africa, and the U.S.S.R. will continue to dominate world reserves and production — is more apparent than real. Dr. Harkin's conclusion assumes a change in the pattern of exploration, while Mrs. Govett tacitly assumes that, at least in the foreseeable future, there will not be much change. One can hope that the more optimistic position will prove to be right — that means will be found to overcome the problems of attracting capital for mining ventures and new exploration in the less developed countries, that the traditional political and social barriers to foreign investment in politically risky countries will be surmounted, and that new arrangements that are acceptable to both the investing countries and the less developed countries will evolve. It is more difficult to be optimistic about agreement between the mineral-producing and the mineral-consuming countries.

While the intensity of conflict between mineral consumers and producers is a relatively modern feature of the minerals scene, removal of minerals from the ground and their subsequent processing has always been in conflict with the goal of preserving man's environment. Concern for damage to the surface of the earth, the atomosphere, and ecological balance, although very topical today, is scarcely new. In 1556 Georgius Agricola wrote: ". . .fields are devasted by mining operations, for which reason formerly Italians were warned by law that no one should dig the earth for metals and so injure their very fertile fields, their vineyards, and their olive groves . . . the woods and groves are cut down, for there is need of an endless amount of wood for timbers, machines, and the smelting of metals. And when the woods and groves are felled, then are exterminated the beasts and birds, very many of which furnish a pleasant and agreeable food for man. Further, when the ores are washed, the water which has been used poisons the brooks and streams, and either destroys the fish or drives them away" (Agricola, 1556, p. 8).

What is new is the vast scale of environmental damage which mining and mineral processing can inflict and a public awareness of the consequences of unchecked exploitation of the earth's surface. Dr. Brooks concludes that conservation of mineral resources and conservation of the environment are inextricably linked, and that the environment may be the true nonrenewable resource.

Dr. Brooks also examines some of the problems of land use and reclamation, problems that will become more severe in the future if Drs. Gentry's and Cohen's conclusion that deposits of ever-lower grade will have to be mined are correct. Conservation measures, including recycling, can reduce the pressure on reserves, but it seems likely that fewer and larger mines using bulk mining methods will characterize the future of mining.

While concern for the effects of environmental protection measures on the mining industry, and particularly on the future of surface mining (which accounts for about 70% of total world mining operations today) has accelerated efforts to improve surface mining techniques, the implications of increasingly restrictive legislation to reduce strip-mining or to stop it altogether, are clear. The economic arguments in favour of applying new, improved mineral processing techniques to increasingly lower-grade surface materials can be expected to pose serious problems.

The discovery and mining of deeper deposits to increase reserves of a number of relatively scarce metals (Govett, Collett, Gentry) poses less serious environmental problems, but challenges the mining engineer. Underground mining is still very labour-intensive, has a lower productivity than surface mining, and incurs higher production costs. Dr. Gentry discusses the techniques now being used for mining at depth and suggests that major improvements in existing techniques are likely to be less important than the development of radical new approaches — hydraulic mining, pneumatic solids transport, in situ leaching (especially for deep, disseminated deposits), and underground in-situ gasification of coal (particularly for extremely deep and currently non-economic deposits).

The process of conversion of a naturally occurring mineral substance to an industrially useful product is commonly ignored in discussions of the adequacy of mineral resources; its importance is gaining recognition through the realization that if extractive processes are available which allow economic production of, for example, metallic copper from an ore containing only 0.1% of the metal, the reserve situation for copper would be dramatically changed. In Chapter 14 Professor Cohen considers the economic and technical complexities of mineral extraction and processing; the present high cost of fuel and environmental restraints demand that conventional process design and operation be reappraised, and that new combinations of processes be used to meet local conditions. Professor Cohen believes that there is considerable scope, even with existing technology, for better recovery, higher-grade products, lower unit costs, and better fuel efficiency.

Availability of energy is essential to any mineral processing technique, and the trend toward processing lower-grade and difficult ores is necessarily accompanied by increased energy consumption per unit of metal extracted. The increasing demands for energy in the minerals industries intensifies the problems of energy supply caused by the general increase in energy consumption throughout the world in the past two decades. In an examination of the availability of conventional energy supplies, Mr. Thomas concludes that crude oil supplies *could* be adequate for many decades; there are excellent prospects of new discoveries, particularly offshore, in many parts of the world. Coal reserves (at only a 50% recovery rate) have an energy equivalent of more than five times the known combined oil and gas reserves, and coal resources are enormous. The resources in the tar sands and oil shales, which high-priced Middle Eastern oil may now make economically exploitable, are several times those of the known reserves in the Middle East. Nevertheless, the rate of growth of energy fuel consumption — and the product mix of demand — will be critical in determining whether known reserves will be adequate.

There are a number of other potential energy sources awaiting development — solar power, hydro power, tidal power, geothermal power and, of course, nuclear power. The world's potential undeveloped hydroelectrical capacity alone is equivalent to three times the total electrical capacity today, but its development is severely restricted by the distribution of resources relative to demand; nuclear power, which now accounts for less than 1% of energy consumption in the world, is just beginning to be developed, but the technical problems are increasingly severe, environmental restraints may make its future uncertain, and development of the breeder reactor is vital to its widespread use. Nevertheless, the development of these alternatives to fossil fuels must be accelerated, not only to meet increased demand but also because fossil fuels are a source of raw materials for many commodities necessary to industrial society (e.g., plastics and synthetic fibres).

Even though there is no absolute shortage of energy fuels, it is clear that mammoth development programmes are required for exploration for petroleum and natural gas and to develop new sources of energy. This same conclusion applies to the development of the truly enormous resources of the oceans and the sea floors. Drs. Wang and McKelvey point out that, if sufficient talent, time, and money are spent on research and development, marine resources could become one of the most important factors in satisfying man's seemingly insatiable appetite for minerals, although in the next few decades technological and economic restraints will limit exploitation to areas of exceptionally favourable circumstances such as offshore deposits in easily accessible coastal waters and recovery of minerals in very short supply. Technological breakthroughs in this decade for deep-sea mining of manganese nodules could eventually permit more extensive marine mining in the world's shelf areas; however, the most serious problem will be in devising means to determine who owns and controls the world's marine resources and to develop them in a way which will avoid conflicts among nations.

In the remaining years of this century the role of the exploration geologist, geochemist, and geophysicist will become increasingly important in the world economy; a major effort is now required to develop better exploration techniques for shallow-buried deposits and entirely new techniques for deeply buried deposits. In the longer-term the development of new extraction and processing techniques will probably be the most important factor in assuring future adequacy of mineral supplies.

Conservation measures can do a great deal to ease the pressures on mineral reserves in the short- and medium-term. Increased world trade and new approaches to investment in exploration and mining in the less developed countries could reduce some of the pressures of the geographic unevenness of resource distribution. High-cost energy may encourage the development of new sources of energy fuels, and the resources of the oceans and sea floors could become an economic source of minerals.

While there is unanimity of opinion that science and technology are inherently capable of the task of providing new sources of mineral supplies, there is far less certainty that the necessary support will be available to conduct basic research and large-scale practical testing of new methods and techniques or that governments are acting in a manner to encourage new exploration and investment in developing mineral supplies. The task is urgent — the average lead-time in mineral exploration is six to ten years; in scientific research it is of the order of ten years. At scientific meetings throughout the world the scientist is continually crying out for more support. Sometimes the amounts needed are small — $50,000 to $200,000 for small-scale instruments, $12,000 to $15,000 annually to support a laboratory technician, approximately $8,000 a year to support a post-graduate student's research — but without this support even the most limited research is not possible.

Research demands money, continuity of operation, and talented people. The needs are global in scope, but realistically the problem will have to be tackled on the national and regional level. Governments must be convinced of the need to finance research; mining companies must be persuaded to overcome their reluctance to spend money on research, and government fiscal policies should encourage such support; exploration should be encouraged rather than discouraged; universities should recognize the importance of both "pure" and "applied" research and acknowledge that the division between the two is neither so sharp nor so important as is often alleged.

Thoughtful scientists in the broad field of mineral resources are generally conscious that there is *potentially* a shortage of world mineral supplies; governments are also becoming increasingly aware that there may be a crisis of supply. The general conclusion of this book is that improvements in technology and development of procedures and methods which are only theoretical concepts or which are at the experimental stage can, conceptually, provide the world with adequate supplies of minerals if research and development are supported. If there is also a concerted effort to conserve known reserves and to reduce the exponential rate of population growth, there is little doubt that the future can be viewed with some optimism. "Resolution of the problem is within the technical capacity of man; what is in doubt is the ability of man to recognize, in time, the existence of the problem" (Govett and Govett, 1972, p. 288).

REFERENCES CITED

Agricola, Georgius, 1556. *De Re Metallica* (translated by H.C. Hoover and L.H. Hoover). Dover, New York, N.Y., 638 pp.

Boyd, J., 1973. Minerals and how we use them. In: E.N. Cameron (Editor), *The Mineral Position of the United States, 1975—2000*. University of Wisconsin Press, Madison, Wisc., pp. 1—8.

Brobst, D.A. and Pratt, W.P. (Editors), 1973. *United States Mineral Resources*. U.S. Geol. Surv., *Prof. Paper 820*, 722 pp.

Brown, H., 1954. *The Challenge of Man's Future*. Viking, New York, N.Y., 290 pp.

Govett, G.J.S. and Govett, M.H., 1972. Mineral resource supplies and the limits of economic growth. *Earth-Sci. Rev.*, 8: 275—290.

Landsberg, H.H., 1964. *Natural Resources for U.S. Growth: A Look Ahead to the Year 2000*. Johns Hopkins, Baltimore, Md., 257 pp.

McKelvey, V.E., 1972. Mineral resource estimates and public policy. *Am. Sci.*, 60: 32—40.

Meadows, D.H., Meadows, D.L., Randers, J. and Behrens, W.W., III, 1972. *The Limits to Growth*. Universe Books, New York, N.Y., 205 pp.

Park, C.F., Jr., 1968. *Affluence in Jeopardy: Minerals and the Political Economy*. W.H. Freeman, San Francisco, Calif., 235 pp.

Skinner, B.J., 1969. *Earth Resources*. Prentice-Hall, Englewood Cliffs, N.J., 150 pp.

PART I

AN ASSESSMENT OF WORLD MINERAL SUPPLIES

Chapter 1

DEFINING AND MEASURING WORLD MINERAL SUPPLIES

M.H. GOVETT and G.J.S. GOVETT

> "The sufficiency of the physical resources in the earth's crust has become a matter of hot debate. Prophecy about the prospective exhaustion of the world's resources is rooted in the shifting sands of statistics on proved reserves."
>
> U.S. National Commission on Materials Policy
> (1973, p. 9-7)

INTRODUCTION

Debate about the adequacy of national and international mineral supplies to meet present and projected world demand is hindered by the ambiguity of the various terms used to describe and measure mineral supplies. In spite of a number of recent attempts to achieve an agreed terminology, there is still a proliferation of definitions. In a recent survey of world iron ore resources the United Nations (1970) appended a glossary of terms listing sixteen separate definitions for the word "reserves" and a further thirteen definitions of terms used exclusively to describe iron ore.

The concept of resources and reserves used by economists and geologists differs significantly. The geologist is concerned with a single ore deposit of limited areal extent; the economist is concerned with regional, national, or international supplies of minerals. A mining company wants to know how many tons of ore of what grade exist in a given deposit to determine whether the deposit can be economically mined. The economist and the politician want to know whether national mineral supplies are adequate for the present, for the next five years, and for the next decade and, if supplies are inadequate, what politically possible and acceptable alternative sources of supply are available.

Studies undertaken in the U.S. during and immediately after World War II were among the first attempts to develop a realistic economic appraisal of national mineral resources (U.S. Bureau of Mines and U.S. Geological Survey, 1947; U.S. President's Materials Advisory Commission, 1952). Earlier Leith (1938) had attempted to relate mineral supplies to costs, and in 1956, in a paper prepared for a committee of the Society of Economic Geologists, Blondel and Lasky (1956) tried to further reconcile the economists' and the geologists' views of mineral supplies. One of the most important points that arose from this work (and subsequent work by Flawn, 1965, 1966; Pruitt, 1966; Skinner, 1969; Lovering, 1969; Hubbert, 1969; McKelvey, 1972, 1973; Zwartendyk, 1972; Brooks, 1973, 1975; and Govett and Govett, 1974, 1976, among others) is that reliable estimates of mineral supplies — whether regional, national, or world-wide — must take into consideration not only the geologic factor of the concentration of ore elements in a mineral deposit, but also the level and pattern of the demand for minerals, the effects of exploration on mineral supplies, and developments in extraction and processing methods as they affect prices and the grade of ore which can be economically exploited. It is becoming increasingly obvious that political factors must be added to the list (Govett and Govett, 1976).

In terms of traditional supply and demand analysis, the shape and position of the demand curve for any mineral will depend on at least some or all of the following factors: population, national income, consumption patterns, manufacturing product "mix", transportation patterns and facilities, the success of conservation measures, and international politics and finance. The supply side of the equation — which is most obviously affected by changes in the rate of exploration and consequent new discoveries, the development of extractive and processing methods, the practical problems of mining at greater depths and developing unconventional sources of supply, and technological advances in utilizing lower-grade ores — also depends on economic factors. If the political situation in a country is unstable, exploration may not be undertaken in otherwise geologically favourable areas; accessibility can be a serious limiting factor in developing a deposit; rising labour costs and labour unrest may change an economic mine into an uneconomic mine.

In the debate about national and international mineral supplies it is clearly necessary to recognize the essentially dynamic character of the minerals industry and to take into account economic as well as geologic factors. The importance of the distinction between estimates of mineral reserves (ore which can be mined economically today) and appraisals of mineral resources (which may become available in the future) is vital.

THE GEOLOGY AND ECONOMICS OF MINERAL SUPPLIES

Geological concepts

In the narrow geological sense, a mineral is a naturally occurring crystalline substance with a specific set of properties and a chemical composition which varies only within well-defined limits. Such a definition, as pointed out by Flawn (1966) in an amusing treatment, is relatively meaningless outside the academic world; minerals such as petroleum are not crystalline, and a number of other minerals, such as coal, do not exactly fit the definition (Behre and Arbiter, 1959). Flawn concluded that "Orderly people who like pigeonholes just big enough but not too big will have to make two compartments for 'mineral' — one might be labeled 'scientific definition' and the other 'economic and legal definition' " (p. 4). Nevertheless, there is a fair amount of agreement that a definition such as that given by Skinner (1969, p. 10) ". . . all nonliving, naturally occurring substances that are useful to man whether they are inorganic or organic" — is acceptable, and that minerals include all natural crystalline solids, fossil fuels, and the waters and gases of the earth and atmosphere.

A mineral deposit, whether it is metallic, non-metallic, or a fuel, is a natural concentration of minerals; in the geochemical sense it is a rare and abnormal event (an anomaly) and the area of the earth's surface where anomalies occur is very small. The natural aggregate of minerals in any rock contains some of the identified hundred-odd chemical elements which appear in the periodic table, but only nine elements account for 99% of the earth's crust (Skinner, 1969); of these, the only economically useful minerals which account for a sizeable portion are iron and aluminium. The content of 100 tonnes of continental crust includes 8.3 tonnes of aluminium and 4.8 tonnes of iron; other economic minerals in the 100 tonnes are present in much smaller amounts (12.0 kg of vanadium, 8.1 kg of zinc, 7.7 kg of chromium, 6.1 kg of nickel, 5.0 kg of copper, and 1.3 kg of lead; from data in Lee Tan and Yao Chi-Lung, 1970). A copper deposit, to be economically mineable today must contain approximately one hundred times the average concentration of copper in the continental crust (Govett and Govett, 1972).

Mineral deposits are fixed in quantity, since, for all practical purposes, no new mineral deposits are being formed today, with the possible exceptions of the ocean-floor manganese nodules and, hypothetically, the Red Sea-type brine deposits; hence the term "non-renewable resources" to describe mineral deposits. While it can be argued that the minerals contained in a deposit are not strictly "non-renewable" since they can be recovered in scrap form and reused, the deposits themselves are certainly "non-renewable". Over time geologists may learn more about *known* mineral deposits, and exploration efforts may locate presently *unknown* deposits; nevertheless, the supply of mineral deposits was essentially fixed in the distant geologic past, and in this sense the supply of minerals is geologically fixed.

Economic concepts

"Ores are rocks and minerals that can be recovered at a profit" (Park and MacDiarmid, 1964, p. 1). An orebody is generally defined as a mineral concentration (a mineral deposit) from which an element or a compound can be *economically* extracted at the present time.

Thus, the geologic concept of a mineral deposit as a natural concentration of minerals is complicated by price, grade, and technology (see Chapter 2). Deposits containing about 13% copper were copper ores in the year 1700; by the year 1900 deposits with between 5 and 2.5% copper were considered to be orebodies; today copper is commonly extracted from ores containing less than 0.5% of the metal. A cupriferous pyrite deposit in Cyprus, containing only 3 million tons of 0.24% copper, is currently being mined because there is a market for this particular ore in Spain and because the infrastructure necessary for the mining and transport of the ore was built earlier in the century (Govett and Govett, 1974). A similar deposit in Canada is not considered an orebody at the present time, nor are a number of high-grade iron ore deposits in the Canadian far north. Uranium and titanium deposits, as well as the ocean-floor manganese nodules, were not considered as ores at the beginning of this century; the deposits of petroleum in oil shales and tar sands of North America are only now reaching the point where they may be considered as "orebodies".

While the world's stock of mineral deposits is geologically fixed, the supply of orebodies is constantly changing. Blondel and Lasky (1956) emphasized that estimates of the productivity of a mineral deposit (or of a geologic region) depend on the limits accepted for the economic possibilities of exploitation. At any given period of time, the estimate of the productivity of a mineral deposit will increase when prices rise (or costs fall), when recovery techniques are improved, when lower-grade ores can be mined, or when ores at greater depths can be recovered. In any appraisal of world mineral supplies it must be emphasized that "resources" cannot be mined until they have been converted into "reserves" (Brobst and Pratt, 1973a).

MINERAL RESERVES AND RESOURCES

Mineral reserves

Reserves can be defined as known deposits from which minerals can be extracted profitably under present economic conditions and with existing technology. In spite of a general agreement on this definition, a mining company announces that "new reserves" of so many million tons could become available in the next few years, headlines announce that "known reserves" of uranium in the free world may not be adequate for the next decade at current prices, and geologists estimate that the sea and the ocean floors hold "vast" reserves of metallic minerals.

The distinction between the geologists' and the economists' concept of reserves is largely a function of the difference between the commercial, short-run point of view and the longer-run, public-policy point of view (Netschert and Lion, 1957). The mining engineer, concerned with management investment decisions on a single ore deposit, normally determines the tons, barrels, or pounds of ore positively available for a projected annual rate of production for 10 to 20 years (see Chapter 2); some companies may actually estimate ore in three categories of decreasing geologic certainty similar to those described by Leith (1938) and discussed at length by Blondel and Lasky (1955, 1956):

Proved or assured ore. Ore blocked out in three dimensions either by drilling or by actual underground mining operations; included are additional minor extensions beyond actual drill holes and openings where geological factors are definitely known and where the chance of the ore not reaching the estimated limits is remote and therefore would not be a factor in the actual planning of mine operations.

Probable or semiproved ore. Adjacent ore where it is likely that the existence of the ore will probably be proved, but where limiting conditions cannot be as precisely defined as in the case for proved ore. Semiproved ore may include ore that has been cut by scattered drill holes too widely spaced to guarantee continuity of the deposit.

Possible ore. Ore located near adjacent orebodies and in geologic structures where the relationship to proved orebodies is such that the presumption of its presence is warranted. Lack of exploration and development work, however, precludes certainty about the actual location or extent of the ore.

These estimates rest on geological appraisals, referred to by Blondel and Lasky (1955) as the "geological indeterminate". Except for proved ore, a margin of error in the estimates is inevitable, since some of the ore is hidden and therefore cannot be measured. Proving an ore deposit is a costly and time-consuming operation, and a company is only willing to undertake the mining tests (trenches, borings, pits, adits) if geological evidence is favourable and economic conditions for mining the grade ore expected are advantageous. Only part of the deposit may be proved since mineral deposits are seldom neat, well-defined bodies; often the entire deposit cannot be developed economically, and a company will choose more or less arbitrarily the boundaries to use in making measurements.

Reserve figures may fail to take into account potential losses in ore as a result of mining operations. For some minerals the recovery ratio is low; for example, for petroleum it is only about one-third, while for other minerals (such as iron ore and copper) the ratio may be as high as 90% (Brooks, 1975). The recovery ratio is a technologic factor which can change with time and can vary from mine to mine. To allow for these factors, reserves are sometimes reported as "recoverable", reflecting the amount of material which will be lost in extraction; even when reserves are not quoted as "recoverable", the estimates may actually have been adjusted to allow for this factor.

Since ore-reserve figures are designed primarily for use within a mining company, there is no certain relationship between a company's published figures and the actual potential of a property (McDivitt, 1957). Companies are often unwilling to publish reserve figures. In those provinces, states, or countries where proved reserves are taxed at a higher rate than other reserves, estimates are unavailable.

Mining company reserves remain constant over long time periods since most companies block out reserves for a specific period relative to anticipated annual production levels. Traditionally companies have blocked out 20 tons of ore for each ton produced annually (in metallic deposits), although recently, as interest rates have risen, only 10 to 15 tons are blocked out for each ton produced annually (see Chapter 2).

Outside of the petroleum industry, the proved-possible-probable terminology has been largely superceded by another classification system (originally proposed by the U.S. Geological Survey and the U.S. Bureau of Mines); this system was recommended by Blondel and Lasky (1955, 1956) for regional and national reserve estimates and is the most widely used today. The categories are:

Measured reserves. Tonnages are compiled from drill holes, trenches, workings, and outcrops; grade is computed from detailed sampling. The mineral content, size, and shape of the deposits are well established within defined limits and the tonnage and grade estimates should not vary by more than 20% from actual figures. In practice, this margin of error has tended to become the rule.

Indicated reserves. Tonnage and grade are computed partly from specific measurements and partly from projections made for reasonable distances; the size and shape of the deposits are not fully outlined nor is the grade established precisely.

Inferred reserves. Tonnages are estimated on the basis of broad geological knowledge and may include concealed deposits where there is enough geological evidence to warrant their inclusion.

The margin of error for indicated and inferred reserves is necessarily larger than in the case of measured reserves; Blondel and Lasky (1955) estimate that if both indicated and inferred reserves are included in a reserve estimate the margin of error is commonly of the order of 50%. This does not mean that the estimate is suspect, but merely that the user should be conscious of the limitations.

Many published reserve estimates combine two or all three of the reserve categories. Blondel and Lasky (1956) themselves suggested that "measured" and "indicated" reserves could be combined into a single figure called "demonstrated reserves", although this practice ignores the differentiation between geologically certain and less certain measurements. In the voluminous *Mineral Facts and Problems* (U.S. Bureau of Mines, 1970), the term "apparent reserves" was used to describe proved reserves (in the case of petroleum), measured plus indicated reserves (the platinum group of metals), and the sum of measured, indicated, and inferred reserves (nickel and aluminium). Other commonly used terms are "marginal", "submarginal", "latent", "ultimate", "identified", and "potential". Blondel and Lasky (1956, p. 692) defined "potential ore" as "... mineral masses which are currently non-exploitable, yet which may be exploitable if conditions become only slightly better." Although the authors pointed out that there might be difficulties in the usage of this term, and a dictionary compiled by the U.S. Bureau of Mines (Thrush, 1968) states that the term is considered obsolete, it has nevertheless entered the literature. Flawn (1966) preferred the term "potential reserves" which is now widely used. The writers share the opinion of Zwartendyk (1972, p. 4) that ". . . a clump of rock may be called a 'potential iron ore', which in turn may be called a 'potential bicycle' " and argue that the adjective "potential" as a qualifier for reserves be used sparingly if at all (Govett and Govett, 1974).

While some qualifying adjectives may be useful in the conceptual sense — a case can be made that the concept of "developed" reserves (Forrester, 1946) to describe that portion of reserves which can be immediately exploited by present mining methods would be useful since it would distinguish that part of reserves for which no additional investment was needed for exploration and which could therefore be mined as long as price was sufficient to cover direct production costs (Brooks, 1975) — the use of such adjectives makes comparisons between different sets of reserve measures, particularly regional and national comparisons, difficult. Uncertainty as to exactly what is being included when reserves are aggregated also makes comparisons over time uncertain.

Recently the U.S. Bureau of Mines and the U.S. Geological Survey have agreed on a new set of definitions to be used in future work (see McKelvey, 1974). Reserves are divided into measured, indicated, and inferred; "demonstrated" is used as a collective term to designate the sum of measured and indicated. Resources are divided into identified resources (including subeconomic, paramarginal, and submarginal) and undiscovered resources (hypothetical and speculative). It is pointed out that this set of classifications is not meant to replace the proved-possible-probable categories used by industry, but is meant to provide a framework for the appraisal of unexplored but known deposits and those which are assumed to exist but are not yet known.

The problems of an agreed terminology are not nearly as severe as the problems of deciding the grade and tonnage of a deposit which should be counted as a reserve and providing a means of revising data over time. Several million tons of 2% copper in the porphyry deposits of Chile and the southwestern U.S. were known in the last century, but since there was no demand for this grade copper in 1900, the deposits were not then classed as reserves (Lowell, 1970). Some of the iron ore deposits of India which were classed as reserves in 1955 are no longer considered to be economically mineable and are excluded from the United Nations (1970) survey of world iron ore reserves.

Any change in mineable grade will immediately change reserves. Since the relationship between grade and reserves is not a simple arithmetic relation, a reduction in mineable grade by one-half may actually result in an increase of reserves by a factor much greater than two (Lasky, 1950; Lovering, 1969; see Chapter 2). The development of techniques which allow economic extraction of ores from greater depths will also increase reserves. Expansion of transport facilities can transform uneconomic deposits in previously remote areas (such as the petroleum deposits in the Arctic) into new reserves. Even political changes, which make exploration and exploitation more attractive in hitherto undeveloped areas, will increase reserves. The effect of price changes on reserves is illustrated by estimates of uranium reserves at different prices: at $8.00 a pound, reserves of U_3O_8 in the U.S. are estimated at 173,000 tonnes; in the price range $8.00—$10.00 an additional 40,500 tonnes are recoverable; at higher prices the amounts rise dramatically, and at a price near $70.00 a pound the 2.4 million tonnes in shales could be recovered (U.S. National Commission on Materials Policy, 1973; see also Chapter 2).

On the other hand, reserves may decrease if rising costs make it uneconomic to mine a given deposit or if prices fall sufficiently to deter investment in development work. If there is no exploration or development work, production from a deposit will inevitably reduce the reserves in the deposit to zero, even if ore still remains in the ground. Substitutes for a mineral may effectively reduce demand; environmental protection measures may make it economically unattractive to develop or exploit a deposit and, in some cases, may make it impossible to continue mining operations.

These factors are largely ignored when the adequacy of reserves of a mineral is evaluated by calculating a "life-index" (also called a reserve/production ratio) of a mineral. A life-index is derived by dividing the *current* reserves of a mineral by the total production of the mineral in a given year (or dividing by a projected growth rate of future annual production). While this simple statistic may be useful for an individual mine, in that it provides a mining company with a measure of the life expectancy of the mine's currently measured reserves, it will not predict how long the mine may be worked, since measured reserves are usually much less than the ultimate production of the mine (see discussion above). The life-index clearly should not be used in the aggregate, even to measure the life span of *current* reserves in a region or country, since an aggregated life-index glosses over large differences in size and grade of individual deposits and the consequent differences in the size and scope of individual mining operations. Most importantly, the life-index fails to take into consideration the flexibility — or inflexibility — of the production capability of a mine in reacting to changing demand situations. Zwartendyk (1974, p. 70) calls the life-index a "statistical mirage", and concludes that ". . . in the final analysis it is not the size of present reserves that counts, but our ability to meet future production requirements" (see also Risser, 1973).

Mineral resources

The U.S. Geological Survey appraisal of mineral supplies (Brobst and Pratt, 1973b) likened reserves to funds in a bank account and other liquid assets, while resources are analogous to "birds in the bush"; resources cannot be mined until they have been converted into reserves. Blondel and Lasky (1956) defined resources as the sum of reserves plus potential ores. Govett and Govett (1974) defined resources as the total amount of an element — both known and unknown — down to some defined grade which is higher than the crustal abundance of the element in the earth's crust but lower than the present economic grade. McKelvey (1972) divided resources into known resources (reserves, paramarginal resources, and submarginal resources) and unknown resources. Brooks (1975) differentiates between low-grade resources in deposits similar to those now being mined and non-conventional resources, which include materials not hitherto exploited (e.g., manganese-bearing slags and ocean-floor manganese nodules).

Whatever definition is adopted, resources must include all reserves, plus all known deposits which are not currently economically or technically recoverable, plus all undiscovered deposits. While in senso stricto it might be argued that there were no uranium reserves one hundred years ago since uranium was not a known, economically useful mineral, clearly uranium has always been a resource; only *known* resources of uranium have increased in the past one hundred years. Germanium was identified in 1886; while the chemistry of its recovery was known shortly thereafter, there was no demand for the metal until the invention of the transistor in the 1940s. Aluminium, beryllium, cadmium, columbium, magnesium, tantalum, titanium, tungsten, uranium, vanadium, and zirconium have all come into use in industry only in this century (Schroeder and Mote, 1959).

While the degree of exploration success and technical factors determine the size and geographic distribution of *known* resources, the only hypothetical upper limit which may be placed on total resources of an element is that represented by the abundance of elements in the earth's crust (see the work of Turekian and Wedepohl, 1961; Vinogradov, 1962; Taylor, 1964; Horn and Adams, 1966; and Lee Tan and Yao Chi-lung, 1970 on crustal abundance; see also Chapter 3). Flawn (1966) uses the concept of crustal abundance in his discussion of a "resource base" of a mineral; Brooks (1975) suggests that a depth of 3 miles (4.8 km) below the land surface would be an appropriate limit, although an arbitrary one, for the limit of calculations of a resource based on *average* crustal abundance to determine an "ultimate" cut-off grade for the resource base of a mineral. Brooks (1975) also points out that the concept could be further refined by estimating the distribution of elements by concentration to determine the quantity of metal available in "high-grade" rocks and how much is contained in successively lower concentrations.

The concept of a "resource base" entirely ignores the role of technology and economics in determining mineral supplies, but if an economic technology was available to process whole rock for its contained elements (cf. Brown, 1954) there would obviously be a virtually limitless supply of any mineral. However, this is not a realistic basis for calculating the limit of resources, even taking into account possible dramatic improvements in mineral extraction technology and unlimited sources of cheap energy; the actual *reasonable* limit must be at a level at which it will be conceivably possible to process earth materials, and this will be well above crustal abundance.

Resources and crustal abundance

Nevertheless, some order of magnitude estimates of total world resources may be derived from examining the relation between reserves and crustal abundance. McKelvey (1960) first showed that there is a fairly close linear relation between reserves in the U.S. and the crustal abundance of elements in the earth's crust; Sekine (1963) showed that the same relation held in Japan; not unexpectedly, it is also generally true on a world-wide basis (Govett and Govett, 1972).

The relation between world reserves and abundance of elements in the continental crust is shown for sixteen common metals in Fig. 1-1. Reserves for silver, mercury, antimony, tungsten, molybdenum, tin, lead, nickel, zinc,

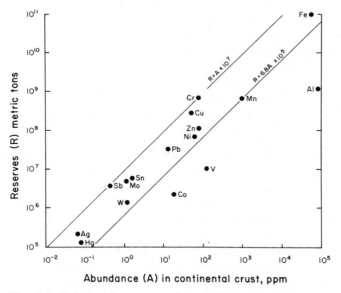

Fig. 1-1. Relation between world reserves and abundance of elements in the continental crust for sixteen common elements.

copper, chromium, manganese, and iron may all be expressed by the relation:

$$R = 6.8A \times 10^5 \text{ to } A \times 10^7$$

where R = reserves in metric tons and A = abundance in the continental crust in parts per million — a remarkably narrow spread of only just more than one order of magnitude. Of the metals included in Fig. 1-1, only cobalt, vanadium, and aluminium fall outside the defined belt; much of the high abundance of aluminium relative to its reserves may be explained by the large amount of aluminium in clay minerals which are not presently regarded as reserves; cobalt and vanadium are largely by-product metals of other ores.

Erickson (1973) points out that in the United States the metals which have been longest sought (lead, molybdenum, copper, silver, and zinc) are close to the upper limit of the relation:

$$R = 2.45A \times 10^6$$

which expresses the reserve relation in the U.S. For the world reserve distribution, it is interesting that chromium and antimony are at the upper limit, while lead, zinc, and silver are mid-way in the defined belt. The closeness of the correlation suggests that undiscovered reserves (i.e., resources) for the metals considered should, in general, bring total resources to at least:

$$R = A \times 10^7$$

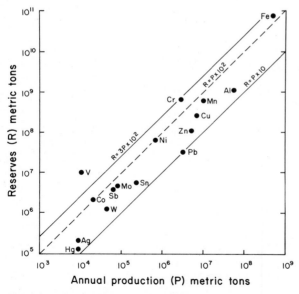

Fig. 1-2. Relation between world reserves and world annual production for sixteen common metals.

The correlation between reserves and abundance is, of course, a function of metal use and, as shown in Fig. 1-2, there is a close relation between reserves and annual production, P:

$$R = P \times 10 \text{ to } 3P \times 10^2$$

This correlation has about the same spread as the reserve-abundance relation; the upper limit of the relation can be reduced to $R = P \times 10^2$ if chromium is allowed to fall outside the belt. Only vanadium falls far outside the defined belt and is considerably underproduced, possibly contributing to its low reserve figure.

A third relation (Fig. 1-3) can be shown relating production to abundance:

$$P = A \times 10^4 \text{ to } 2.8A \times 10^5$$

From this relation it may be deduced that cobalt, vanadium, and aluminium are underproduced in relation to their abundance, while lead, copper, tin, antimony, silver, and mercury are being produced at the upper limit of their abundance relation.

These correlation diagrams seem to show that industrial society, on the whole, has found uses for metals approximately proportional to their natural abundance, and that those metals which are apparently under-used (cobalt, vanadium, and aluminium, Fig. 1-3) also have anomalously low reserves (Fig. 1-1) which may perhaps increase with added exploration. Since mining

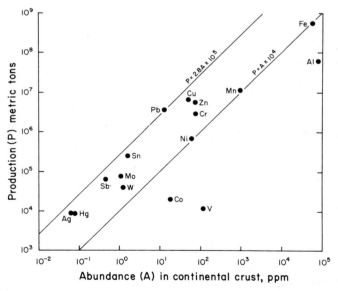

Fig. 1-3. Relation between world annual production and abundance of elements in the continental crust for sixteen common metals.

companies commonly block out reserves for some time period based on current production rates (see above), it is not surprising that there is a close relation between reserves and production.

None of these correlations can indicate the upper limit of resources; however, if it may be assumed that the general slope of the reserve-abundance correlation will be maintained, the current upper limit of the correlation factor indicates the minimum resources which may be expected to be found. It suggests that exploration should reveal considerable additional supplies of those metals which fall in the lower range of the relation:

$$R = 6.8A \times 10^5 \text{ to } A \times 10^7$$

A corrolary is that if the upper limit of the relation is raised by new discoveries of one or more metals, it should give impetus for renewed exploration for other metals.

Conceptual models of mineral resources

A number of attempts have been made recently to illustrate the relation between reserves and resources and to define the various degrees of economic and geologic certainty in resource estimates. McKelvey (1972) divided resources into two main categories on the basis of their "feasibility of recovery": (1) identified resources include proved, probable, and possible reserves, plus a category of "recoverable" reserves which would be mineable if economic and technical conditions were only slightly more favourable; (2) undiscovered resources are divided into "paramarginal" (those recoverable at prices one and one-half times current prices) and "submarginal" resources (those of a still lower grade). The classification system adopted by the U.S. Geological Survey (Brobst and Pratt, 1973b) places greater emphasis on the geologic evaluation of resources and the definition of varying degrees of geologic certainty; this reflects their emphasis on the concept that there is no economic availability if there is no geologic availability. In defining the various degrees of geologic certainty, resources are divided into two main categories:

Identified resources: specific mineral deposits whose existence and location are known. If the deposits have been evaluated as to extent and grade they are considered to be reserves; otherwise, they are considered as "conditional resources", i.e., resources that may become reserves under changed economic and technical conditions.

Undiscovered resources: divided into hypothetical and speculative resources. *Hypothetical* resources are those undiscovered mineral deposits which may reasonably be expected to be found in known districts, while *speculative* resources include conventional deposits in areas where there have not yet been discoveries, and unconventional resources which have either been only recently recognized or which have yet to be recognized. Geologic estimates of hypothetical resources would include concealed copper porphyry

26

deposits in known copper districts in the southwestern U.S. and Asia; speculative copper deposits would be estimated on the basis of recent discoveries (such as those in eastern North America) that allow "speculation" that new regions nearby or in a similar environment may have a potential for discovery (see next section).

Zwartendyk (1972) devised a conceptual model for the Canadian government, dividing resources into "reasonably assured" and "additional" resources, on the basis of both the degree of geologic certainty and economic recovery. Brooks (1975) designed a matrix, using McKelvey's classifications, to illustrate the degree of geologic uncertainty and the cost-price factor. In an attempt to emphasize the dynamic character of reserve and resource estimates Govett and Govett (1974) presented a triangular model (Fig. 1-4) in which changes in grade, changes in technology, and exploration success rates are shown on the three axes. The entire area of triangle *ABC* represents 100% of total resources of a mineral on a national or global scale; known resources are represented by the smaller triangle *ADC* (a function of exploration success); the division between non-exploitable deposits and reserves is dictated *at any given time* by the position of line *EF* (a function of grade

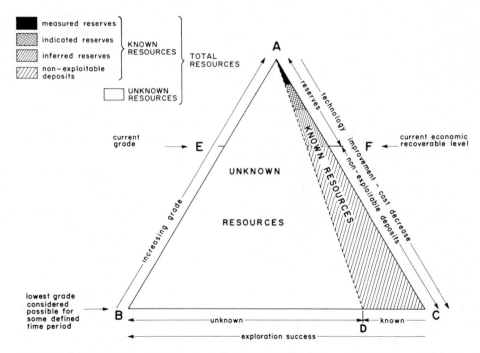

Fig. 1-4. Conceptual model showing relation between reserves and resources as a function of grade, technology, cost, and exploration (after Govett and Govett, 1974; redrawn with permission, *Resources Policy*, Vol. 1, p. 53).

and technology). The model could be further subdivided into the various categories of resources used by McKelvey (see Fig. 2-1) and others, but the authors have argued in favour of a simple terminology (known and unknown resources) on the grounds that other distinctions — hypothetical, paramarginal, conditional — change constantly as economic and technical factors change.

In any conceptual model reserves are necessarily relatively fixed in the short-run by consumption and technology; in the longer run, reserves are determined by the rate at which resources are converted into reserves. In the final analysis, both reserves and resources are changing concepts closely associated with the dynamics of economic change.

MEASUREMENT OF RESERVES AND RESOURCES

Recent reserve and resource estimates

Table 1-I, compiled for illustrative purposes, gives some recent estimates of world reserves, identified resources, and the U.S. Geological Survey (Brobst and Pratt, 1973b) estimates of hypothetical and speculative resources for a number of industrially important minerals. The reserve data (compiled from a number of sources) includes measured, indicated, and inferred reserves; identified resources are known mineral deposits which may or may not have been evaluated as to extent and grade, and include minerals which may not be economically recoverable at present; hypothetical and speculative resources were defined in the last section.

Estimates of resources for the fuel minerals uranium and petroleum are notably absent in Table 1-I. Data for uranium are simply not available since, in most countries, uranium is classed as a strategic material. McCulloh (1973), in discussing petroleum, was unwilling to give an estimate of resources for the U.S. Geological Survey resource appraisal since the range of published data is so enormous that any single figure could be criticized as too high or too low (see below).

Govett and Govett (1976), using similar data, noted that the ratio between identified resources and reserves was remarkably low. In Table 1-I reserves are included in the estimates of "identified resources", so that the ratios are somewhat misleading, and are probably lower than indicated; even with these data, the ratio is greater than five for only lead, zinc, and petroleum.

However, a more extraordinary feature of the data in Table 1-I is that, where estimates were available, the ratio between identified resources and hypothetical resources is not nearly as great as might have been expected; for most of the minerals, except zinc, the ratio of hypothetical resources to identified resources is less than or not much greater than one. In the case of copper and tin, where a ratio of speculative to hypothetical resources can be

TABLE 1-I

Estimates of world reserves, identified resources, hypothetical resources, and speculative resources of some industrially important minerals (million metric tons unless otherwise specified)

Mineral	Reserves	Identified resources	Hypothetical resources	Speculative resources	Ratios identified resources/ reserves	Ratios hypothetical resources/ identified resources
Bauxite	5,900	12,000–15,000	>9,600		2.0–2.5	>0.8–0.6*
Copper	273	312	363	290	1.1	1.2
Chromium	1,785	4,390	>3,650		2.5	>0.8*
Iron ore	251,000	779,000	enormous		3.1	
Lead	141	~1,500	n.e.	n.e.	10.6	
Manganese	~6,500	~7,700	~10,000	n.e.	1.2	1.3
Mercury (thousand 76-1b flasks)	7,000	17,080	n.e.	n.e.	2.4	
Petroleum (billion barrels)	408	2,385	n.e.	n.e.	5.8	
Phosphate	n.a.	6×10^9	9×10^9	n.e.		
Silver (million troy oz.)	6,800	20,400	n.e.	n.e.	3.0	1.5
Tin (thousand)	10,100	10,400	9,400	7,630	1.0	0.9
Titanium	n.a.	1,785	1,450	very large		0.8
Uranium	944	1,455	n.e.	n.e.	1.5	
Vanadium (thousand)	4,200	18,100	large	large	4.3	
Zinc	235	1,510	3,575		6.4	2.4

* Ratio hypothetical + speculative resources/identified resources.

n.e. = not estimated; n.a. = not available.

Source. Reserves: bauxite and petroleum, U.S. Geological Survey (1968); copper, U.S. Bureau of Mines (1970); iron ore, United Nations (1970); mercury, U.S. Geological Survey (1968); silver, U.S. Bureau of Mines (1970); vanadium, U.S. Geological Survey (1968); uranium, U.S. Bureau of Mines (1973). Other data based on Brobst and Pratt (1973b).

calculated, the ratio is less than one. McKelvey (1973) pointed out that the distinction between speculative and hypothetical resources has forced the compilers of resource data to realize that speculative resources for some minerals are, in fact, relatively low, and that, from a world-wide geologic perspective, regions that have a significant potential for discovery of certain minerals are already largely known; thus a significant amount of undiscovered resources necessarily come under the heading of "hypothetical" rather than "speculative" resources. He concluded that "For some commodities this may simply reflect insufficient knowledge of regional geology to identify other promising areas, but for others it is an expression of confidence that the favourable terranes have all been identified" (pp. 72—73).

It is expected that many of the estimates in Table 1-I will change significantly within this decade as new information becomes available and better methods for making resources estimates are developed. Reserve estimates, since they are most directly affected by economic factors, will change more rapidly and frequently than estimates of undiscovered deposits; however, revision of hypothetical and speculative resource estimates should be important in focusing research and directing exploration efforts.

Problems in reserve estimation

Estimates of reserves depend directly on the past work of mining companies in discovering, measuring, and exploiting orebodies. Techniques for estimating reserves in individual deposits are quite well developed; those for metallic minerals are more or less geometric, and those for fluid hydrocarbons depend on characteristics of fluid or pore space (American Gas Association et al., 1972). There is a wide range of accuracy in the estimates; while proved reserve figures should, theoretically, be accurate within 20% (as noted above), the margin of error rises to plus or minus 50% when indicated and inferred reserves are added to proved reserves. McKelvey (1972) estimated that the error in proved reserve figures is of the order of 25%, but, to illustrate how wide the margin of error can be in practice, he cited the case of a mining geologist who arbitrarily tripled his calculations of the amount of ore in a district; twice the amount of this tripled estimate was actually mined in the next twenty years.

Most regional and national reserve estimates are subject to a very wide degree of error. As Zimmerman (1951, p. 443) pointed out: "A corporation knows reasonably well what its ore reserves are . . . National reserves . . . are largely a matter of conjecture." Calculations such as those currently being made by the Canadian Department of Energy, Mines and Resources, which estimate ultimate reserves for each mine in Canada by combining current proved reserves with estimates of likely future extensions (Zwartendyk, 1974), could improve national reserve estimates, as could the development of new techniques for estimating ore tonnages (e.g., Sinclair, 1974), although the problem of defining an economic grade for reserves is difficult (see Chapter 2).

The reliability of national reserve estimates for a given mineral can be reasonably well assessed since most compilers of reserve data are careful to indicate the degree of certainty inherent in their estimates. On the other hand, estimates of world reserves, in part because of simple unavailability of data, must be used with caution.

All reserve estimates, whether regional, national, or international, must be carefully defined as to what is being measured and should be constantly revised. New information can radically change reserve figures; for example, a recent revision of uranium reserve estimates for the free world, based on re-assessments of U.S., Australian, and southwest African deposits, resulted in an increase of nearly 50% in reserves in the five years between 1968 and 1973 (U.S. Bureau of Mines, 1973); the revised estimates are still incomplete since most countries do not release data on uranium deposits. Inclusion of the large lateritic nickel deposits in tropical and semi-tropical areas of the world has made a significant difference in world nickel reserves in the last decade; reserves in New Caledonia and Cuba each doubled between 1965 and 1970, and reserves in the Philippines, Indonesia, the Dominican Republic, and Guatemala in 1970 were twice the amount reported for the non-communist world (excluding Canada) in 1965 (Govett and Govett, 1974). Between 1947 and 1973 world iron ore reserves increased thirteen-fold, chromium reserves increased seven-fold, bauxite and manganese reserves increased approximately four-fold, and reserves of copper, silver, lead, and zinc increased from two to three times (Govett and Govett, 1974). In some of these cases the revisions were the result of new information rather than re-evaluations; the thirteen-fold increase in world iron ore reserves was, in part, a result of adding reserves in the U.S.S.R. which were excluded from earlier estimates (the U.S.S.R. currently accounts for 44% of world iron ore reserves; United Nations, 1970).

Techniques for resource estimates

While it is difficult to estimate reserves with accuracy, errors in estimating incompletely explored deposits (let alone in predicting the content of deposits in unexplored areas and of unconventional mineral deposits) are several orders of magnitude greater. The accuracy of the estimates will be determined by the amount of geological knowledge available and the methods by which this knowledge can be applied. Since appropriate geological information is generally woefully lacking and the level of sophistication of geological deduction is inadequate, resource appraisals must be largely conjectural and, as pointed out in Chapter 2, useless as working figures. However, to the degree that resource estimates focus attention on the inadequacy of geological information and attempt to define geographic zones of interest, they are useful exercises.

Most resource estimates are derived by analyzing unexplored areas on the basis of analogy to geologically similar areas which have been explored. Extrapolation has been most widely used in the petroleum industry where the estimates vary widely depending on which region was used as the basis for the extrapolation and which statistical techniques were applied to the data. The most important improvement in petroleum extrapolation would be the development of a sound measure of the population distribution of the productivity of hydrocarbon per unit of rock volume based on a large number of explored basins (McCulloh, 1973). Theobald et al. (1972) found that estimates made between 1968 and 1972 of the total petroleum liquids in the U.S. ranged from a low of 660—670 billion barrels (Moore and Hubbert) to a high of 3,033 billion barrels (U.S. Geological Survey). McCulloh (1973) plotted the more authoritative estimates of petroleum resources as a function of their date of publication, together with a curve of production plus cumulative discoveries. The different projections show a great deal of scatter (as would be expected), but they tend to increase more or less with the production and discovery curve.

The extrapolation method is also used in estimates of metallic mineral resources. Lowell (1970) derived estimates of copper resources by examining the relation between reserves found by drilling and the areal extent of the favourable geological environment for copper. Using this approach he estimated the number of copper porphyry deposits awaiting discovery in a number of favourable regions. In Chile and Peru he found that about one-half of the geologically favourable belt has been explored and twelve deposits found; on this basis he assumes that another twelve deposits await discovery in the unprospected part of the belt. Similarly, since thirteen deposits have been found in the 10—15% of the geologically favourable region that has been explored in British Columbia (Canada), he deduces that an additional one hundred deposits can be expected. Olson and Overstreet (1964) and Nolan (1950), among others, have used a similar geographic analogy approach to extrapolate areal and size distributions of known metallic mineral deposits to adjacent unexplored and possibly concealed deposits, while McAllister (Chapter 2) discusses the unreliability of this approach.

Most extrapolation methods have the inherent drawback that they are based on reserve data, since extrapolations are made from information on known deposits. Also, many of the geographic analogies are restricted to relatively well-explored areas. Most of the estimates of tonnages assume that mining operations will continue at approximately the current mining depth, although exploration techniques for more deeply buried deposits are being developed and mines are being exploited at increasing depths. It has been calculated (Grossling, 1970) that present reserves of many minerals could be increased by as much as 50—100 times if more deeply buried deposits were mined, although a lower limit has been suggested by evidence that most hydrothermal-type deposits are restricted to the 2 or 3 km of the earth's upper crust (Skinner and Barton, 1973). Furthermore, the assumed relationship between grade and tonnage noted above — that there is an inverse nonlinear relation between grade and tonnage, so that if the economic grade is lowered by some factor the tonnage increases by a much larger factor — is now being questioned (initially for copper deposits; see Singer et al., 1975).

Geochemical studies of grade-frequency distributions of elements have been made (e.g., DeGeoffroy and Wu, 1970); in the past few years geologists have also developed techniques to use the relation of tectonism and volcanism to moving crustal plates to locate favourable geological areas for exploration (Brobst and Pratt, 1973a). The tendency for metal deposits to be "grouped" in geographic belts (metallogenic provinces) has led to considerable interest in compiling maps of these provinces (e.g., Mayakyan, 1960), studying their geologic history, and determining the ore-forming factors (Bilibin, 1968). A great deal has been published recently in North America, the U.S.S.R., and Japan on specific mineral belts and on metals in various regions, often in the form of maps (e.g., the mapping of copper and molybdenum porphyry deposit-type mines; Argall and Wylie, 1973, see also Chapter 3).

Techniques for resource appraisals of nonconventional deposits and minerals are considerably less developed than those for conventional sources. Little more than a decade ago it was difficult to find estimates of the potentials of the seas and the ocean floors, although recently there has been a great deal published on the subject (see Chapter 8). The U.S. Geological Survey resource appraisals (Brobst and Pratt, 1973b) included the resources in the ocean-floor manganese nodules and in the metal-bearing muds of the Red Sea as "identified resources" on the ground that they are now known both as to existence and location, even though they have not been fully evaluated. While the estimates are admittedly arbitrary, they are supported by recent evidence that U.S. mining companies expect to exploit the resources in the ocean-floor nodules by the end of this decade (Faltermayer, 1972).

CONCLUSIONS

In any resource appraisal it is essential to make some assumptions about economic factors when considering geological evidence. The current energy "crisis" is based, in part, on mistaken predictions about energy prices and supplies in the last decade, especially in the U.S. Fears that uranium supplies may be inadequate before the end of this decade are intensifed by a recognition of the long lead-time from the decision to undertake exploration to actual production. While world uranium resources are very large, and the development of a breeder reactor would significantly reduce the amount of uranium needed for energy generation, in the next decade the world may well face a uranium *"reserve crisis"*. Shortages of such critical minerals emphasize that, while the geologic factor may be the most critical determinant of mineral supplies in the long run, the economic "indeterminant" cannot be ignored in the present.

There is a serious danger implicit in ignoring the distinction between reserves and resources; the unbridled optimism of some observers about the future adequacy of world (or national) mineral supplies is often based on a failure to recognize that while mineral resources are vast, mineral reserves at any given time are limited. The decision to exploit an ore deposit will depend on such factors as the grade of the ore, the tonnage of ore contained in the deposit, the percent recovery expected, the market price for the ore, the tons that can be mined per day, and the operating and capital costs per ton of ore.

In the current energy debate arguments that new oil resources will be exploited in the next few decades fail to take into account the enormous investment necessary to exploit these new sources and to convert the resources into reserves. There is no great technical problem in extracting oil from shales or from the Canadian tar sands, but to meet a U.S. projected target of 1 million tons of oil from shales per day by 1985 (which would produce only 10% of the projected U.S. oil demand in 1985), 2 million tons of rock would have to be mined, moved, and dumped *every* day — such an enterprise would be equivalent to the entire U.S. mining industry's operations at present (*Observer*, 1973); the potential environmental impact of such an operation is frightening to contemplate.

On the other hand, the pessimism which arises from figures that state that there are only x number of years supply of a mineral remaining (based on dividing current reserves by current annual production) should also be tempered by the realization that new resources are constantly being discovered and converted into reserves; even if few new reserves are added to the world's stock, there would not be an abrupt halt to the mining industry, but rather a slow running down of production of certain minerals and an increase in their price. Substitutes would be sought, reclamation and recycling would be increased, conservation measures would be introduced, and research and development would be increased to develop new methods of exploration and exploitation.

Discovery of new mineral deposits will be largely a function of increased geological knowledge and the amount invested in exploration; exploitation of new deposits will be determined by economic and political factors and the development of new extraction and processing techniques. Future mineral supplies depend, to a large degree, on the research and exploration which is currently under way; whether there should be optimism or pessimism about future mineral supplies will depend on the support which these activities receive.

34

ACKNOWLEDGEMENTS

We would like to thank Dr. D.B. Brooks for making available a pre-print of his chapter "Mineral Supply as a Stock" to be published in the forthcoming AIME volume on mineral economics.

REFERENCES CITED

American Gas Association, American Petroleum Institute, and Canadian Petroleum Association, 1972. *Reserves of Crude Oil, Natural Gas Liquids and Natural Gas in the United States and Canada and United States Productive Capacity as of December 31, 1971.* Washington, D.C., 248 pp.
Argall, G.O., Jr. and Wylie, R.J.M., 1973. Open pit copper mining. *World Min.*, 26(8): 34—37.
Behre, C.H., Jr. and Arbiter, N., 1959. Distinctive features of the minerals industries. In: E.H. Robie (Editor), *Economics of the Minerals Industries*. AIME, New York, N.Y., pp. 43—79.
Bilibin, Yu.A., 1968. *Metallogenic Provinces and Metallogenic Epochs.* Queens College, Flushing, N.Y., 35 pp.
Blondel, F. and Lasky, S.G., 1955. Concepts of mineral reserves and resources. In: United Nations, *Survey of World Iron Ore Resources.* New York, N.Y., pp. 169—174. Reprinted in: United Nations, *Survey of World Iron Ore Resources, 1970.* pp. 53—61.
Blondel, F. and Lasky, S.G., 1956. Mineral reserves and mineral resources. *Econ. Geol.*, 51: 686—697.
Brobst, D.A. and Pratt, W.P., 1973a, Introduction. In: D.A. Brobst and W.P. Pratt (Editors), *United States Mineral Resources. U.S. Geol. Surv., Prof. Paper* 820, pp. 1—8.
Brobst, D.A. and Pratt, W.P. (Editors), 1973b. *United States Mineral Resources. U.S. Geol. Surv., Prof. Paper* 820, 722 pp.
Brooks, D.B., 1973. Minerals: an expanding or a dwindling resource? *Can. Dep. Energy, Mines Resour., Miner. Bull.*, MR 134, 17 pp.
Brooks, D.B., 1975. Mineral supply as a stock. In: W.A. Vogeley (Editor), *The Economics of the Mineral Industries.* AIME, New York, N.Y., in press.
Brown, H., 1954. *The Challenge of Man's Future.* Viking, New York, N.Y., 290 pp.
DeGeoffroy, J. and Wu, S.M., 1970. A statistical study of ore occurrences in the greenstone belts of the Canadian Shield. *Econ. Geol.*, 65: 496—504.
Erickson, R.L., 1973. Crustal abundance of elements, and mineral reserves and resources. In: D.A. Brobst and W.P. Pratt (Editors), *United States Mineral Resources. U.S. Geol. Surv., Prof. Paper* 820, pp. 21—26.
Faltermayer, E., 1972. Metals: the warning signals are up. *Fortune*, October, beginning p. 109.
Flawn, P.T., 1965. Minerals: a final harvest or an endless crop? *Eng. Min. J.*, 166: 106—107.
Flawn, P.T., 1966. *Mineral Resources.* Rand McNally, Chicago, Ill., 406 pp.
Forrester, J.D., 1946. *Principles of Field and Mining Geology.* John Wiley and Sons, New York, N.Y., 647 pp.
Govett, G.J.S. and Govett, M.H., 1972. Mineral resource supplies and the limits of economic growth. *Earth-Sci. Rev.*, 8: 275—290.
Govett, G.J.S. and Govett, M.H., 1974. The concept and measurement of mineral reserves and resources. *Resour. Policy*, 1: 46—55.

Govett, M.H. and Govett, G.J.S., 1976. The problems of energy and mineral resources. In: F.R. Siegel (Editor), *Review of Research on Modern Problems in Geochemistry.* UNESCO, Paris, in press.

Grossling, B.F., 1970. Future mineral supply. *Econ. Geol.*, 65: 348—354.

Horn, M.K. and Adams, J.A.S., 1966. Computer-derived geochemical balances and element abundances. *Geochim. Cosmochim. Acta*, 30: 279—297.

Hubbert, M.K., 1969. Energy resources. In: National Academy of Sciences and National Research Council, *Resources and Man.* W.H. Freeman, San Francisco, Calif., pp. 157—242.

Lasky, S.G., 1950. How tonnage-grade relations help predict ore reserves. *Eng. Min. J.*, 151: 81—85.

Lee Tan and Yao Chi-Lung, 1970. Abundance of chemical elements in the earth's crust and its major tectonic units. *Int. Geol. Rev.*, 12: 778—786.

Leith, C.K., 1938. Mineral valuations of the future. In: C.K. Leith (Chairman), Elements of a National Mineral Policy. AIME, New York, N.Y., pp. 47—48.

Lovering, T.S., 1969. Mineral resources from the land. In: National Academy of Sciences and National Research Council, *Resources and Man.* W.H. Freeman, San Francisco, Calif., pp. 109—133.

Lowell, J.D., 1970. Copper resources in 1970. *Min. Eng. (N.Y.)*, 22: 67—73.

Mayakyan, I.G., 1960. A metallogenetic map of the world. *Int. Geol. Rev.*, 2: 489—497.

McCulloh, T.H., 1973. Oil and gas. In: D.A. Brobst and W.P. Pratt (Editors), *United States Mineral Resources. U.S. Geol. Surv.*, Prof. Paper 820, pp. 477—496.

McDivitt, J.F., 1957. Mineral resources survey: concepts and problems. *Can. Min. J.*, 78(5): 104—107.

McKelvey, V.E., 1960. Relation of reserves of the elements to their crustal abundance. *Am. J. Sci.*, 258A: 234—241.

McKelvey, V.E., 1972. Mineral resource estimates and public policy. *Am. Sci.*, 60: 32—40.

McKelvey, V.E., 1973. Mineral potential of the United States. In: E.N. Cameron (Editor), *The Mineral Position of the United States, 1975—2000.* University of Wisconsin Press, Madison, Wisc., pp. 67—82.

McKelvey, V.E., 1974. Potential mineral resources. *Resour. Policy*, 1: 75—81.

Netschert, B.C. and Lion, D.M., 1957. Comment on mineral reserves and mineral resources. *Econ. Geol.*, 52: 588—590.

Nolan, T.B., 1950. The search for new mining districts. *Econ. Geol.*, 45: 601—608.

Observer, 1973. London, 9 December.

Olson, J.C. and Overstreet, W.C., 1964. Geologic distribution and resources of thorium. *U.S. Geol. Surv. Bull.*, 1204, 61 pp.

Park, C.F., Jr. and MacDiarmid, R.A., 1964. *Ore Deposits.* W.H. Freeman, San Francisco, Calif., 475 pp.

Pruitt, R.G., Jr., 1966. Mineral terms — some problems in their use and definition. *11th Annual Rocky Mountain Mineral Law Institute.* Matthew Bender, New York, N.Y., pp. 1—34.

Risser, H.E., 1973. The U.S. energy dilemma. *Ill. State Geol. Surv., Environ. Geol. Notes* 64, 64 pp.

Schroeder, N.C. and Mote, R.H., 1959. Dimensions and changing patterns of supply and demand. In: E.H. Robie (Editor), *Economics of the Minerals Industries.* AIME, New York, N.Y., pp. 351—392.

Sekine, Y., 1963. On the concept of concentration of ore-forming elements and the relationship of their frequency in the earth's crust. *Int. Geol. Rev.*, 5: 505—515.

Sinclair, A.J., 1974. Probability graphs of ore tonnages in mining camps — a guide to exploration. *Can. Inst. Min. Metall. Bull.*, 67: 71—75.

Singer, D.A., Cox, D.P. and Drew, L.J., 1975. Grade and tonnage relationship among copper deposits. *U.S. Geol. Surv., Prof. Paper* 907-A, pp. A1—A11.

Skinner, B.J., 1969. *Earth Resources*. Prentice Hall, Englewood Cliffs, N.J., 150 pp.

Skinner, B.J. and Barton, P.B., 1973. Genesis of mineral deposits. *Annu. Rev. Earth Planet. Sci.*, 1: 183—211.

Taylor, S.R., 1964. Abundance of chemical elements in the continental crust: a new table. *Geochim. Cosmochim. Acta*, 28: 1273—1285.

Theobald, P.K., Schweinfurth, S.P. and Duncan, D.C., 1972. Energy resources of the United States. *U.S. Geol. Surv., Circ.* 650, 27 pp.

Thrush, P.W. (Editor), 1968. A Dictionary of Mining, Mineral and Related Terms. U.S. Government Printing Office, Washington, D.C., 1269 pp.

Turekian, K.K. and Wedepohl, K.H., 1961. Distribution of the elements in some major units of the earth's crust. *Geol. Soc. Am. Bull.*, 72: 175—192.

United Nations, 1970. *Survey of World Iron Ore Resources*. New York, N.Y., 479 pp.

U.S. Bureau of Mines, 1970. *Mineral Facts and Problems*. U.S. Government Printing Office, Washington, D.C., 1291 pp.

U.S. Bureau of Mines, 1973. *Minerals Yearbook, 1971*. U.S. Government Printing Office, Washington, D.C., Vol. 1, 1303 pp.

U.S. Bureau of Mines and U.S. Geological Survey, 1947. *Mineral Position of the United States*. Appendix to Hearing before Subcommittee of the Committee on Public Lands, U.S. Senate, 80th Congr., 1st Sess., May, 1947. Republished as *Mineral Resources of the United States, 1948*. Public Affairs, Washington, D.C., 212 pp.

U.S. Geological Survey, 1968. Geological Survey Research, 1968, Chapter A. *U.S. Geol. Surv., Prof. Paper* 600-A, 371 pp.

U.S. National Commission on Materials Policy, 1973. *Materials Needs and the Environment Today and Tomorrow*. U.S. Government Printing Office, Washington, D.C., paged by chapters.

U.S. President's Materials Policy Commission, 1952. *Resources for Freedom*. U.S. Government Printing Office, Washington, D.C., 5 volumes.

Vinogradov, A.P., 1962. Average contents of chemical elements in the principal types of igneous rocks of the earth's crust. *Geochemistry*, 7: 641—664.

Zimmerman, E.W., 1951. *World Resources and Industries*. Harper, New York, N.Y., revised ed., 832 pp.

Zwartendyk, J., 1972. What is "mineral endowment" and how should we measure it? *Can. Dep. Energy, Mines Resour., Miner. Bull.*, MR 126, 17 pp.

Zwartendyk, J., 1974. The life-index of mineral reserves — a statistical mirage. *Can. Inst. Min. Metall. Bull.*, 67: 67—70.

Chapter 2

PRICE, TECHNOLOGY, AND ORE RESERVES

A.L. McALLISTER

INTRODUCTION

Data on ore reserves are generally published at regular intervals, most commonly annually, by mining companies. Companies estimate the magnitude and grade of ore in a deposit by a variety of engineering and geological techniques applicable to each particular commodity and orebody to the level of reliability deemed necessary to determine type and size of plant, financing requirements, environmental control, labour needs, and the optimum rate and nature of production. The degree of sophistication necessary to establish reserves varies widely, depending on the size, shape, and overall grade of the orebody, the consistency with which elements and minerals are distributed throughout the host rock, the nature and complexity of the wall rock, and the cost of data acquisition. Most of these factors are highly variable, and their evaluation and treatment is a very subjective procedure.

There is an optimum rate at which the ore from a given mine should be withdrawn in order that maximum present value of future profits can be realized. This rate is dependent upon the margin of profit per unit of ore, availability of a market at a given price, current interest rates, taxation and other government policies, and, above all, the size and shape of the orebody. It follows that before a large amount of money can be obtained or spent for construction of mining and treatment facilities of a given capacity, reasonably precise estimates of ore reserve must be made. It should, therefore, be emphasized that ore-reserve data are obtained primarily for "local" consumption and for application in operation control relative to a particular and probably unique orebody. The level of reliability of reserve estimates is established to the extent necessary for efficient operation; operators who have been involved in the development process are generally well aware of the degree of reliability with which each orebody or segment of an orebody is defined. From an operational point of view there is little incentive to engage in what must, at times, appear a somewhat academic process of attempting to classify ore reserves according to categories which have consistently defied unambiguous definition (see Chapter 1).

Difficulties arise when it becomes necessary to communicate ore-reserve estimates to those who are unfamiliar with the local scene, since the engineering or geological "experience" factor inherent in the estimates cannot be conveyed. For example, experience would show that reserves in many flat-lying coal seams may be established by drill holes on a grid of some hundreds of metres interval, but such a grid would be inadequate to the point of being meaningless in many gold deposits in which the gold is erratically distributed. How close should the holes be placed in the case of a gold deposit? The answer to this question will not come from simple statistical evaluations alone, but must include a knowledge of geological controls which may not be expressable quantitatively.

Some companies are obviously cautious and conservative in their published estimates; experience shows that others, even those with the utmost integrity, are too optimistic and strain the meaning of the various defined categories of ore reserves to the limit. The subjective element inherent in reporting ore reserves of individual mines becomes part and parcel of data compilations which are based on company reports — a fact which should be fully realized when using reserves estimates on a local, national, or international level.

The even more nebulous concept of "resources" (discussed in Chapter 1) and their availability at the national and international level has been added to the long-standing concern over the meaning of ore-reserve data at a local or national level. This inevitably has brought into focus the complex relation between reserves and resources — the factors influencing this relationship are the subject of this chapter.

FACTORS DETERMINING ORE RESERVES

The problem of defining "reserves" and "resources" has been discussed at length in Chapter 1. A diagrammatic classification proposed by McKelvey (1974) depicts the normal upgrading of resources to reserves and reserves from one category of reserves to the next higher category by "increasing the degree of geological certainty" (see Fig. 2-1). Drilling, mapping, sampling, and exposure by surface and underground workings (commonly part of a development programme at operating mines or prospects under various stages of exploration) are important factors in upgrading reserves in the various categories of "proved", "probable", and "possible". It should be noted, however, that while attempts at defining the various categories of reserves are necessary for discussions of a conceptual nature, and that defined categories are necessary for statistical compilations at the national level, the impact of these types of estimates at the operational level is, at best, limited; it is common practice for most mining companies to simply report data under the unqualified term "reserves", although such data commonly (although not universally) are interpreted to include "proved" and "probable" ore. While some companies do report "proved" and "probable" ore separately, few report on inferred, possible, or potential ore, let alone on "resources".

Fig. 2-1 implies, by means of relative area, that the quantity of material

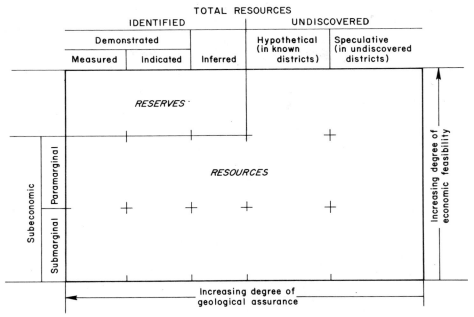

Fig. 2-1. A schematic classification of mineral resources (redrawn with permission from McKelvey, 1974, *Resources Policy*, Vol. 1, p. 76).

currently classifiable as "reserves" is much less than the total quantity of resources; there is an intuitive implication that the quantity of reserves will become progressively smaller as more and more material is withdrawn through mining operations *unless* an equal amount of material is transferred from the "resources" area into the category of reserves. This may be accomplished in three ways: by normal development procedures, by technological advances, and by price increases. In fact, the transfer of resources to the reserve category is generally the result of the interplay of these three factors.

Development procedure

As more and more people who are unfamiliar with mining company practice become involved with the problems of estimating long-range supplies of minerals, the misleading practice of assuming that ore reserves are a measure of the size of an orebody becomes increasingly common. The life of a mine is determined by dividing the size of the orebody, *not* reserves, by the average annual production. In practice, ore reserves are established to allow efficient planning and operational control, and, while plans for the immediate future must be definite and precise, demanding an accurate picture of the nature and dimensions of ore available for immediate mining, planning for years which are in the progressively more remote future can be more general. Thus, ore which will not be mined for a number of years is not delineated with the same care as ore which will be mined immediately. Expenditures at any given time for the purpose of establishing the general dimensions of an ore zone are a necessary function of company planning; expenditures to derive detailed definitions of ore reserves are postponed until such data are actually required.

Of equal importance to the development of ore reserves is the necessity of reducing, for evaluation purposes, the profits to be derived from a tonne of ore to their present value. The present value of a profit of one dollar to be received one year from now is determined by reducing the one dollar by one year's interest, i.e.:

Pv of $1.00 due in 1 year $= 1/(1 + r)$

where Pv = present value and r = interest rate. If $r = 12\%$, $Pv = \$0.89$; similarly, $1.00 due in 20 years has a present value of only $0.10. Thus, there is a natural limit beyond which the ratio of development cost to present value becomes economically prohibitive.

Commonly mining companies with large orebodies maintain reserves for a given number of years at some anticipated rate of production; it is the task of mine geologists to maintain a continual programme of exploration and development to find and upgrade sufficient quantities of ore from one category of reserves to another to replace what has been mined. In cases where production increases, it may be necessary not only to develop a single year's increased production, but to find a much more substantial amount in order to maintain reserves for a given number of years. Failure to understand the implications of this can lead to serious misunderstanding. The controversial *Report on Natural Resources Policy in Manitoba* by Keirans (1973, p. 35) reads in part:

"The companies (INCO) refuse to disclose the value of their reserves beyond a 10—15 year horizon. In their annual reports they note modest increases in reserves sufficient to cover the years production figures, designed more to justify exploration outlays and to ensure investors that the time-horizon of the firm is unimpaired than to provide factual information. Indeed, it would not be in their interest to do so for full disclosure of the real asset values underlying their operations could only increase public concern and invite appropriate tax measures."

While the final conclusion as stated may or may not be correct, to imply ulterior motives to a company for following what is commonly accepted mining practice is neither warranted nor conducive to confidence in such potentially important documents. The actual figures referred to are given in Table 2-I. Aside from the fact that the "10—15 year horizon" is obviously not true, it seems reasonable to assume that the statement of Keirans is

TABLE 2-I

Production and reserves for International Nickel Company * (data from *Northern Miner Handbook of Mines*, various issues)

Year	Reserves (10^6 tonnes)	Production of ore (10^6 tonnes)	Ratio of reserves/ annual production
1954	237	13.2	18
1959	240	13.9	17
1965	276	17.9	15
1966	276	15.9	17
1967	295	18.5	16
1968	325	22.6	15
1969	336	17.1	20
1970	345	25.7	13
1971	348	25.0	14
1972	351	17.4	20
1973	353	17.9	20
1974	362	—	—

* Includes all mines in the Sudbury district plus all Manitoba mines after 1964.

based on a lack of understanding of mining company procedure and practice at the operation level; moreover, except for variations in the reserve/production ratio due to year to year fluctuations in the market, it is obvious that the company, like many others, maintains reserves for about 18 years.

In this case, as in most cases, the creation of reserves is brought about neither by price nor technological changes, even though substantial changes in both have occurred over the 20-year period depicted in Table 2-I. The only relation between published ore-reserve data and mine life is that ore reserves represent a *minimum* quantity of remaining ore.

The relation of price and technology

The fact that a higher selling price for a mineral commodity and an increasingly sophisticated technology make it possible to mine lower and lower grade material, thus increasing the amount available, needs no elaboration. It is only when an attempt is made to introduce a quantitative aspect into the concept that difficulties arise. Fig. 2-2 illustrates the problem. If man possesses an infinite capacity to improve his technology, the reserve field may be extended by shifting the line *BC* an infinite distance to the right until the size of the reserve field encompasses the entire crust of the earth — at no increase in cost. Similarly, if price is without limit, *AB* moves infinitely upward to encompass the crust — and ore reserves are defined by crustal averages for the metals (see also discussion in Chapter 1).

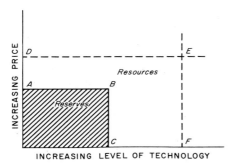

Fig. 2-2. Effect of price and technology on development of ore reserves from resources.

Both suppositions are exercises in absurdity because they envision a time of consumption far beyond our ability to plan and because practical limitations exist. A striking example of such limitations is presented by Chapman (1974) who analyses the consumption/production ratios of nuclear generating plants using uranium fuel in different types of reactors, showing that for uranium ore below a certain grade level, the plants become net *consumers* of energy.

There is, therefore, a practical grade limit — in the above example, for uranium — which is a currently undefined cut-off line, *EF*, lying between the grade of ore now mineable and the average abundance in crustal rocks. Similarly, price limitations must ultimately be established as a line, *DE*, which represents the proportion of total Gross National Product (GNP) we are willing to devote to one particular commodity relative to another commodity. For any given level of price or technology, lines *DE* and *EF* have some unique position, and their combined positions define the quantity of material available under those conditions. The principle is easy, but practical results arising out of attempts to identify available resources at projected levels of price and technology have met with very limited success.

It has been said that the sum total of scientific knowledge in the world doubles every 10 years. The quantitative basis for such a statement may not be as easy to establish as growth curves for consumption of mineral resources, but there can be little doubt that man's efficiency in finding and producing mineral materials in almost all cases has outstripped demand for minerals, resulting in progressively lower mineral prices. There can be no more striking example than the case of copper, a metal mined since antiquity. Copper was smelted in Egypt over 5,000 years ago; it was being mined in North America at about the same time from small surface pits. As indicated in Table 2-II, the price of copper has fallen dramatically since 1,840 B.C., except during the past 30—40 years; this reversal of the long-term price trend may or may not be a permanent reversal.

The long-term price decline is primarily a function of the ability of man to

43

TABLE 2-II

Historical price and grade of copper mined (modified from Lowell, 1970)

Place	Date	Grade mined (%)	Price (1970 U.S. $/lb)
Lake Superior	3000 B.C.	15	—
Middle East	1840 B.C.	—	25.00
Middle Europe	1540	8	10.00
Middle Europe	1890	6	00.45
U.S. (Bingham)	1906	2	—
U.S.	1910	—	00.31
U.S.	1945	1.3	00.26
North America	1974	0.5	
		(open pit)	00.58
		0.65	
		(block caving)	—

mine large quantities of low-grade material, simultaneously reducing costs and making available large quantities of material which would not otherwise be ore. Price reductions have been accompanied, therefore, not only by increased consumption, but also by steadily increasing reserves.

The quantity of minerals available in the earth's crust is, for all purposes, infinite providing that technological advance continues (see Chapter 3); unfortunately, even the most optimistic observer intuitively feels that there must be some limit to technology, even though a look at achievements from an historical perspective inspires confidence. The records show that the mineral resource industry has been innovative, productive, and resourceful in maintaining a continually expanding source of raw materials both through "instant breakthroughs" and, more commonly, through small, incremental improvements in exploration, mining, milling, smelting, and product development.

Technical advances

Technological advances do not always result in the ability to mine lower-grade material, thereby increasing the reserves; the opposite is equally possible. It is necessary to view mineral products as a package in which technological advance in one area may act to the detriment of another. Fortunately substitution of one mineral for another does not always have a dramatic impact. Copper producers have lived in constant apprehension of possible incursions of aluminium, a good conductor of electricity, into traditional copper markets; however, despite 80 years of growth and development of the aluminium industry, the copper industry is still strong and growing. Somewhat more dramatic has been the effect of petroleum and natural gas on the coal mining industry and the spectacular rise in importance of nuclear energy as a competitor for both coal and petroleum. An extreme example of the effect of substitution was that of the cutting off of Chilean nitrate deposits from Germany during World War I; the Haber process of producing nitrogen from the atmosphere was the direct result and proved disasterous to the natural nitrate industry. On the other hand, remarkable achievements in the use of plastics, conservation of metals through high-strength alloys, and increased recycling appear to be acting only as a brake on the accelerating use of a variety of metals.

The marvelously simple invention of the horseshoe is an example of the effect of an "instant breakthrough" in technology. Few in our modern horse-less society are even aware that such a mundane bit of iron could have had a major impact on resource technology; it is even more surprising that it should have taken nearly 1,000 years after the birth of Christ for it to be invented. Prior to that time, the use of horses in rubble strewn mines was prohibited by rapid wear of hoofs; clad in his new footwear, the horse pro-vided cheaper energy than manual labour and relieved man, for the first time in his history, of the excruciating tasks involved in moving heavy ore and waste and started him on his way of managing non-human energy. The much later use of explosives in mining must also be regarded as a major breakthrough, as was the development of pumps.

Few technological achievements have affected the mining industry as dramatically as the process of flotation; it is difficult to envisage a viable modern mining industry without it. As first utilized over 100 years ago few could have realized its ultimate importance in making possible the use of ore from hundreds of currently productive complex sulphide bodies and allow-ing the use of billions of tons of low-grade iron ores. From the first clumsy, expensive processes of bulk oil flotation (patented by Haynes in 1860) through the skin flotation techniques of 1880—1910 and froth flotation (as established by Froment in Italy in 1902 and Ballot in Australia) to the com-plex differential flotation used in 1975, the process has been continually improved until today it is in standard use for concentrating ores of a variety of metallic and non-metallic minerals. Considering the very low cost of the flotation process, it is remarkably efficient; indeed, many orebodies with vast reserves exist because of its development. This should not, however, obscure the fact that roughly 10% of the world's mineral production regularly goes to the tailings piles where it is not only lost but also contributes to problems of environmental control; this loss constitutes a challenge for better recovery processes and for expansion of the practices now used in the copper industry for recovery of material by leaching processes.

The iron ore industry illustrates the degree to which technological achieve-ment can directly affect mineral supplies. When it became apparent after 1940 that the high-grade ores were rapidly being depleted in the U.S., new sources in Canada and South America were sought and developed; at the same time massive efforts were made to find ways of treating the low-grade taconites which were generally much more accessible. The resulting technological revolution in treating taconites has swept not only the American industry but is now world-wide in its application. High-grade direct shipping ore has been largely displaced by pellets made from low-grade material (grades have dropped from about 50% to as low as 20% iron). Within a decade this technological assault changed the iron ore picture from one of looming shortage to what can only be described as enormous available reserves and resources.

Other striking changes in technology (which have directly affected the availability of mineral resources) are in the areas of energy resources and in exploration and mining techniques; these are briefly reviewed below.

Energy resources

A high level of industrial civilization was made possible by man's mastery of energy fuels which began about nine centuries ago when coal was first mined as a continuous enterprise in northern England; exploitation of petroleum began in Rumania in 1857 and in the U.S. in 1859. Harnessing of the energy in atomic nucleii began in 1942. It now appears that research into nuclear technology may be the one hope of maintaining or achieving the comforts of a civilization dependent largely on fossil hydrocarbons.

Certainly the impact of nuclear technology is by far the most outstanding achievement of the 20th century. Not only has it created a new and growing demand for raw materials, which in turn created a whole new family of orebodies out of previously worthless material, but it suggests that adequate sources of energy may be made available for the processing of the huge quantities of lower and lower grade materials which will be required with increasing urgency as the demand for minerals increases throughout the world. If the development of nuclear forms of energy can be achieved within the limits of environmental constraints and at a reasonable cost, it may well be that our overwhelming dependence upon petroleum products will have lasted less than a century — which may well be too long in consideration of their value for non-energy applications and possible long-term uses for purposes where nuclear energy cannot be substituted easily.

Exploration techniques

The maintenance of mineral reserves is, in the first instance, governed by the presence of potentially exploitable deposits and by the success or failure of the mineral industry to meet consumption demands by finding deposits at a non-prohibitive price. Fig. 2-3 is an impressive representation of the effect of advances in exploration techniques on total U.S. petroleum resources. It is not intended to imply that each indicated technique has had a profound impact on the search for petroleum; it would be difficult to equate the importance of geochemistry with seismic or gravity techniques in finding oil

46

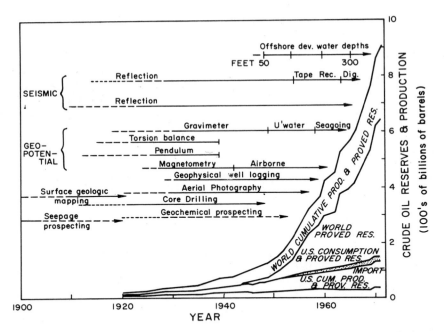

Fig. 2-3. U.S. and world crude oil production and proved reserves in relation to selected important developments in petroleum exploration technology (redrawn with permission from McCulloh, 1973, *U.S. Geological Survey, Professional Paper* 820, p. 492, fig. 61).

reserves. However, similar diagrams for copper or other metallic minerals would probably indicate a reverse order of importance for geochemistry and seismic or gravity techniques (see discussion in Chapter 10).

The progressively changing costs of replacing a ton of metal reserves may be one of the best indications of the technological ability to maintain present or projected rates of mineral production; similarly, the rate of discovery relative to production should be a measure of exploration ability and incentive. The measurement of these parameters should be a prime objective in assessing overall future mineral supply. However, data on exploration are extremely difficult to assemble and assess — costs of unsuccessful ventures are often not recorded; data on many discoveries (especially of low-grade material) become available only years after the initial find; and statistics are seldom directly comparable. Based on work in Canada, Derry (1968, 1970, 1972) and Cranston and Martin (1973) have provided a good summary of the problems in assembling exploration cost and performance data (this topic is also discussed in Chapter 11; the general conclusions are similar, although the absolute figures differ). On the assumption that Canada provides an acceptable model of exploration technology for the period 1946—1971 and that similar patterns in metal exploration have or will be developed in other countries, the conclusions reached by Cranston and Martin (1973) are summarized below:

(1) Exploration expenditures (in 1971 dollars) have increased by a factor of five times in the period 1946—1971.

(2) The aggregate value of discoveries rose sharply during the period 1951—1955, but has remained relatively constant since then despite increasing exploration expenditures.

(3) The value of metal discoveries per dollar of exploration expenditure reached a peak in the period 1951—1955.

(4) The number of discoveries reached a maximum in the period 1951—1955, while the average expenditure per discovery was at a minimum in 1951 and has increased sharply since.

The average cost of a metal discovery in Canada was about $2 million in the period 1946—1955; it rose to $6 million in 1956—1965; by 1966—1970 it was $14 to $15 million (Cranston and Martin, 1973; Govett, in Chapter 11, derived a figure of $2.7 million for the period 1951—1953 and $27 million for the period 1967—1969). Regardless of which cost figures are used, the "value" of the average discovery increased significantly in the past 20 years; Cranston and Martin (1973) calculated that the value of an average discovery rose from an estimated $245 million in Canada in 1946—1951 to at least $711 million in 1970, and that over the 25-year period the value of minerals discovered has exceeded the value of minerals mined by a factor of more than three. This increase in value is probably strongly influenced by the development of large porphyry copper deposits in British Columbia made possible by the ability to mine large tonnages of 0.5% copper and not necessarily as a result of new exploration technology; also, while the Canadian data may be meaningful for continental exploration elsewhere, they are unrealistic for offshore and marine resources.

The increasing costs of finding deposits suggests that exploration techniques, despite the introduction of geophysics and geochemistry into exploration, are in need of new "breakthroughs". If the figures presented by Cranston and Martin (1973) are reasonably correct, it cost 1.7 billion exploration-dollars to find 115 billion dollars worth of ore — about 1.5% of the total value of the ore discovered. Relative costs of finding 1 ton of metal in 1966—1971 were about double what they were in 1951—1956 (expressed in constant dollars). When one considers that only about 4% of the world's GNP is in the form of mineral products and that the 1970 cost of finding ore is only 1.5% of the ultimate value of the ore, the world would do well to expand exploration programmes and research into exploration technology many fold.

Mining technology

Chapter 13 deals at length with new advances in mining technology. The importance of mining techniques in increasing ore reserves is a direct one, particularly as they affect the handling of large tonnages of low-grade material. For example in 1906 the grade of ore mined in the open pit at Bingham Canyon (U.S.) was about 2.0% copper, considerably lower than competing underground operations; by 1946 the grade of open-pit copper deposits in the U.S. had dropped to 1.3%; currently several operations on porphyry copper deposits are based on material of less than 0.5% grade. This was achieved primarily by using heavy equipment, lower-cost ammonium-nitrate fuel oil explosives, improved fragmentation techniques, automation, and conveyor systems.

The net result of advances in mining technology in the past few decades has been the gradual transfer of interest from high-grade vein deposits to low-grade surface deposits capable of being mined by open pit operations or low-cost underground methods such as block caving or large-scale cut and fill. However, the progressive transfer to energy-consuming machines may in itself eventually impose limitations on the grade of material mineable if energy costs continue to increase. Environmental constraints also make the future development of large-scale surface operations questionable.

Price changes

Fig. 2-2 indicates that an increase in price makes available additional quantities of material which can be mined at a profit. It also follows that an increase in price should provide incentives to press the search for new deposits, some of which may be high-grade and economically mineable even at old price levels. As a general concept both of these points are valid; in practice the resource industries have responded to price changes in the search for and establishment of ore reserves.

At the practical level the treatment of mineral prices is far from simple, and predictions of future price levels is the most hazardous part of the procedure of evaluating and estimating ore reserves. Price predictions involve an assessment of short-term and sometimes violent price fluctuations, the effects of inflation, the increasing burdens of taxation and royalties, political events, competing international cartels, and nationalistic movements.

Copper prices are a case in point. In North America many major primary producers are integrated through to the semi-fabricated production level (West, 1974); the net result of this integration is a relatively uniform price. Elsewhere prices on the London Metal Exchange provide the basis for sales; these prices are subject to much wider fluctuations than North American prices, especially during times of shortage or impending shortage. Such fluctuations impose hardships, especially on small producers operating on marginal-grade orebodies who may see their ore reserves wiped out in a relatively short period (alternatively, they may operate at a handsome profit during high points of the cycle). The opening and closing of many small mines is a reflection of the characteristic cyclic price curve; contraction

TABLE 2-III

Copper prices relative to the prices of other commodities

Year	Price (U.S. $/lb)	GNPI * (Canada) index 1961 = 100	Price adjusted to 1961 level (U.S. $/lb)
1956	0.40	93.0	0.43
1958	0.27	96.3	0.28
1970	0.57	133.6	0.43

* GNPI = gross national product implicit price index.

and expansion of ore reserves is the inevitable result.

The effect of double-digit inflation on costs and selling price cannot be ignored by the mining industry; the possible complex effects are examined by Heath et al. (1974) who point out that a failure to estimate the effects of inflation will certainly lead to erroneous estimates of capital cost, misleading conclusions on profitability of an operation, and creation of ore reserves where no such reserves exist. The data in Table 2-III illustrate the effect of inflation on copper prices. The drop in price from $0.40 in 1956 to $0.27 in 1958 was, in fact, a real reduction of 35% in the purchasing power of copper and signified a real slump in the metals industry. The increase to $0.57 in 1970 only represented a return to the $0.43 level in terms of 1961 dollars; the purchasing power of a pound of copper at $0.57 per pound was exactly the same as $0.43 copper in 1956. Similar effects arise from the increasing frequency of parity adjustments between nations, making long-term evaluation of marginal-grade reserves particularly difficult.

Obviously it is a major concern to the mineral industry that some price stability be maintained so that a semblance of confidence can be maintained in feasibility studies. The fact that an orebody, by definition, is a body of rock from which one or more minerals or elements can be extracted at a profit, is commonly misconstrued to mean that any orebody capable of yielding a profit will be developed and mined. Unless price levels are such that anticipated profits exceed the current "safe" interest rates of 10—15% capital will not be invested in the development of mineral resources. The current trend in taxation policy in countries such as Canada is such that, to maintain reasonable after-tax profits, a very high pre-tax profit must be achieved. Brant (1974) calculates that the effect of newly introduced tax policies in British Columbia on a typical low-grade porphyry copper operation are such that 0.5% copper is less profitable to mine than 0.3% copper was prior to the new legislation — in effect, all copper ore between 0.3 and 0.5% automatically becomes waste. He further comments that should such stringent policies be applied on a world-wide basis, a full 50% of copper in deposits above 0.3% grade would become useless.

Clearly it is difficult if not impossible to predict future prices for metals;

in the light of current government attitudes in some countries it is becoming increasingly difficult to translate prices into profits. But just as clearly, any attempt to define reserves or to assess the effect of prices on reserves demands an attempt to predict future prices. The various approaches to price estimates include extrapolation of historical data, the use of computer-based long-term economic models, the assumption of an historically established margin between cost and selling price, and calculation of break-even prices for the majority of current producers (assuming that prices cannot fall below that level in constant dollars). All of these methods are used, with varying success.

It can be shown that the real price of most metal products has declined over the last century, despite a vast increase in consumption and greatly increased total reserves. Whether these trends — which the energy fuels are not following — will continue in the future, remains to be seen.

THE EFFECT OF PRICE AND TECHNOLOGY ON RESERVES OF SPECIFIC COMMODITIES

The real concern about long-range supply of non-renewable resources arises from the accelerating rate of consumption since World War II; data on trends in price, rates of production, and growth of reserves prior to that date are of doubtful relevance. Typical relationships of these factors for the four most common non-ferrous metals over the critical last quarter of a century are shown in Figs. 2-4 and 2-5.

Production and reserves of all four metals (aluminium, lead, zinc, and copper) have increased substantially. Reserves of aluminium have increased particularly rapidly, with the discovery of large deposits at a variety of localities around the world, especially in Australia (see Chapter 4). The price of aluminium has decreased steadily in terms of constant dollars up to the mid-1940s but since that time has remained relatively constant — despite the marked increase in consumption. The grade of aluminium ore is not indicated in Fig. 2-4 because of the similarity of deposits, and the relatively large influence of other factors such as transportation, metallurgical quality, and stripping ratios. Australian production from deposits averaging only 30—35% aluminium is nevertheless a noteworthy achievement.

Lead and zinc generally occur together in polymetallic deposits along with small quantities of silver, cadmium, bismuth, and other metals. Because it is difficult to establish production grades individually (see Chapter 3), grade curves are not shown; it is reasonable to assume that mineable grades have not dropped as dramatically as they have for copper because the lead-zinc industry has not profited from large-scale open-pit operations (copper is dealt with in more detail below).

Although the prices of all four metals have progressively fallen over the

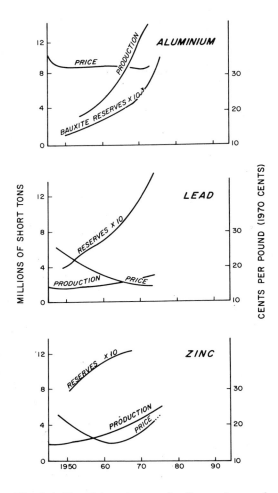

Fig. 2-4. Trend in mine production, price, and reserves of aluminium, lead, and zinc over the past 25 years.

past century, significant variations are noted in Figs. 2-4 and 2-5. Copper has moved steadily upward in the past 40 years, zinc has now recovered from a price slump in the late 1950s and (in constant dollars) is selling at roughly its 1950 price. The price of lead has continued downward.

Similar diagrams could be drawn for other commodities, and in most cases similar trend relationships prevail. The assembly of such data is a necessary first step in the establishment of viable predictions as to future prices, production, cut-off grades, and reserves. To illustrate some of the problems and the dilemmas in assessing reserves and resources three widely different resource materials are considered below — oil shales, copper, and uranium.

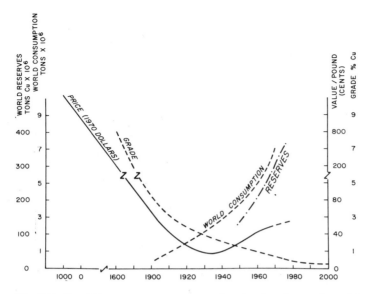

Fig. 2-5. Trend in copper reserves, grade mined, price, and world consumption.

Oil shales

The drastic increase in oil prices in 1973—1974 has provided ample political as well as economic support for the development of alternate energy supplies; the alternate sources most actively pursued and most likely to provide immediate and significant relief include oil from tar sands, and oil and possibly gas from oil shales. There are substantial supplies of these materials in the world. Tar sands are discussed in some detail in Chapter 7; oil shales are discussed here.

Oil shales are sedimentary rocks rich in organic material resulting from deposition in fresh-water lakes. The quantity of oil which can be derived from a unit of shale is related to the organic content, which in turn is related to the environment in which the rock formed. Normal procedures of studying various facies of sedimentary rocks are applicable and, particularly in a shale environment, characteristics such as thickness and quality may be assumed to continue over distances not normally applicable in metallic orebodies. Reliable estimates as to grade and tonnage can be made with relatively widespread samples. Distribution of individual beds on surface and at depth are determined with confidence from structural and stratigraphic maps, or, in some cases, by geophysical methods (especially in areas covered by much younger beds). Areas of potential production can be easily identified and, of equal importance, large areas of no potential may be identified and eliminated.

In the case of most metallic minerals uncertainty centres about the availability of reserves. In oil shales large quantities of defined grade are known to be available and major uncertainties involve primarily probable price levels, various aspects of technology (such as water consumption), and the environmental impact of the necessary large-scale operations.

Despite the vast quantities of oil in shale beds throughout the world, production has generally declined for a number of years simply because of the over-supply of oil from conventional sources which was maintained by unusually low prices until recently. In 1970 the National Petroleum Council of the U.S. estimated that the production of oil from high-yield shale (30—35 gallons/ton or 125—146 litres/tonne) in beds greater than 10 m thick and less than 450 m below the surface by underground methods would cost between $4.30 and $5.30 per barrel (this compared to the then existing price of about $3.50 for conventional oil). However, there has been no commercial production of oil from shales in North America for over 100 years.

Pilot plants and industrial-size testing in western Colorado have confirmed the feasibility of synthetic oil production from oil shales; it is estimated that by 1980 the U.S. could have a sizeable oil-shale industry (see Chapter 7 and Chapter 13 for more detail). The high price of imported oil is stimulating a re-examination of both the production time-table for oil from oil shales and the estimates of the quantity economically recoverable both in the short-term and the medium-term.

Table 2-IV gives estimates of world resources of shale oil. The 678 billion barrels of identified resources of material containing 25—100 gallons/ton (104—417 litres/tonne) is of the same order of magnitude as world petroleum

TABLE 2-IV

World shale oil resources (billions of barrels/ton; compiled from Culbertson and Pitman, 1973)

Continents	Identified resources		Hypothetical resources		Speculative resources	
	25—100 gal/ton	10—25 gal/ton	25—100 gal/ton	10—25 gal/ton	25—100 gal/ton	10—25 gal/ton
Africa	100	small	n.e.	n.e.	4,000	80,000
Asia	90	14	2	3,700	5,400	110,000
Australia/ New Zealand	small	1	n.e.	n.e.	1,000	20,000
Europe	70	6	100	200	1,200	26,000
N. America	418	1,600	350	1,700	1,600	46,000
S. America	small	800	n.e.	3,200	2,000	36,000
Total	>678	>2,421	>452	>8,800	15,200	318,000

n.e. = not estimated.

reserves. Speculative resources are listed at an almost unbelievable 318 trillion (318×10^{12}) barrels, but include large volumes from low-grade black shales which will surely not be useable during any realistic forecast period.

As in the case of other minerals, any estimate of the economically recoverable quantity of shale oil in the short- to intermediate-forecast periods is dependent on the future price structure and improvements in extraction and processing technology. Any estimates, including those in Table 2-IV, are necessarily order-of-magnitude figures, becoming increasingly speculative as lower and lower grade material, or material in areas of difficult access, is included. Shramm (1970) suggests that of an estimated 160 billion barrels of high-grade oil (30—35 gallons/ton; 125—146 litres/tonne) lying above 300 m in beds more than 8 m thick, some 80 billion barrels is recoverable using present technology (and presumably at a 1970 price of $3.00 per barrel); he also estimates that if 90% of the 600 billion barrels of +25-gallons/ton shale (104 litres/tonne) were recovered, along with one-half of the 10—25-gallon/ton shale (42—104 litres/tonne) the ultimate amount recoverable would be about 1,300 billion barrels (pilot projects by the U.S. Bureau of Mines utilizing room and pillar mining techniques are particularly applicable to thick flat-lying beds, but they generally allow less than 50% extraction).

Using Shramm's (1970) data (which he warns are highly speculative) and recalculating it to conform with international data, allowing for cost inflation of 100% since 1968 (which is deliberately high to allow for the runaway capital costs of 1974—1975), and adjusting for escalated oil prices, Fig. 2-6 gives an order-of-magnitude figure for the availability of shale oil, first using current methods, and secondly using various technological improvements in the conceptual or experimental stage. Perhaps the most important aspect of Fig. 2-6 is the wide disparity between the curves — an honest reflection of the degree of uncertainty as to quantities which might become available, and a striking example of the importance of possible future technical developments. The difference in availability at $12 per barrel of oil is roughly 1 trillion barrels (for a gross value of $12 trillion) and is a significant incentive

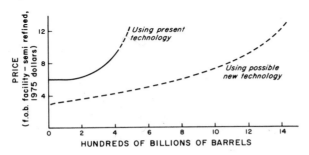

Fig. 2-6. World availability of oil from shales at different price levels under conditions of current and envisaged technology (order-of-magnitude values only). (After Shramm, 1970.)

for continuing research into such oil-shale technology. Cost-reducing technology must be directed toward cheap methods of large-scale mining, process-water conservation, waste disposal, land conservation, and the extraction and utilization of a wide range of by-products. On the other hand, the current pre-occupation with the need for more oil could lead to an over-emphasis on shale-oil; national policy on the relative costs and benefits of shale oil versus other energy sources (including coal) is clearly an immediate need.

Copper

Copper provides the most interesting case study of the effect of price and technology on ore grade and reserves. It has been in constant use since pre-historic times and is vital to modern technology; since it is difficult to envisage the effects of an extreme shortage of copper on modern industry, some assessment of its future supply is essential.

Although copper is mined from a variety of geological environments, the bulk of its production comes from deposits in which by-products are sufficiently small to allow effective estimation of mining costs per unit of copper. The data on world copper resources are shown in Table 2-V (see Chapter 1 for definitions and comparable data for other minerals).

Current evidence suggests that the bulk of copper production in the next few decades will come from porphyry coppers, and the assessment of well-defined belts (subduction zones of the plate-tectonic hypothesis) where copper porphyries occur assumes special significance. Smaller, but important, volumes of copper are mined in sedimentary stratiform deposits of the central African types and from shales of the White Pine Michigan type. Large conditional resources in many sedimentary beds are known and increased prices and technological advances could create large reserves from these beds.

Fig. 2-5 illustrates a number of important points about copper reserves, consumption, price, and grade. The consumption of copper has increased over the past two decades at a very rapid rate. On a world-wide basis the rate of consumption is about equal to the rate at which ore reserves are growing, and the ratio of reserves to production remains about 50/1. However, the

TABLE 2-V

World copper resources (from Cox et al., 1973)

	Million tons	Tonnes
Identified resources	344	312×10^6
Conditional resources	381	346×10^6
Speculative resources	320	290×10^6
Hypothetical resources	400	363×10^6

"reserves" curve does not take into account "potential" ore — and it almost certainly occurs in many of the deposits now being worked. Also, the fact that ore reserves are being maintained suggests that the level of exploration technology is advancing at a rate sufficient to maintain discovery levels; however, this would only be so if discovery costs per tonne of metal remained constant and the Canadian experience (see above) indicates that exploration costs are increasing rapidly (although it could be argued that discovery costs are a relatively small factor in metal costs). While the impact of technology on price and mineable grade is as impressive as it is obvious, projections to . the year 2000 — although they may be necessary for social and economic planning — are hazardous.

The 1974 break-even point for large-scale open-pit operations which mine approximately 0.50% copper has been estimated as lying within a range of $0.70 to $0.75 per lb ($1.54 to $1.65 per kg); this would be equivalent to about $0.50 per lb ($1.10 per kg) in 1970 dollars; below that point mining would be uneconomical for a number of current operations. A continuation of trends established over the past three decades suggests a price of about $1.00 per lb ($2.20 per kg) by the year 2000 (expressed in 1970 dollars), which is almost double the current price. This is not exhorbitant if one considers the long historical downtrend of copper prices. The projected price, along with continued technical advances, would mean a substantial decrease in the grade required to make ore, probably to the often suggested cut-off grade of 0.25% by the year 2000. Price, or grade, or both could fall if there are technological advances; on the other hand, grade could be further reduced if prices increase and there is little technical change. It should be noted that the price-grade relationship is not necessarily linear. Mining lower-grade ore, in many cases, means mining larger orebodies and a consequent lower stripping ratio (see Chapter 14 for a discussion of the effects of size on operations and price). Milling costs are relatively constant per tonne of ore mined, but smelting and refining are more closely related to grade of concentrate and quantity of metals. Recovery is marginally lower in low-grade deposits.

This leads to the real point at issue — what would be the effect on reserves if the cut-off grade were reduced to 0.25% copper? Firstly, it would mean that mining material with a grade in excess of 0.5% copper would become highly profitable, and this increased profitability could encourage the search for new orebodies with grades of 0.25% or higher. Secondly, it would allow the mining of known and newly discovered bodies of low-grade material.

Those who hold optimistic views about the adequacy of world mineral supplies imply that there are vast quantities of such low-grade material available for addition to reserves as the cut-off grade of ore is reduced, and there are numerous, if isolated, examples to support this optimism. One of these is Gaspé Copper Mines Ltd., a skarn-porphyry copper deposit in eastern Canada with reported reserves of 57 million tons (51.7×10^6 tonnes) in 1970, averaging 1.08% copper; in 1971 the company reported 290 million tons

(263×10^6 tonnes) of 0.59% copper to be mined by large-scale open-pit operations. Thus, ore reserves were increased by a factor of about five, and available copper metal increased by a factor greater than two. But how typical is the Gaspé example? We simply do not know with any degree of confidence the rate at which grade drops off from the margins outwards away from most existing copper deposits; consequently the available volume of low-grade material is unknown.

Various workers have ventured estimates on the effect of reducing cut-off grades on ore reserves. Brant (1974) suggests that there is as much copper in grades between 0.3 and 0.5% as there is in richer deposits. Carman (1972), presenting views similar to those of Lasky (1950), has suggested that reducing grades by one-half would increase resources by some factor much greater than doubling, say ten. Cox et al. (1973) recognize that there is actually no sound basis for such estimates; however, for the purpose of estimating they *assume* a doubling of available copper if grade is reduced by one-half (the matter of grade-reserve relations is also discussed briefly in Chapter 1). More research is clearly needed on the whole subject of grade-tonnage relations (see Singer et al., 1975).

Aside from the effect which reducing grade may have on future reserves, the amount of copper (and other minerals) contained in the sea-floor nodules constitutes an enormous "resource". Estimates vary over a wide range, and the actual total may well exceed all estimates.

The simple fact remains that we do not know how much copper can be made available as ore reserves in the future, even with a grade of between 0.25 and 0.5%. It is probable that a good deal of relevant data exists; many companies have information on low-grade material which is marginal to existing ore and data on depth, grade, and thickness of many sedimentary copper occurrences are probably available. Meaningful estimates of future reserves of copper — or any other mineral — at any level of price and technology will be made only by serious attempts to recover and expand pertinent field data on depth, grade, and mining characteristics of low-grade deposits — not in conceptual papers or in more detailed studies of already published data.

Uranium

The problem of estimating uranium reserves is even more urgent than that of copper or other metallic minerals. Uranium reserves, at least outside of the communist countries, are at best dangerously low and are probably downright inadequate relative to anticipated demand. Despite an imminent shortage of uranium, both demand and price are low, and there is little incentive for exploration. The sharp increases in the price of petroleum will hopefully shock the industrialized world into action; however, results are not yet apparent (see Chapters 5 and 7).

During the period 1945—1960 uranium exploration led to the discovery of many new deposits and reserves were developed at relatively low cost. However, as readily detectable surface deposits were located the discovery rates fell and relative costs increased; between 1965 and 1970 discovery costs in North America rose from $0.30 to $0.80 per lb ($0.60 to $1.76 per kg) of uranium (Barnes, 1972).

Data on uranium reserves compiled periodically by the OECD Nuclear Energy Agency and the International Atomic Energy Agency (1973) indicate non-communist reserves of 1,126 thousand tons (1.0×10^6 tonnes) and estimated additional reserves of 1,191 thousand tons (1.1×10^6 tonnes) available at less than $10.00 per lb ($22 per kg); there are 680 thousand tons (617×10^3 tonnes) of reserves and 821 thousand tons (745×10^3 tonnes) of estimated additional reserves at a price of $10.00 to $15.00 per lb ($22 to $33 per kg).

Figures for 1972 indicate a reserves/annual production ratio of close to 90 and a discovery/consumption ratio of nearly three to one. For most metals such figures would indicate a very comfortable supply situation, and it is not surprising that uranium exploration has not been intensive. Nevertheless, it is hard to draw a parallel between the supplies of uranium and other metals because of the expected very high growth rate of the requirements for nuclear power in the future. Between 1970 and 1981 annual production of uranium is estimated at 67,000 tons (61,000 tonnes) of U_3O_8 per annum, and by 1985 it is expected to be double that amount. Under such conditions the 1973 reserves of less than $10.00/lb uranium would be used up by 1987 and the $10.00 to $15.00/lb uranium would be exhausted before the turn of the century. These estimates do not include the almost certain acceleration in conversion to nuclear power resulting from the increase in the cost of oil (see OECD, 1974); they do, however, involve a "best guess" as to the mix of reactor types apt to be in use in the immediate future (in the longer-term the development of the breeder reactor could significantly change the picture; see Chapter 7).

Fig. 2-7 gives the U.S. Bureau of Mines estimates of quantities of uranium available at different price levels. It is noteworthy that an increase in price by a factor of ten increases resources by about the same factor. Uranium could probably be recovered from seawater for $70.00 per lb ($154 per kg) according to one observer; however, it has also been suggested that it could be recovered for one-third that figure (Davis, 1972). There are similar discrepancies between the U.S. Bureau of Mines estimates and those of the International Atomic Energy Agency; however, these different estimates all indicate that the effect of price on available reserves is extremely important and that high-cost reserves are very large. The fuel component in a nuclear station employing enriched uranium is 20—25% of the total cost of generation and about one-third of this fueling cost (i.e., 7 or 8% of the electricity cost) is attributable to uranium ore; therefore, a quadrupling of present uranium

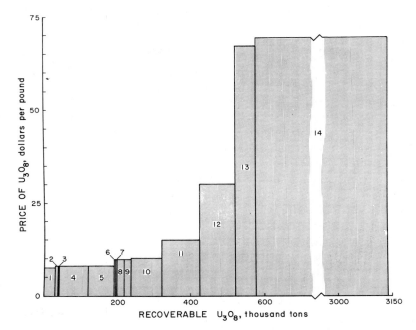

Fig. 2-7. Availability of uranium at different price levels (redrawn with permission from Bieniewski et al., 1971, *U.S. Bureau of Mines, Information Circular* 8501, p. 14). 1 = copper leach solutions; 2 = uranium-vanadium ore; 3 = uraniferous lignite ore; 4 = other uranium ore mineable by open pit; 5 = other uranium ore mineable by underground methods; 6 = uranium-vanadium deposits; 7 = uraniferous lignite deposits; 8 = other uranium deposits mineable by open pit; 9 = other uranium deposits mineable by underground methods; 10 = wet-process phosphoric acid; 11 and 12 = high-cost conventional deposits; 13 = Florida phosphate rock leached zone; 14 = Chattanooga shale.

price would increase power costs by only about 30% (Davis, 1972); with breeder reactors the increase would be much less. In view of the rising costs of petroleum, a substantial increase in uranium price would seem to be bearable. Thus, in consideration of the high capital cost of nuclear generating stations and the need to assure long-term uranium supplies, serious consideration should be given to stimulating exploration and to developing high-cost sources of uranium.

CONCLUSIONS

The simple laws of economics require companies to establish reserves for the period of time necessary to maintain optimum control of operations — this is seldom more than 20 years. The need for society to extend its vision beyond that time period is apparent, and concern has been increasingly directed toward "potential" ore and the quantities of material which might become available at different prices and levels of technology.

Companies' ore-reserve estimates vary according to the nature of the commodity, the disparity between current cost and selling prices of the mineral concerned, and the length of time of projections. The most common method of estimating mineral reserves at the national or international level utilizes intuitive guesses based on company reports and geological knowledge. Such guesses are quoted again and again and eventually are accepted with an unwarranted confidence; unfortunately they are often higher than the limited data justify; for example, the common assumption that a vast increase in reserves would result from decreasing cut-off grade for ore by one-half has not been substantiated.

The most reliable estimates are based on adding known occurrences of low-grade material to existing reported reserves and estimating the proportions which would become available under different prices. While this provides a reliable minimum estimate, it fails to account for what must be a substantial number of low-grade deposits which have been explored only in a preliminary fashion and are not necessarily recorded.

The popular method of estimating possible future reserves on a volumetric basis — initiated three decades ago for predicting oil content of sedimentary basins by comparing them with the relatively well-explored basins of the southwestern U.S. under the tacit assumption that the inevitable laws of statistics were on the side of the estimator if only he compared crustal units of sufficient size — provided the petroleum industry with a very dubious measure of actual reserves (see Chapter 1). The fact that the comfort derived from such predictions has now evaporated would be of little significance if the estimates had not had such an important influence on the profligate use of reserves, lulling governments as well as the public into a false sense of security. Fortunately the oilfield example was adopted only to a limited extent for other mineral resources. However, where it was used the assumption was made that, for a given area, the number of future mines and the expected magnitude of production could be estimated by comparison with well-explored areas of similar size and similar gross geological features (e.g., Lowell, 1970; see Chapter 1). While such assumptions are as difficult to disprove and as they are to prove, there seem to be a number of fallacies in the argument. For example, the Cordilleran belt of British Columbia, the southwest U.S., Peru, and Chile all have numerous occurrences of porphyry copper and should be directly comparable; but, in Canada, known deposits average about 0.49% copper, in the U.S. they average about 1.0% and in South America they are about 2.0%. It is quite possible that different portions of the earth's crust might be fundamentally different either in metal content, or have differed in some process of concentration not yet understood, despite the apparent similarity of environment.

The uncertainty factor in resource estimates is emphasized here because of its direct bearing on the transfer of resources to reserve category and to dispel any undue confidence that may be engendered by basically unfounded and over-optimistic resource estimates. The point is well illustrated by the following quotation from North (1974, p. 25).

"It has been established by unchallengeable experience that volumetric estimations of 'potential' reserves, made prior to their discovery, are meaningless. Between 1955 and 1962, whilst the concept still held attractions for some exploration companies (especially for those under government control), at least three countries large enough to be compared with Canada (Brazil, India and Australia) were subjected to blanket exploration with volumetric forecasts acknowledged to be the spur. In all three cases, the forecasts were totally demolished by results. Every large basin, contributing major volumes of sedimentary rock to the calculations proved to be completely or virtually unpetroliferous — Amazonas, Parana, Maranhao, Indo-Gangetic, Bengal, Cucla, Great Artesian, Fitzroy-Canning. In each country the only important production accrued from tiny coastal grabens, contributing insignificant volumes of sedimentary rock to the calculations (Bahia Reconcavo, Sergipe-Alagoas, Camban, Gippsland)."

Coal, oil shale, many non-metallic minerals, some copper deposits, and iron ore deposits occur as sedimentary strata; such deposits have good continuity and their existence at considerable depths or in broad unexposed areas can be assumed with confidence. Coal beds, for example, can be identified with relative ease. Their occurrence as sedimentary beds allows extrapolation over reasonable distance, correlation of beds from one occurrence to another, and allows reasonable estimates to be made of the continuity of thickness and quality. Furthermore, relatively large tracts of land can be safely dismissed as being completely unproductive as far as coal is concerned. As a result we probably have a much better estimate of our coal resources than we have of either hydrocarbons or uranium. Similarly, estimates of enormous potential reserves of sedimentary iron ore, along with a variety of sedimentary non-metallic minerals, may be accepted with some confidence. It is also possible to make reasonable predictions as to amenability of ore to treatment, quality, technical problems, mechanical characteristics of wall rock, etc. Thus technological problems and feasible price levels can be estimated and available potential reserves predicted for different levels of price and technology.

The very real efforts to evaluate world mineral resources over the past decade have succeeded mainly in highlighting our inability to provide meaningful data for the future beyond the time of exhaustion of existing proven and probable reserves. Current estimates at any level are, however, a necessary first step. As in the case of estimates of ore reserves at the mine level the degree of reliability is generally well appreciated by those doing the compilation, but unfortunately the data are mainly for consumption on the policy-

62

making level by economists, social planners, and politicians who may not be *really* aware of the risk factors inherent in the minerals industry.

The historical pattern has been for consistent improvements in technology to provide mineral products at progressively lower prices, and, at the same time, to maintain reserves at more or less standard operational levels of 15—20 years, except for those such as iron ore, bauxite, oil shales, and some porphyry copper deposits in which reserves for a very long time are established at a very low cost. There is no reason to believe that technological advances are slowing down, and several areas of probable advances are recognizable. Geologists, geophysicists, and geochemists will probe deeper and deeper in their search for ore, and engineers will improve recovery techniques (especially for coal, petroleum, and potash), reduce energy consumption, and continue to improve materials handling in low-grade ores. Improved design, substitution, and recycling may, in some measure, alleviate the pressures of demand. If all else fails, society may be required, at long last, to suffer a reversal of the long-established trends and find it necessary to pay more for non-renewable resources.

REFERENCES CITED

Barnes, F.Q., 1972. Uranium exploration costs. In: S.H.U. Bowie, M. Davis and D. Ostle (Editors), *Uranium Prospecting Handbook.* Institution of Mining and Metallurgy, London, pp. 79—94.
Bieniewski, C.L., Persse, F.H. and Brauch, E.F., 1971. Availability of uranium at various prices from resources in the United States. *U.S. Bur. Min. Inf. Circ.* 8501, 92 pp.
Brant, A., 1974. Increasing taxes turn low-grade ore into barren rock. *North. Miner,* June 6, p. 56.
Carman, J., 1972. Comments on the report entitled: "The Limits to Growth". Address at University of New Brunswick, Mimeogr., 19 pp.
Chapman, P., 1974. The ins and outs of nuclear power. *New Sci.,* 64(928): 866—869.
Cox, D.P., Schmidt, R.G., Vine, J.D., Kirkemo, H., Tourtelot, E.B. and Fleischer, M., 1973. Copper. In: D.A. Brobst and W.P. Pratt (Editors), *United States Mineral Resources. U.S. Geol. Surv., Prof. Paper* 820, pp. 163—190.
Cranston, D.A. and Martin, H.C., 1973. Are ore discovery costs increasing? *Can. Min. J.,* 94: 53—64.
Culbertson, W.C. and Pitman, J.K., 1973. Oil shale. In: D.A. Brobst and W.P. Pratt (Editors), *United States Mineral Resources. U.S. Geol. Surv., Prof. Paper* 820, pp. 497—504.
Davis, M., 1972. Uranium supply and demand. In: S.H.U. Bowie, M. Davis and D. Ostle (Editors), *Uranium Prospecting Handbook.* Institution of Mining and Metallurgy, London, pp. 17—32.
Derry, D.R., 1968. Exploration. *Can. Inst. Min. Metall. Bull.,* 61(670): 200—205.
Derry, D.R., 1970. Exploration expenditure, discovery rate and methods. *Can. Inst. Min. Metall. Bull.,* 63(695): 362—366.
Derry, D.R., 1972. Exploration statistics show despite funds spent barely enough new ore is being found for future needs. *North. Miner,* April 6, p. 14.
Heath, K.C.G., Kalcov, G.D. and Inns, G.S., 1974. Treatment of information in mine evaluation. *Trans. Inst. Min. Metall.,* 83: A20—A33.

Keirans, E., 1973. *Report on Natural Resources Policy in Manitoba.* Unpublished Report for Planning and Priorities Committee of the Cabinet, Government of Manitoba, Winnipeg, Man., 19 pp.

Lasky, S.G., 1950. How tonnage grade relations help predict ore reserves. *Eng. Min. J.*, 151: 81—85.

Lowell, J.D., 1970. Copper resources in 1970. *Min. Eng. (N.Y.)*, 22(4): 67—73.

McCulloh, T.H., 1973. Oil and gas. In: D.A. Brobst and W.P. Pratt (Editors), *United States Mineral Resources. U.S. Geol. Surv., Prof. Paper* 820, pp. 477—496.

McKelvey, V.E., 1974. Potential mineral reserves. *Resour. Policy*, 1: 75—81.

North, F.K., 1974. Canada's oil and gas resources. *Geosci. Can.*, 1(1): 24—34.

Northern Miner Handbook of Mines, various issues.

OECD Nuclear Energy Agency and International Atomic Energy Agency, 1973. *Uranium: Resources, Production, and Demand.* OECD, Paris, 140 pp.

OECD, 1974. *Energy Prospects to 1985.* Paris, 2 volumes.

Shramm, L.W., 1970. Shale oil. In: U.S. Bureau of Mines, *Mineral Facts and Problems.* U.S. Government Printing Office, Washington, D.C., pp. 183—202.

Singer, D.A., Cox, D.P., and Drew, L.J., 1975. Grade and tonnage relationship among copper deposits. *U.S. Geol. Surv., Prof. Paper* 907-A, pp. A1—A11.

West, M.J., 1974. Price stability for copper — prospects and problems. *Trans. Inst. Min. Metall.*, 83: A13—A19.

Chapter 3

THE ABUNDANCE AND AVAILABILITY OF MINERAL RESOURCES

F.M. VOKES

INTRODUCTION

This chapter will consider two questions of prime importance for the assessment of the mineral resources of the earth. Firstly, how much of each element of economic interest is present in those shells or zones of the earth which are ever likely to be accessible to man? Secondly, how much of the quantities present are likely to be available for man's use?

A survey will be made of the abundance and availability of elements of industrial importance in the various outer shells of the earth. It will be argued that, with the exception of a very few, though important, commodities, the solid crust of the earth is, and is likely to remain, the source of mineral commodities for the foreseeable future; the earth's oceanic crust does not appear to enter the picture as a supplier of mineral raw materials to any great extent — with the important exception of deposits lying at the ocean/crust interface (see Chapter 8).

The composition of the continental crust of the earth — as a repository and supplier of industrially important elements — will be presented, using published data from various sources. The relation between abundance and availability will necessarily be treated at length, and the reasons for the non-correlation between these two parameters will be discussed in the light of the geochemical, mineralogical, and technological-economic characteristics of the elements concerned.

The existence of metallogenic provinces lies behind the uneven geographical distribution of mineral resources between the nations of the world and the fact that some nations will always have excesses of resources while others will have deficiencies; improved recognition of the factors controlling metallogenic provinces will be a powerful tool in future exploration for, and assessment of, mineral resources. Thus the question of the non-uniform distribution of mineral deposits in the earth's crust will be discussed, and various authorities' views on the relation between metallogenic provinces and the major units of crustal structure will be presented.

70

from 0.02 to 4% by weight) plays a very important role in regulating climatic conditions on the surface of the earth. Sulphur compounds, present in variable quantities, are of the nature of contaminants rather than normal constituents; contamination may be due to geological processes (e.g., volcanicity) or to industrial processes (e.g., combustion of coal). Sodium chloride is another important temporary constituent of the atmosphere; nearly all of it represents salt taken up by vapour from the surface of the oceans, and its concentration is greatest near the sea, diminishing rapidly away from the coasts. Other halogens, such as fluorine, bromine, and iodine have also been detected in the atmosphere; however, fluorine is mostly due to industrial contamination, and bromine and iodine are to a great extent taken up from the oceans.

From an economic point of view nitrogen is the most important gas, due to its role as a basic constituent of fertilizers. The other abundant atmospheric constituents, oxygen and argon, together with the rare gases — xenon, neon, and krypton — are also recovered, but on a relatively small scale.

Skinner (1969) has pointed out that the uses to which gases are put do not permanently remove them from the atmosphere, and that we can thus classify them as reusable resources. In addition, the amounts used are relatively so small that their temporary removal has no observable effect on atmospheric composition. Thus, as is the case with the hydrosphere, the atmosphere provides an essentially limitless source of the few economically recoverable elements concentrated in it; their exploitation does not lead to noticeable deleterious effects on the medium from which they are removed. This is very much in contrast with the situation in respect of exploitation of mineral raw materials from the third outer shell or zone of the earth — the crust.

The earth's crust — oceanic and continental

The crust, the zone of the solid earth above the Mohorovičić discontinuity, comprises only about 0.4% of the total mass of the earth; the continental crust accounts for only 0.29% of the mass of the earth. However, the crust has an importance very much greater than its bulk alone would justify since it is the only part of the earth which is directly accessible for sampling and since it contains all the economically exploitable concentrations of mineral raw materials available to mankind.

The oceanic crust
The oceanic crust, relatively speaking, is very much less well-sampled than the continental crust, primarily due to the difficulties of obtaining samples from under the deep oceans (although considerable progress has been made in recent years). The results of sampling to date seem to show that the oceanic crust is relatively constant in composition, approaching that of a basalt. Wedepohl (1969), for example, suggested that the oceanic crust is

composed primarily of material which has a composition approximating that of primitive oceanic tholeiitic basalt; Taylor (1964a,b) also indicated that the composition of average basalt can be used as a first approximation of the composition of oceanic crust.

All indications so far are that the oceanic crust of the earth is of much less importance as a repository of potential mineral resources than is the continental crust, even though the absolute abundances of many elements of economic interest are higher in the oceanic crust. On the one hand, there is little evidence suggesting that the oceanic crust contains any significant concentrations of minerals of economic interest, nor does there seem to be much reason to expect them. Guild (1974) pointed out that the products of the tholeiitic magmas within the oceanic part of earth plates are noteworthy for their lack of endogenic mineral deposits, and that the likelihood of any conventional mineral deposits having been formed there seems remote.

On the other hand, Erickson (1973) suggests that resource target maps should be made showing the areas of the world where segments of oceanic crust are within reasonable exploration depths. In recent years other workers have pointed to the mid-ocean ridges (spreading ridges) as sites of formation of massive sulphide deposits. The famous deposits on the island of Cyprus have been pointed to as examples of such mid-ocean ridge sulphide accumulations (Pereira and Dixon, 1971). However, recent investigations of the trace element geochemistry of the Troodos volcanics of Cyprus (Miyashiro, 1973; Pearce, 1975) have thrown doubts on the reality of their postulated mid-oceanic origin and consequently on the reality of the "type example" of mid-ocean ridge sulphide deposits. None of the recent sampling of mid-oceanic ridges has resulted in any indication of sulphide or other mineral accumulations in the oceanic crust*; it seems reasonable to conclude that there is an exceedingly small probability of it containing mineral resources sufficiently important to warrant their recovery. This conclusion, of course, relates to the oceanic crust as such, not to the sediments covering it, since these are sites of mineral deposits of considerable interest in the form of the so-called manganese nodules as well as of base metal-bearing, iron-rich sediments of the Red Sea type (see Chapter 8 for more details).

The continental crust

One of the most obvious and important points regarding the distribution of elements in the continental crust, as compared with the atmosphere and hydrosphere, is its great heterogeneity (Skinner, 1969). The elements are present as discrete crystalline minerals or other solid substances having specific and rather simple compositions. These solid substances are not randomly distributed throughout the crust, but are relatively concentrated

---

* Editors' note added in proof: Dr. P.A. Rona has drawn our attention to the discovery of the Trans-Atlantic Geotraverse (TAG) Hydrothermal Field where metal oxides and possibly sulphides are being deposited on the Mid-Atlantic Ridge at 26°N (Scott et al., 1974).

in particular rock types and mineral deposits. This means that even those elements with low overall abundances in the crust are sometimes found in relatively or absolutely high local concentrations.

It is the existence of such concentrations (which we call mineral deposits or ore deposits) that has enabled mankind to exploit the earth's mineral resources from earliest times. It is these deposits to which we will still have to look in the foreseeable future for our continuing supplies of mineral products, even though our ideas of the richness of workable concentrations are continually changing as technology and economics make it possible to exploit lower and lower absolute concentrations and as exhaustion of accessible rich concentrations make this exploitation necessary.

ABUNDANCE OF ELEMENTS IN THE CONTINENTAL CRUST

Major elements

Many discussions of the composition of the continental crust have been published in an attempt to arrive at the best figures for the abundances of

TABLE 3-IV

Published estimates of the major element composition of Precambrian shields (in wt. %)

Element composition as oxides	Canadian Precambrian Shield (Shaw et al., 1967)	Finnish Precambrian Shield (Sederholm, 1925)	Norwegian Precambrian, average (Barth, 1961)	Australian Shield, average (Lambert and Heier, 1968)	Scottish Lewisian, average (Holland and Lambert, 1972)	Average crystalline shield (Poldervaart, 1955)
SiO_2	64.93	67.45	69.7	67.0	64.0	66.4
TiO_2	0.52	0.41	0.7	0.5	0.6	0.6
Al_2O_3	14.63	14.63	14.1	14.5	15.4	15.5
Fe_2O_3	1.36	1.27	1.4	1.5		1.8
FeO	2.75	3.13	2.6	3.0		2.8
Fe_2O_3*					6.1	
MnO	0.07	0.04	n.e.	0.2	0.1	0.1
MgO	2.24	1.69	1.7	2.5	3.1	2.0
CaO	4.12	3.39	2.8	4.0	5.3	3.8
Na_2O	3.46	3.06	3.3	2.5	3.7	3.5
K_2O	3.10	3.55	3.6	3.0	1.7	3.3
P_2O_5	0.15	0.11	0.2	0.2	n.e.	0.2
Total	97.33	98.73	100.1	98.9	100.0	100.0

* Total Fe as Fe_2O_3.
n.e. = not estimated.

the various chemical elements. Early estimates of the composition of the crust were based on one of three approaches: surface sampling and averaging of rock compositions over surface areas; average igneous rock types; and material balance compositions. Most of the estimates refer to, at best, the composition of an upper crust of less than 10 km in thickness, which is by far the most important part to be considered. According to Heier (1973) the properties of the deeper parts of the crust are best represented by high-pressure granulite facies rocks.

Heier (1973) has reviewed the various estimates made of the composition of Precambrian shield surface rocks; a selection of these estimates are given in Table 3-IV. They show a general agreement on the composition of the rocks from different parts of the world, and, according to Heier (1973) the compositions are (with some allowance for a general increase in basicity with depth) considered not only typical of shield areas, but of the entire shield crust.

Other approaches attempt to estimate the composition of average igneous rocks based on the reasoning that igneous rocks are the ultimate origin of the material building up the continents, and that all sedimentary and metamorphic rocks are derived from this material (see Table 3-V). Among these estimates are those of Clarke and Washington (1924) which are based on a

TABLE 3-V

Estimates of the major element composition of the "average igneous rock" (after Heier, 1973)

Element composition as oxide	Average 5,159 analyses (Clarke and Washington, 1924)	Average two Norwegian Caledonian opdalites (Goldschmidt, 1954)	1:1 granite/basalt mixture (Taylor, 1964a)	1:1 felsic/mafic mixture (Vinogradov, 1962)
SiO_2	59.12	61.94	60.3	63.11
TiO_2	1.05	0.95	1.0	0.75
Al_2O_3	15.34	15.30	15.6	15.21
Fe_2O_3	3.08	0.94		6.65*
FeO	3.80	4.56	7.2*	
MnO	0.12	0.07	n.e.	0.13
MgO	3.49	4.10	3.9	3.10
CaO	5.08	4.66	5.8	4.14
Na_2O	3.84	3.42	3.2	3.37
K_2O	3.13	3.37	2.5	3.01
P_2O_5	0.30	0.16	n.e.	0.21
Total	98.37	99.50	99.5	99.68

* Total Fe.
n.e. = not estimated.

74

compilation of 5,129 "superior analyses"; these estimates are still among the most commonly quoted. Goldschmidt (1954) pointed out the similarity between average igneous rock and granodiorite, as exemplified by the opdalite of the Norwegian Caledonides. More recently, Taylor (1964a) used what Heier (1973) refers to as a "brilliant approach" to arrive at an average composition of igneous rock based on the lanthanum-normalized rare earth element distribution pattern in granite, basalt, and sedimentary rocks. Taylor calculated that a mix of one part granite to one part basalt would explain the sedimentary pattern and he used this mix to calculate the abundances of most elements in the crust; these calculations are widely used as reference data, although Heier (1973) questions their use. The approach of Vinogradov (1962) is similar to that of Taylor (1964a). Several other estimates have also been made recently, including those of Turekian and Wedepohl (1961), Lee Tan and Yao Chi-Lung (1965), Horn and Adams (1966), and Wedepohl (1969).

In spite of the variations in the estimates, the main features of crustal abundances brought out by Mason (1966) some years ago seem to have been little affected. Mason pointed out that eight elements — oxygen, silicon, aluminium, iron, calcium, sodium, potassium, and magnesium — make up nearly 99% by weight of the total composition of the earth's crust. Of these, oxygen is absolutely predominant; this predominance is even more marked when the figures are recalculated to atom percent and volume percent (see Table 3-VI). Oxygen atoms make up over 60% of the total number of atoms in the earth's crust; on a volume basis, oxygen makes up more than 90% of the total occupied by all the elements. Mason (1966) concludes that the crust of the earth may be regarded as, essentially, a packing of oxygen anions bonded by silicon and ions of the common metals; this agrees with Gold-

TABLE 3-VI

The eight most abundant elements in the earth's crust (from Mason, 1966, reproduced with permission from *Principles of Geochemistry*, John Wiley and Sons, p. 48)

Element	Weight percent	Atom percent	Radius (Å)	Volume percent
O	46.60	62.55	1.40	93.77
Si	27.72	21.22	0.42	0.86
Al	8.13	6.47	0.51	0.47
Fe	5.00	1.92	0.74	0.43
Mg	2.09	1.84	0.66	0.29
Ca	3.63	1.94	0.99	1.03
Na	2.83	2.64	0.97	1.32
K	2.59	1.42	1.33	1.83

schmidt's (1954) conclusion that the lithosphere might equally well be termed the oxysphere.

The crustal abundances of several economically important elements — including iron, aluminium, silicon, magnesium, sodium, and potassium — are in excess of 2% by weight. This means that there are truly enormous absolute quantities of these elements present in that part of the solid earth which is accessible to us. Using Mason's (1966) figures for the volumes and masses of earth shells (see Table 3-I) and the crustal abundance of, say, iron, the total mass of this element in the crust can be calculated; using the figure of 5% for the abundance of iron, the calculation shows some 12×10^{17} tonnes of iron; in 1960 the total world proven reserves of iron were 129×10^9 tonnes, and world production of iron in 1971 was 760×10^6 tonnes of metal (Klemic et al., 1973). Similar calculations can be made for the other major elements in the crust, showing similar large absolute amounts.

Minor and trace elements

While there are differences in the order of several tenths of a percent, in some cases, between the abundance figures given by various workers for the major elements, from the point of view of mineral resource abundance the uncertainties are of comparatively small importance since we are dealing with such large absolute quantities. In the case of the minor and trace elements, uncertainties about the true crustal abundances are more serious and are compounded by difficulties of analytical techniques.

Several of the authorities cited with regard to major element abundances have also produced estimates of the abundance of the minor and trace elements (Turekian and Wedepohl, 1961; Vinogradov, 1962; Taylor, 1964a; Lee Tan and Yao Chi-lung, 1965; Mason, 1966; Horn and Adams, 1966; Wedepohl, 1969); there are considerable differences between these various estimates. There are also differences in estimates made for more restricted geographical areas (see Eade and Fahrig, 1973, whose trace element data for parts of the Precambrian Shield of Canada differ considerably from earlier published data).

Table 3-VII compares the abundances of some of the economically more important minor and trace elements. Disregarding the absolute discrepancies in the various estimates, the data illustrate the tremendous variation in the levels of concentration of the elements in the crust as a whole and in its two component parts. Many of the elements which play an important part in our daily life and which have long been known and used by man are actually very rare in the crust. As Mason (1966) has pointed out, copper is less abundant than zirconium (not shown in Table 3-VII), while lead is comparable in abundance with a "rare" element such as gallium; mercury is, in fact, rarer than any of the so-called "rare earth" elements. On the other hand, many unfamiliar elements are relatively abundant; Mason (1966) has shown that

TABLE 3-VII

Abundances of some minor and trace elements in the earth's crust (in ppm)

Atomic No.	Element	Mason (1966)[1]	Taylor (1964a)[2]	Lee Tan and Yao Chi-Lung (1965)[2]	Taylor (1964a)[3]	Lee Tan and Yao Chi-Lung (1965)[4]
9	F	625	625	470	400	420
15	P	1,050	1,050	1,200	1,400	1,400
22	Ti	4,400	5,700	5,300	9,000	8,100
23	V	135	135	120	250	170
24	Cr	100	100	77	200	160
25	Mn	950	950	1,000	1,500	1,800
27	Co	25	25	18	48	37
28	Ni	75	75	61	150	140
29	Cu	55	55	50	100	85
30	Zn	70	70	81	100	120
41	Nb	20	20	20	20	18
42	Mo	1.5	1.5	1.1	1	1.5
47	Ag	0.07	0.07	0.065	0.1	0.091
48	Cd	0.2	0.2	0.15	0.2	0.16
50	Sn	2	2	1.6	1	1.9
56	Ba	425	425	400	250	370
74	W	1.5	1.5	1.2	1	0.94
78	Pt	0.01	—	—	0.028	0.075
79	Au	0.004	0.004	0.0035	0.004	0.0035
80	Hg	0.08	0.08	0.08	0.08	0.11
82	Pb	13	12.5	13	5	10
91	Th	7.2	9.6	6.8	2.2	4.2
92	U	1.8	2.7	2.2	0.6	1

[1] In earth's crust.
[2] In continental crust.
[3] In average basalt (equated with oceanic crust).
[4] In oceanic crust.

rubidium is present in amounts comparable with nickel, while vanadium is more abundant than tin, and scandium is more abundant than arsenic. The relative industrial or technological importance of an element (or its availability) is not necessarily proportional to its absolute abundance in the rocks of the crust; this is dealt with in more detail in the next section.

The data in Table 3-VII also show the differences in the relative abundance of the elements in the oceanic and the continental crust. Relative to the oceanic crust, the continental crust is depleted in phosphorus, titanium, vanadium, chromium, manganese, cobalt, nickel, copper, zinc, silver, and platinum; it is enriched in fluorine, barium, tungsten, lead, thorium and uranium.

AVAILABILITY OF ELEMENTS IN THE EARTH'S CRUST

As indicated above the availability of any element in the earth's crust may not necessarily be proportional to its absolute abundance. Some of the elements with low absolute abundances are among those which have long been available to mankind and have been extracted and used from earliest times; others, present in equal or greater abundances in the crustal rocks, have either not been used or have had to await the development of advanced technology before they could be extracted in sufficient quantities to become readily available for use. Iron and aluminium illustrate this point; the absolute abundances of the two elements are, respectively, approximately 5.0% and 8.1% (Mason, 1966). Iron, the less abundant of the two elements, is a metal which man was able to extract and use at a very early stage of his development; aluminium, on the other hand, was only isolated in the 1820s and it was not until 1886 that an industrial reduction process was invented (see below). The element with the lowest crustal abundance (Table 3-VII) is gold, an element which has played a considerable economic role throughout the history of man; its value is based both on its intrinsic properties and on its scarcity.

Factors affecting availability

In order to be available to man, an element should occur in the rocks in chemical combinations or minerals, which are often referred to as ore minerals or economic minerals, in which it is the main or at least a major constituent. This enables the element to be physically separated from the unwanted constituents (gangue minerals) of the rock (or deposit) in which it occurs as the first step in its extraction and recovery. The majority of the commonly used elements are those forming the major constituents of easily recognized minerals. Extreme examples of this are the metals gold, silver, and copper, which can and do occur predominantly or to a certain degree as native elements. These are minerals which are easily recognized and easily extracted from their natural occurrence, even with quite primitive technology. Other well-known and long-used elements occur in minerals of comparatively simple composition, such as oxides and sulphides which can be easily reduced to metals by simple smelting methods.

The elements which seldom or never occur in minerals of their own, even though they may show quite high crustal abundances, are called "dispersed elements" since they are systematically dispersed throughout common rock-forming minerals of other major elements. Examples are rubidium, which is dispersed in potassium minerals; gallium, in aluminium minerals; hafnium, in zircons; and rhenium, in molybdenite. Even when the element forms a discrete mineral in which it is a major constituent, this mineral may be widely dispersed in small quantities throughout some or many of the common rocks of the crust, e.g., titanium, widely dispersed as rutile, titanite, and ilmenite.

The frequency with which the minerals of the various elements occur in special concentrations of their own (mineral deposits or ore deposits) has had a very great influence on the availability of the elements, even though modern technology is becoming increasingly efficient in separating out smaller amounts of ore minerals from rocks in which they occur. It is evident that both economically and from an environmental point of view, there is an ultimate limit to which technological improvements can be carried; thus, in the foreseeable future, the factor of mineral concentrations in deposits — what Brobst and Pratt (1973) call "geologic availability" — will continue to be a governing factor in determining the availability of an element. The concept of metallogenic provinces — areas of the earth's crust where the frequency of occurrence of mineral or ore deposits is high compared with adjacent or surrounding areas — will be discussed later.

Problems of availability of some specific elements

Availability of elements which do form their own minerals is highly dependent on their exact nature; elements which occur in complex minerals from which their extraction is technologically difficult or costly have only recently been used or have yet to be used on a scale commensurate with their crustal abundances. Aluminium, titanium, and magnesium (described below) are some examples. A different problem is presented in the recovery of some minor or rare elements which are recovered as a byproduct of other elements (e.g., cobalt, vanadium, silver, tellurium).

Aluminium

Because of the difficulties connected with its reduction from its ore minerals, aluminium has only become generally available at a relatively late stage in man's history. Even now the proportion of aluminium minerals that are ·technologically available for the extraction of the metal is very small in comparison to their number and total amounts in crustal rocks. The "ore minerals" of aluminium are chiefly hydroxides of the metal contained in the ore type "bauxite", which is the only economic source of aluminium with present-day "western" technology. The amounts of aluminium "locked-up" in aluminous minerals, such as the clay minerals, feldspars, feldspathoids, sillimanite group of minerals, dawsonite, staurolite, etc., are much greater. The aluminium in these minerals will only become available with advances in technology which could be accelerated in some places by near exhaustion of bauxite ores. Nepheline is mined in the U.S.S.R. as a source of aluminium, and aluminous shales have been exploited in Poland; on a pilot scale, alumina has been extracted from anorthosite in the U.S. (see Pratt and Brobst, 1974).

Titanium

Titanium has a relatively high crustal abudance, and it occurs in discrete minerals in which it is a major constituent (rutile, ilmenite). In places these minerals are found concentrated in deposits of such size and grade that their recovery is technically and economically feasible. However, the industrial uses to which titanium metal has been put are so far rather limited, in spite of its many excellent industrial properties, due to the costly metallurgical process which is necessary to produce pure titanium, as a result of its high fusion point and its strong affinity for nitrogen, carbon, and oxygen. Thus, titanium is a good example of an element where technological difficulties outweigh "geologic availability" in making it widely available for use.

Magnesium

Magnesium is also one of the more common elements in the earth's crust, with an abundance of 2.1%. It is a prominent constituent of many rock-forming minerals which can be found in rather large concentrations of great purity, e.g., dolomite with 13% magnesium and olivine with about 30%. Yet the bulk of the magnesium used today is extracted from seawater (in which it occurs to the extent of only about 0.13%) due to the ease and cheapness of the method of extraction from seawater and the unlimited "reserves" available.

Minor by-product elements

There are many less common elements where availability is conditioned by the fact that they are solely, or to a large extent, produced as by-products of the processing of ores in which they occur for the more common elements. The rare elements are seldom present in sufficient amounts as to be decisive in determining whether a deposit is workable, although they can obviously add to the profitability of an operation. Silver is one such minor by-product element; the bulk of silver mined comes from ores which are worked primarily for their lead, zinc, or copper contents (e.g., the Kidd Creek Mine in Ontario, Canada, where about 20% of the silver mined in Canada comes from). When lead, zinc, and copper production is high the production of silver is also high.

The recovery of the ferroalloy element vanadium, which is chiefly present in technologically available form and amounts in certain magnetite-ilmenite deposits, is governed primarily by the iron and titanium grades represented in the magnetite and ilmenite, even though the magnetite concentrate itself generally commands a considerable premium due to its vanadium content. Cobalt is also a by-product element; only in Morocco is it produced as a major product (Vhay et al., 1973). Cobalt is a by-product of nickel-copper ores in Canada, and is produced from massive pyritic ores in Uganda and Finland; more than one-half of the world's production of cobalt comes from the copper deposits of Zambia and Katanga. The only commercial source of rhenium is the mineral molybdenite, in which it occurs in solid solution in small amounts.

Another aspect of by-product availability (noted by Mason, 1966) is that certain elements which possess no minerals of their own, are often produced as a result of the processing of ores of more abundant elements; they become available with the discovery of new industrial applications. Tellurium is a case in point; sufficient quantities can be made available during the electrolytic refining of copper to encourage the development of industrial uses of tellurium. Goldschmidt (1954) has pointed out that if there were a demand for gallium or germanium large amounts of the elements (1,000 tonnes or so annually) could be extracted from the ashes of certain coals; the demand for germanium for industrial purposes has, in fact, resulted in the practical application of Goldschmidt's suggestion (see Mason, 1966).

RELATIONSHIP OF CRUSTAL ABUNDANCE AND AVAILABILITY

It is obvious that a mineral, a mineral deposit, and a metallogenic province represent *concentrations* of one or more elements compared to their average crustal abundance as expressed in Table 3-VII. The term "clarke" (after the eminent American geochemist F.W. Clarke) was introduced by the Russian geochemist A.E. Fersman to denote the average percentage (abundance) of an element in the earth's crust, and to quantify the concentration of an element in any particular geological body; the term "clarke of concentration" or "concentration clarke" was introduced by V.I. Vernadsky (it is also called a "concentration factor" or a "ratio of concentration").

The "concentration clarke" shows how many times an element is concentrated in a particular mineral, deposit, or province compared with its average crustal abundance ("clarke"). Thus, if the clarke of copper is 0.0055, the concentration clarke in chalcopyrite is 34.5/0.0055 or 6,270; in bornite it is 11,450; in chalcocite with 79.9% copper it is 14,520. An economic mineral deposit (or ore deposit) senso stricto is simply a deposit in which the clarke of concentration of the element of interest has a value sufficient to make its extraction profitable. Table 3-VIII shows some of the concentration clarkes necessary to form orebodies of the commoner metals under existing economic conditons and in deposits of the requisite tonnage.

The concept of the concentration clarke may also be used to express relative concentrations of an element or elements within specific subdivisions of the earth's crust, e.g., within metallogenic provinces as compared with crustal average or with other metallogenic provinces, within the continental crust as compared with the crust as a whole, within the shield areas compared with the crust as a whole, or within the fold mountain belts (Lee Tan and Yao Chi-lung, 1965). The problem with this approach lies in the extreme difficulty of arriving at a meaningful figure for the average content of any element in that part of the crust under consideration. Aside from the sampling problems, there is an added problem of defining the limits of the

TABLE 3-VIII

Concentration clarkes for orebodies of the common metals (partly after Mason, 1966)

Metal	Clarke	Minimum extractable grade (%)	Clarke of concentration for orebody
Aluminium	8.13	30	4
Iron	5.00	30	6
Manganese	0.10	35	350
Chromium	0.01	30	3,000
Copper	0.0055	0.5	90
Nickel	0.0075	0.5	66
Zinc	0.0070	3	430
Tin	0.0002	1	5,000
Lead	0.0013	4	3,075
Uranium	0.0002	0.1	500
Gold	0.0000004	0.001	2,500

province. In dealing with this concept Lee Tan and Yao Chi-lung (1965, p. 786) have pointed out that "... a concentrated (or dispersed) element as compared with the crustal background may not be concentrated (or dispersed) compared with the regional background, or the degrees of concentration (or dispersion) may be far apart. This is because the regional background values may be higher (or lower) than the crustal background values".

A metallogenic province represents an area where the abundance of some elements is considerably higher, and of other elements is considerably lower perhaps, than their clarke values. Friedensburg (1957) has estimated that only about one-millionth of the metals and other elements of economic importance contained in the rocks of the earth's crust down to a depth of 2,000 m from the surface may be concentrated into workable deposits. The proportion of any element concentrated in these workable (available) deposits will vary considerably from element to element. The factors involved will include the geochemical-mineralogical character of the elements, the geological environments in which the element was concentrated, and the genetic conditions of concentration. Economic-technological factors will also act to determine the lower levels of workability (cut-off grades) of the various elements.

Given all these factors, it would appear at first sight to be unreasonable to expect any general relationship between crustal abundance and availability of elements. However, there have been several recent studies which have investigated the relation between crustal abundance (clarkes) of various elements of economic importance and their availability as expressed by estimates of annual production and present reserves. These studies do seem to show that, for many of the common metals at least, there is a rather close linear relationship between clarke and estimated reserves. McKelvey (1960), Sekine (1963), Ovchinnikov (1971), Govett and Govett (1972), and Erickson (1973) have shown a linear relationship for limited areas of the earth's crust, such as the U.S. and Japan, and for the land surface of the earth as a whole. The problem has also been discussed in Chapter 1, where the linear relationship is confirmed.

The existence of a linear relationship between crustal abundance and ore reserves would appear to conflict with much of what has been argued previously in this chapter — that a high crustal abundance does not necessarily imply a high availability. However, when one examines the elements (metals) dealt with in the studies cited above, a number of interesting facts emerge. Govett and Govett (Chapter 1) show that the three elements which fall noticeably outside (below) the limits of the linear relationship shown by the majority of the elements investigated — aluminium, cobalt, and vanadium — are just three of the elements which earlier in this chapter were singled out as presenting difficulties in regard to availability due to their manner of occurrence in nature. The bulk of the crustal content of aluminium is "locked-up" in minerals which are not amenable to treatment at present and therefore are not ore minerals and do not figure in ore reserves. Cobalt and vanadium are typical by-product elements which are recovered from ores of other metals; the "reserves" of these two elements are expressions of the quantities of the major element ores available.

An earlier study by Govett and Govett (1972) showed that world reserves of titanium and tungsten are considerably less than might be expected from the crustal abundance figures of these elements. As we have seen, much of the crustal content of titanium is present in a "dispersed" form as discrete titanium minerals such as rutile, sphene, and ilmenite which occur in accessory amounts in many common rock types; the proportion of titanium minerals concentrated into mineral deposits is much less than in the case of minerals of many of the other, relatively less abundant, elements. In the case of tungsten (wolfram), Govett and Govett (1972 p. 284) suggest that its low reserve category is ". . . partly due to lack of information about . . . relatively strategic metals". Tungsten is a metal which appears to occur mainly in minerals in which it is a major constituent, for all practical purposes as the two tungstates scheelite and wolframite. As far as the present writer is aware, there appears to be little evidence that wolfram minerals occur dispersed in common rocks in accessory amounts, as is the case with titanium minerals; they seem to be typically present in the crust as "mineral deposits", albeit of very varying grades. It would seem, on the basis of present evidence, that the data presented by Govett and Govett (1972) indicate that present-day reserves of tungsten are indeed underestimated and that additional reserves remain to be discovered (see also Chapter 4).

The situation with tungsten can be compared with that for molybdenum and tin, elements which have clarke values almost identical to that of tungsten (see Table 3-VII). Tungsten, molybdenum, and tin are also similar geochemically and occur in ore deposits of the same general genetic types, even though mixed deposits tend to be the exception rather than the rule. This would seem to reinforce the conclusion that the world's reserves of tungsten are either not fully disclosed or are underestimated (see below).

Ovchinnikov (1971), in a study covering a far wider range of elements than those discussed above, concluded that in comparison with their clarke values, deposits of cesium, tantalum, beryllium, and thorium have been clearly underinvestigated. He also noted the fact that only the aluminium present in bauxite ores had been taken into consideration; similarly, only the fluorine in fluorite deposits is reported in ore-reserve figures, although it is obvious that they constitute only a fraction of the fluorine in the earth's crust.

Ovchinnikov (1971) and McKelvey (1960) found that those elements which deviate furthest from the average line towards higher average concentrations in the crust are those elements which are important "petrogenetically" and which are relatively evenly distributed at all depths in the crust (e.g., vanadium, titanium, aluminium, zirconium, fluorine, barium). Most of the other elements showing little or no variation are those which do not normally play a role as rock-forming minerals, but are concentrated by ore-forming processes into deposits of their own.

In discussing the relationship between crustal abundance and ore reserves in the U.S. and Japan, Sekine (1963) showed that the elements which have a fairly clear linear relationship between reserves and abundance were such chalcophile elements as copper, lead, zinc, antimony, and mercury, the first three of which have a long history of utilization by man. The elements which have small ore reserves and the greatest deviation from the line are mostly lithophile or siderophile in geochemical character and are mostly of recent exploitation and utilization.

These relationships could help in predicting cases where reserves of a particular metal can be augmented. Ovchinnikov (1971) predicted a marked increase in the known reserves (at the accepted current minimum grade) of beryllium, tantalum, and some other metals including tungsten, zirconium, bismuth, platinum, and thorium, although large increases would depend not only on discoveries of new deposits but also on some reduction in the minimum acceptable commercial concentration.

Erickson (1973) calculated (in relation to lead and molybdenum as unity) the ratio between the resource potential and the actual reserves of many elements of economic interest both for the crust as a whole and for the U.S. (to a 1 km depth); these figures for a selected number of metals are reproduced in Table 3-IX. Regarding U.S. resources, he makes the point that those metals which show known reserves most closely approaching their potential recoverable resources are those which have been most diligently

84

TABLE 3-IX

Reserves, resource potentials, and ratios of some selected elements in the crust of the earth (modified after Erickson, 1973)

Element	U.S. crust			World crust (excluding U.S.)		
	reserve *	recoverable resource potential *	ratio of potential to reserve	reserve*	recoverable resource potential *	ratio of potential to reserve
Lead	31.8	31.8	1	54.0	550	10
Molybdenum	2.83	2.7	1	2	46.6	233
Copper	77.8	122	1.6	200	2,120	10
Silver	0.05	0.16	3.2	0.16	2.75	17
Gold	0.002	0.0086	4.3	0.011	0.15	14
Zinc	31.6	198	6.3	81	3,400	42
Antimony	0.10	1.1	11	3.6	19	5
Mercury	0.013—0.028	0.2	15—7.1	0.11	3.4	31
Uranium	0.27	5.4	20	0.83	93	112
Thorium	0.54	16.7	31	1	288	288
Borium	306	980	32	76.4	17×10^3	222
Tungsten	0.079	2.9	37	1.2	51	42
Beryllium	0.073	3.7	50	0.016	64	4,000
Iron	1,800	118×10^3	65	87×10^3	$2,035 \times 10^3$	23
Chromium	1.8	189	105	696	3,260	4.7
Fluorine	4.9	1,151	235	35	2,000	571
Titanium	25	13×10^3	520	117	225×10^3	1,920
Platinum	0.00012	0.07	580	0.009	1.2	133
Nickel	0.18	149	830	68	2,590	38
Vanadium	0.115	294	2,560	10	5,100	510
Manganese	1	2,450	2,450	630	42×10^3	67
Tin	—	3.9	large	5.8	68	12
Aluminium	8.1	203×10^3	25×10^3	1,160	$3,519 \times 10^3$	3,000

* In millions of metric tons.

sought for the longest time, e.g., lead, molybdenum, copper, silver, gold, and zinc. These metals are very often found in sulphidic combinations (chalcophile elements), whereas the higher the ratio becomes the greater is the tendency for the element to form oxide or silicate minerals.

Comparing Erickson's (1973) data for the U.S. crust and the world crust, the ratios of potential to reserve for the world are generally much higher than those for the U.S. (probably due to lower intensity of prospecting work outside of the U.S.). However, there are certain elements which have distinctly lower ratios; these are chiefly nickel, tin, antimony, vanadium, aluminium, and manganese. At least part of the reason for this is that these metals are among those which are very unevenly distributed on a world basis; the "classical" nickel, tin, and antimony provinces of the world are, respectively, in Canada and Australia; Malaysia-Indonesia, Bolivia, and Cornwall; and China. The large vandium-bearing magnetite-ilmenite deposits of Canada, Scandinavia, South Africa, and Australia do not have their parallels in the U.S. The differences in the ratios for aluminium are due to geographic-climatological environmental factors; in the case of manganese, the differences are considered to be due to the different grades of reserves in the U.S. and in the rest of the world (most of the reserves outside the U.S. are in deposits of great purity, while in the U.S. large tonnages of manganese are contained in subeconomic resources with less than 35% manganese).

On the basis of his calculations, Erickson (1973, p. 24) concluded that "... if we search for an element hard enough, we find it in about the quantities we might expect". Resources of less abundant elements cannot be of the same magnitude as those of more abundant ones; "Thus, in a crude way, because copper is about 20,000 times more abundant than gold in the earth's crust, we can expect that recoverable resources of copper should be about 20,000 times greater than recoverable resources of gold" (Erickson, 1973, p. 25). This simple relationship is, of course, modified by the inherent geochemical nature of each element; furthermore, the relationship does little to tell us where the resources will be found.

MINERAL DEPOSITS AND METALLOGENIC PROVINCES

The concept of metallogenic provinces

In the preceding sections attention has been drawn to the great heterogeneity in the composition of the earth's crust and to the fact that minerals containing elements of economic interest are not scattered uniformly throughout the crust but are relatively concentrated in certain rock types and in the relatively and absolutely high local concentrations we call mineral or ore deposits. A number of geological events must have coincided to effect these concentrations. As Stanton (1972) has remarked, most ore deposits appear to be closely related to their geological surroundings, different kinds of ore showing different kinds of environments and each being a characteristic part of a particular kind of environment. Since these geological environments vary during different stages of the evolution of the continental crust, it follows that particular ores, i.e., particular elements, should have been conspicuously concentrated in certain places at certain times.

That some regions of the crust — metallogenic provinces — are abundantly supplied with workable orebodies or accessible concentrations of elements, while others contain few or none has long been recognized. Among early workers who developed the concept of metallogenic provinces are De Launay (1913), Finlayson (1910), Gregory (1922), and Spurr (1923); Lindgren (1909, 1933) also considered the problem of metallogenic epochs, periods in the evolution of the earth when ore formation appears to have been at a maximum. More recently Bilibin (1955), Turneaure (1955), Smirnov (1959), Radkevich (1961), Petraschek (1965), Dunham (1973), and Guild (1974) have made important contributions.

Turneaure (1955) used the term "metallogenetic province" to refer to strongly mineralized areas or regions which contain ore deposits of a specific type or groups of deposits which have features suggesting a genetic relationship; a metallogenic "epoch" refers to a period during which the deposition of metals was especially pronounced. In Petraschek's (1965) view the term metallogenic provinces should be used for regions which extend at least· 1,000 km in one direction; Park and MacDiarmid (1970), following Turneaure (1955), use the term metallogenic provinces for areas ranging in size from a single mining district to regions extending hundreds or even thousands of kilometres.

A number of subdivisions of provinces have been proposed (Shatalov, 1961; Michel et al., 1966; Dunham, 1973) but there is considerable disagreement on the subject. Shatalov (1961), according to Dunham (1973), recognized several orders of magnitude, ranging from the planetary metallogenic belt, to a linear metallogenic belt and a non-linear metallogenic province, further subdivided into zones, ore districts, centres, and fields. These terms are loosely used by a number of workers without agreement on their exact definition.

The idea of what constitutes a province is also open to debate. Stanton (1972, p.657) for example, concluded: "To some geologists it is a region in which notable ore formation took place during a particular geological period, i.e., metallogenic epoch, while to others it is a region in which notable ore formation has taken place during one or more metallogenic epochs". Petraschek (1965), following Smirnov (1959), has applied the term "polycyclic metallogen" to regions where ore formation has taken place during more than one epoch.

Whatever view one takes of the definitions of provinces and their subdivision, the reality of metallogenic provinces cannot be doubted. Some of the considerable literature concerning their classification, their relation to the major tectonic features of the crust, and their ultimate origin is referred to briefly below.

Relation to major crustal tectonic features

Several schemes have appeared recently which relate metallogenic provinces to the major tectonic features of the crust. Turneaure (1955) distinguished provinces occurring in the Precambrian Shields, the stable regions, and the mountain belts, respectively; Petraschek (1965) referred to provinces in the metamorphic shields, the stable mantle regions, and orogenic belts. Dunham (1973) classified the settings of what he termed "megaprovinces", as follows:

cratonic or shield-platform areas (which form exposed nuclei of the continents and provide the setting for a number of important provinces); epi-platform regions (where surface rocks are little-distorted Phanerozoic and Precambrian shelf or continental sediments lying on a platform eroded across a crustal basement); and mobile belts (which include not only the fold moun-

TABLE 3-X

Plate-tectonic classification of ore deposits and metallogenic provinces (after Guild, 1974)

Main type	Character	Sub-types		
Formed at or near plate margins	Orientation of deposits, districts, provinces generally parallel to plate margins	*Accreting plate margins*: e.g., Red Sea, Southeast Pacific Rise metal-bearing muds; possibly podiform chromites; certain cupriferous massive sulphides?	*Transform plate margins*: e.g., Boleo, Baja California copper-manganese deposits; certain podiform chromites	*Subducting plate margins*: Island-arc or continental (Andean) type; e.g., volcanogenic massive sulphides — Kuroko ores; polymetallic sulphides in Philippines, western North America, Appalachian-Caledonian belt, Urals; copper and copper-molybdenum deposits of American, southwest Pacific, Tethyian belts
Formed within plates	Districts and provinces tend to be equidimensional and less oriented; or along traverse lineaments	*Oceanic plates*: e.g., deep-sea manganese nodules; evaporites in small or newly opened basins	*Trailing continental margins (Atlantic type)*: e.g., heavy sands (titanium, zirconium, iron, etc.); phosphorite on shelf; Laisvall-type lead?	*Continental plates*: e.g., Bushveld-type stratiform chromium; copper-nickel-platinum; iron-titanium-vanadium; carbonatites; kimberlites; Kiruna-type iron-phosphorus; Precambrian and Paleozoic banded iron; gold-uranium conglomerates; red-bed copper; Kupferschiefer; Zambian-Katangan copper; Mississippi Valley lead-zinc; evaporites

tains but also the depressions represented by oceanic deeps and flysch-and-molasse-accumulation troughs and other deep receptacles of sediment and effusives which are not much distorted).

Recent discussion has centred on both the distribution of certain provinces with respect to present lithospheric plates and the ditribution of older provinces with regard to the position and nature of former plate boundaries. Guild (1974) classifies ore deposits and the metallogenic provinces in which they appear into two main classes: those formed at or near plate margins, and those formed within plates. These two main classes are then subdivided according to the types of situation envisaged within each; the classification is summarized in Table 3-X.

It should be noted that not all present-day metallogenicists are convinced of the reality of the connection between metallogenic (or metal) provinces and plate tectonics. An example of a recent voice of opposition to the ideas championed by Guild and others is that of Noble (1974).

Some examples of metallogenic provinces

The relative concentration of ore or mineral deposits in metallogenic provinces, subprovinces, and districts means, from a resource point of view, that certain portions of the earth's crust, and thus certain countries, are well-endowed with minerals while others are not so well-endowed or are very deficient in them. It means that several of the more important mineral commodities needed by man are restricted to a relatively few primary producing countries, while other countries which are important consumers of these commodities have to import all, or a large portion, of their supplies (this is dealt with at greater length in Chapter 4).

From an element availability point of view, metallogenic provinces can be considered as regions or areas of specific metal enrichment, so that they may be referred to as, for example, "copper provinces" or "copper belts" and "lead provinces" or "lead belts". To illustrate provinces of importance to the mineral-consuming world, a short review is made of the major metallogenic provinces of some important mineral resources below.

Aluminium

World resources of aluminium in the form of the hydroxide-rich bauxite ores are limited and very erratically distributed. Bauxite ores occur almost exclusively as residual deposits resulting from tropical weathering processes operating on a wide variety of source rocks. The controlling conditions for bauxite formation are thus largely climatological rather than geological-lithological, although long periods of tectonic stability, permitting deep weathering and preservation of land surfaces, are also vital prerequisites for optimum bauxite development.

Because of these factors, while bauxite is widespread throughout the world, it is primarily found in tropical countries. Even where bauxite is found in presently temperate countries it is apparent that the climate was more tropical at the time of its formation. Because of the vulnerability to erosive processes, most of the known bauxite deposits are geologically young; more than 90% of them formed no earlier than the Cretaceous period, and the largest were formed in modern tropical regions during the last 25 million years (Skinner, 1969).

The metallogenic provinces for aluminium are thus restricted to a broad belt on either side of the equator, with the greatest resources located in the tropical regions of Australia, Africa, the Caribbean, and South America. Lesser subprovinces or districts occur in China, southern Europe, Central America, and southern North America.

Chromium

All known available resources of chromium are associated with ultramafic plutonic rocks (such as dunite, peridotite, pyroxenite, and their varieties) and with closely related anorthositic rocks. There are two major geological types: stratiform (layered), and pod-shaped (usually referred to as Bushveld-type and Alpine-type deposits, respectively). Of these two types, the stratiform type contains more than 98% of the world's chromite resources (Thayer, 1973).

Because of the limited number of stratiform igneous complexes of the ultramafic-mafic type, stratiform chromite resources are limited to relatively few geographical areas. These include the Bushveld Complex of South Africa and the Great Dyke of Rhodesia, as well as the Stillwater Complex of Montana, the Bird River Complex of Manitoba, and the Fiskenæsset anorthositic complex of Greenland.

Podiform chromite deposits are individually much smaller in size than the stratiform types; production comes mostly from bodies containing 100,000 tons or more; less than a dozen deposits originally containing more than 1 million tons of ore are known (Thayer, 1973). The metallogenic provinces for podiform chromite deposits are typically the mobile belts of the earth, in contrast to those of the stratiform types which are to be found within the continental parts of plates or the stable shield regions. They occur in irregular peridotite masses or peridotite-gabbro complexes of the Alpine type in mobile belts such as the Appalachians and the Urals, the Tethyian mountain chain in the Balkans, Turkey, and Iran, and in belts surrounding the Pacific Ocean. The southern Ural mountains appear to include the greatest known concentration of large podiform deposits in the world, although figures on the size of the deposits are unavailable.

Copper

Copper is available in a variety of mineralogical and deposit types in a wide spectrum of geological environments typical of several stages of crustal evolution. Copper metallogenic provinces are present in all of the three major crustal subdivisions, on practically all continents, and in a great many countries. However, about 70% of the world's identified copper resources are present in a rather limited number of deposit types in a limited number of geographical locations (Cox et al., 1973). Both the sedimentary copper deposits and the porphyry-type deposits are centred in the great mobile zone represented by the circum-Pacific planetary metallogenic belt and by provinces in the sedimentary covers of several epiplatformal or stable regions.

Gold

Gold — a very rare element — is found in so great a variety of rocks as to preclude most generalizations regarding the association of gold with a particular rock type (Simons and Prinz, 1973); nevertheless, gold deposits occur most commonly with felsic or intermediate igneous rocks and in siliceous or aluminous sedimentary metamorphic rocks. Because of gold's inertness to chemical weathering and its resistance to mechanical attrition processes during transport, it can either be residually concentrated in situ or transported by water and concentrated along with other "resistates" in placer deposits. Simons and Prinz (1973) group gold deposits into seven broad categories: gold-quartz lodes, epithermal ("bonanza") deposits, young placers, ancient (fossil) placers, marine placers, deposits of disseminated gold, and by-product gold.

The distribution of economically important gold deposits, both geographically and by deposit-type, is extremely uneven. South Africa dominates both the world production and reserve picture with one single metallogenic subprovince or district in the Transvaal and Orange Free State where the gold-bearing conglomerates (ancient or fossil placers) of the Precambrian Witwatersrand system have produced gold since the 1880s. It is possible that other ancient shield areas — such as the Canadian, Brazil-Guyana, or African-Arabian shield areas — may contain extensive gold-bearing conglomerates (Simons and Prinz, 1973).

Next to the ancient conglomerates of the Witwatersrand type, the gold-quartz lode type of deposit, which is also characteristic of the ancient shield areas of the crust, is the most productive type of gold deposit. This type of deposit is exemplified by the Homestake, Porcupine, Kirkland Lake—Larder Lake, Kalgoorlie, Kolar, and probably the Morro Velho deposits.

Iron

Vast quantities of iron ore resources are known to exist throughout the world (Klemic et al., 1973); nevertheless, deposits of certain geological types and certain geological ages account for a large proportion of these resources. Bedded sedimentary deposits of iron minerals in Precambrian banded iron formations containing 30% or more iron are by far the principal sources of iron ore. These banded iron formations are commonly exposed as steeply to moderately dipping belts, a few kilometres to several tens of kilometres long. A few, such as those in the Labrador area of Canada, are in belts several hundreds of kilometres long. Within each province, the tonnages of ore-grade material can be very large, ranging from hundreds to thousands of millions of tonnes. Such ores are well developed on all the world's shield areas — the Lake Superior region of the U.S.; the Michipocoten, Lake Albanel—Misstassini and Labrador areas of Canada; the Quadrilatero Ferrifero of Brazil; the Krivoi Rog and Kursk areas of the U.S.S.R.; the Hammersley and other ranges of Australia, to mention only well-known ones. Thus the presence of an "iron province" is far from being an exclusive feature of any particular continent. The ores do, however, show quite a remarkable restriction in time; they are mainly restricted to the Precambrian, mainly to an age close to 2,200 million years ago — the great "iron epoch" of the earth's crust.

Compared with the Precambrian iron formations, other iron ore resources are of modest dimensions, although they are of great local or regional importance in many parts of the world. The "large" to "medium" resources reported by Klemic et al. (1973) include:

(1) Phanerozoic sedimentary deposits — "ironstones" — mainly of Lower Palaeozoic (Newfoundland, Alabama) or Jurassic (England, France, Luxembourg, and Germany) age.

(2) Laterites formed by deep chemical weathering under conditions of alternating wet and dry seasons in many tropical countries. Iron-rich laterites most commonly are derived from serpentinites, as in Cuba and the Philippines. Although there is only minor production from such deposits at present, their resource potential is rated as "large" on a world scale (Klemic et al., 1973). Nickel, cobalt, and possibly chromium are potential by-products of the treatment of such ores in the future.

(3) Secondary enrichments of low-grade iron deposits. Rich residual deposits formed by the near-surface leaching of pre-existing iron formations have produced a very large proportion of the total amount of iron ore so far used in the world. Although their relative importance is greatly diminished, large resources still exist in many of the Precambrian banded iron-formation provinces and districts of the world — Lake Superior, Brazil, Venezuela, Australia — and in some provinces of later age.

(4) Magmatic segregations. Deposits of both titaniferous and titanium-free types are included here; the class includes some very large individual bodies. The northern Swedish iron ores of the Kiruna type are normally included in this class, although recent research has thrown doubt on the hitherto accepted origin of these deposits (see Parak, 1975).

Nickel

The presently available nickel reserves and resources are contained in two main deposit types; the primary or sulphide type, and the secondary, or nickel-laterite type. The sulphide deposits are by far the most important present source of nickel, both as regards the quantity of metal and the number of deposits. Dominating production from this type of deposit has been one very restricted subprovince or district in the Precambrian Shield of Canada, the Sudbury area, which has accounted for more than one-half of the world's total nickel supply since 1905. More recently other nickel provinces and districts have been discovered in the Precambrian Shield areas, reducing the dominance of the Sudbury area to a marked extent; of these new provinces and districts the principal ones are the Thompson belt of Manitoba (Canada), and western Australia and South Africa, as well as two large but very low-grade deposits in the U.S. (Cornwall, 1973).

Deposits of increasing economic importance are found in the lateritic mantles formed by the tropical to subtropical weathering of ultramafic rocks containing small amounts of primary, silicate-bound nickel — such as dunite, pyroxenite, peridotite, and their serpentinized equivalents. Important large deposits of nickel-silicate laterites, developed on fresh ultramafics (with more than 1.5% nickel) occur in New Caledonia; lesser deposits are found in Indonesia, Venezuela, Brazil, and the U.S. (Oregon). Nickeliferous laterite deposits with 0.9—1.4% nickel, resulting from the weathering of serpentinites, are important in Cuba, the Philippines, Indonesia, the western U.S., Guatemala and elsewhere.

Small quantities (0.2-0.4%) of nickel which are universally present in silicate combination in ultramafic rocks (and which form the source for the lateritic deposits) constitute a subeconomic resource "several orders of magnitude greater" than the identified resources in sulphide and laterite deposits (Cornwall, 1973). The same can be said for the quantity of nickel contained in the deep-sea manganese nodules. Considerable technological advances will be necessary, however, to recover nickel economically from these sources.

Molybdenum

Molybdenum is analogous to nickel in its primary distribution in metallogenic provinces of restricted geographical location and number. Like nickel, the production history of molybdenum has been dominated by one single deposit — Climax, Colorado — which has supplied at least half of the world's production for the last 50 years. In addition, probably more than 95% of the world's supply of molybdenum has come from one deposit type, the so-called porphyry molybdenum and porphyry copper-molybdenum deposits. By far the greatest proportion (about three-quarters) of the world's resources of the metal seem to be located in this type of deposit in the eastern sector of the circum-Pacific planetary metallogenic belt in the Cordillera of North and South America (King et al., 1973). Within this belt a limited area in Colorado is estimated to contain more than 30% of identified world resources; the U.S.S.R. has about 5% of the world's resources, mostly in the form of disseminated and porphyry-type deposits in the Ural and Altai mountains.

The predominance of the western American Cordilleran belt in the resource picture of molybdenum is unlikely to be challenged in the foreseeable future. Within this belt the resources are likely to be divided between deposits that contain molybdenum as the primary metal and those in which molybdenum can be expected as a co-product or by-product. Porphyry molybdenum and porphyry copper-molybdenum deposits are expected to contain the bulk of molybdenum resources (King et al., 1973).

Metallogenic provinces and mineral supplies

The fact that mineral deposits of a particular type are known to occur in a particular geological environment in a particular area has proved of considerable assistance in the search for further deposits of the same type in that area. The expectation of more deposits, of roughly the same type, in established producing areas or districts lies behind a good deal of ore prospecting philosophy — the best place to find new mines is near old mines. Even if we cannot explain the patterns of metal provinces, we can use them in the search for new ore deposits.

Noble (1974) illustrates this philosophy by showing that of the eighteen "first-order" mines that have been discovered in the western U.S. in the past 30 years, eleven are located within the very well-defined metallogenic subprovince of the southern Arizona "Copper Quadrilateral". The overall rate of discovery is nearly one mine per year, and there is no indication that the rate of discovery is slowing down. In the western U.S. as a whole, Noble (1974) noted that the newly discovered mines have, with one exception, not brought about any changes in the boundaries of known metal provinces of the region.

Noble (1970, 1974) appears to advocate a deep-seated (upper mantle) origin for metallogenic (metal) provinces and to doubt that the crust of the earth could in any way be the source of the metals in ore deposits; this view is shared by many writers. On the other hand, the role of crustal, as well of mantle, processes in the determination and perpetuation of metallogenic provinces has been emphasized recently, especially by French writers (e.g., Routhier et al., 1973) who emphasize the concept of "heritage" and of "remanence" or "permanence" of metals within the prism of the earth's crust representing the province. Certain metals occur repeatedly in this prism at different periods and in different ore types, and each concentration inherits from a "geochemical stock" more or less dispersed or concentrated at earlier periods. Routhier et al. (1973) therefore consider the "prism" as a geochemical province within which, at various periods, numerous concentrating phenomena and metallotects give rise to ore deposits which together constitute a more or less polymetallic mineral province. In the view of these workers not very much new material which is likely to be concentrated into mineral deposits has arrived in the crust since about 2,500 million years ago.

Exceptions to this are some mafic and ultramafic bodies (e.g., Bushveld, Sudbury, and some serpentinite belts) and very weakly contaminated volcanisms of deep-seated origin.

Whether the tenets of the theory are accepted or not, the practical conclusions are well worth serious consideration from the point of view of mineral resource assessment and ore prospecting. Routhier et al. (1973) emphasize the view that the recognition of an important metallogenic epoch, with ores of a well-defined type, should not blind us to the possibility of the existence of concentrations formed at other times — earlier or later — in the same province. The concept of heritage leads to the general conclusion that when deposits of a certain metal are recognized in a segment of the crust, this segment should then be treated as a mineral and geochemical province, and other types of deposits of the same metal which may have been formed at different periods should be searched for. As an example of deposits where this heritage approach can be applied, Routhier et al. (1973) cite the tin deposits of Nigeria (Jurassic and Precambrian) and of Bolivia (Tertiary, Cretaceous, Jurassic, Devonian), the lead-zinc deposits of Sierra Cartagena in Spain (Pavillon, 1969), and the molybdenum-tungsten deposits of the Colorado mineral belt (Tertiary and Precambrian; Tweto, 1960).

CONCLUSIONS

As a very broad generalization, elements in the earth's crust are available in workable concentrations in proportion to their abundance. This is generally true of elements which have extremely low abundance, such as gold, but which are readily concentrated because of their chemically inert nature; it is also true of elements of high abundance, such as iron which occurs concentrated in a form which is amenable to ready extraction and processing. Some extremely abundant elements, such as aluminium, are not as readily available because the mineralogical form in which they occur makes them economically unavailable for extraction with current technology. The availability of other elements, such as cobalt and vanadium, is dictated by their recovery as by-products during the extraction of other elements. Generally, all deposits of elements can become ores only when they are present in concentrations considerably exceeding their crustal abundance in a mineralogical form amenable to economic extraction.

The concentrations of elements to form ore deposits in quite well demarcated geological zones of the earth (metallogenic provinces) is well recognized as an important factor in mineral exploration and exploitation. It is not for the writer to discuss the pros and cons of the various conflicting views on the origin of metallogenic provinces — whether they are of crustal or mantle derivation, or whether they are related to plate tectonics or not; a detailed discussion of the fundamental reasons for the localization of metallogenic provinces has been deliberately avoided since the problem is extremely complex and the subject of considerable debate among metallogenicists. It is, however, not without importance in the present context; an understanding of the underlying causes of the relative concentration of mineral deposits is of the utmost importance both in the assessment of resources and for exploration work which seeks to convert resources into reserves.

The recognition of provinces for various metals or deposit types is fundamental to both the concept of hypothetical and speculative resources and to resource assessments (Brobst and Pratt, 1973; see also Chapters 1 and 4). More work needs to be done in identifying metallogenic provinces and in assessing and discovering resources; a sound genetic theory would obviously put this work on a much firmer basis and is an objective which must be worked for with all possible means at our disposal.

REFERENCES CITED

Barth, T.F.W., 1961. Abundance of the elements, areal averages and geochemical cycles. *Geochim. Cosmochim, Acta*, 23: 1—8.

Bilibin, Yu.A., 1955. *Metallogenetic Provinces and Metallogenetic Epochs*. Gosgeol-tekhizdat, Moscow (English transl., Secr. of State, Canada, 1960, 122 pp.).

Brobst, D.A. and Pratt, W.P. (Editors), 1973. *United States Mineral Resources. U.S. Geol. Surv., Prof. Paper* 820, 722 pp.

Clarke, F.M. and Washington, H.S., 1924. The composition of the Earth's crust. *U.S. Geol. Surv., Prof. Paper* 127, 117 pp.

Cornwall, H.R., 1973. Nickel. In: D.A. Brobst and W.P. Pratt (Editors), *United States Mineral Resources. U.S. Geol. Surv., Prof. Paper* 820, pp. 437—442.

Cox, D.P., Schmidt, R.D., Vine, J.D., Kirkemo, H., Tourtelot, E.G. and Fleischer, M., 1973. Copper. In: D.A. Brobst and W.P. Pratt (Editors), *United States Mineral Resources. U.S. Geol. Surv., Prof. Paper* 820, pp. 163—190.

De Launay, L., 1913. *Traité de Metallogenie*. Libr. Polytechn. C. Beranger, Paris and Liege, 3 volumes.

Dunham, K.C., 1973. Geological controls of metallogenic provinces. In: N.H. Fisher (Editor), Metallogenic Provinces and Mineral Deposits in the Southwest Pacific. *Bull. Aust. Bur. Miner. Resour., Geol. Geophys.*, 114: 1—12.

Eade, K.E. and Fahrig, W.F., 1973. Regional, lithological and temporal variation in the abundances of some trace elements in the Canadian Shield. *Geol. Surv. Can., Paper* 72—46, 46 pp.

Erickson, R.L., 1973. Crustal abundance of elements, and mineral reserves and resources. In: D.A. Brobst and W.P. Pratt (Editors), *United States Mineral Resources. U.S. Geol. Surv., Prof. Paper* 820, pp. 21—25.

Finlayson, A.M., 1910. The metallogeny of the British Isles. *Q. J. Geol. Soc. London*, 66(262): 281—298.

Friedensburg, F., 1957. Die Zukunftvorräte der Metalle. *Erzmetall*, 10(12): 573—576.

Goldschmidt, V.M., 1954. *Geochemistry*. Clarendon, Oxford, 730 pp.

Govett, G.J.S. and Govett, M.H., 1972. Mineral resources and the limits of economic growth. *Earth-Sci. Rev.*, 8: 275—290.

Gregory, J.W., 1922. Ore deposits and their genesis in relation to geographical distribution; *J. Chem. Soc.*, 121/122(705): 750—772.

Guild, P.W., 1974. Distribution of metallogenic provinces in relation to major earth features. *Schriftenr. Erdwiss. Komm. Oesterr. Akad. Wiss.*, 1: 10—24.

Heier, K.S., 1973. A model for the composition of the deep continental crust. *Fortschr. Miner.*, 50: 174—187.

Holland, J.D. and Lambert, R.St.J., 1972. Major element composition of the shields and the continental crust. *Geochim. Cosmochim. Acta*, 36: 673—683.

Horn, M.K. and Adams, J.A.S., 1966. Computer-derived geochemical balances and element abundances. *Geochim. Cosmochim. Acta*, 30: 279—297.

King, R.V., Shawe, D.R. and MacKevett, E.M., Jr., 1973. Molybdenum. In: D.A. Brobst and W.P. Pratt (Editors), *United States Mineral Resources. U.S. Geol. Surv., Prof. Paper* 820, pp. 425—435.

Klemic, H., James, H.L. and Eberlein, G.D., 1973. Iron. In: D.A. Brobst and W.P. Pratt (Editors), *United States Mineral Resources. U.S. Geol. Surv., Prof. Paper* 820, pp. 291—306.

Lambert, I.B. and Heier, K.S., 1968. Geochemical investigations of deep-seated rocks in the Australian Shield. *Lithos*, 1: 30—53.

Lindgren, W., 1909. Metallogenic epochs. *Econ. Geol.*, 4: 409—420.

Lindgren, W., 1933. *Mineral Deposits*. McGraw-Hill, New York, N.Y., 4th ed., 930 pp.

Lee Tan and Yao Chi-lung, 1965. Abundance of chemical elements in the earth's crust and its major tectonic units. *Acta Geol. Sinica*, 45(1): 82—91. English transl. in *Int. Geol. Rev.*, 1970, pp. 778—786.

Mason, B., 1966. *Principles of Geochemistry*. John Wiley and Sons, New York, N.Y., 3rd ed., 329 pp.

McKelvey, V.E., 1960. Relation of reserves of the elements to their crustal abundance. *Am. J. Sci.*, 258A: 234—241.

Mero, J.L., 1965. *The Mineral Resources of the Sea*. Elsevier, Amsterdam, 312 pp.

Michel, H., Pelissonier, H., Permingeat, F. and Routhier, P., 1966 Propositions concernant la definition des unies métallifères. *Proc. Int. Geol. Congr., 22nd Sess., India, 1974, Sci. Comm. Geol. Map World*, pp. 149—153.

Miyashiro, A., 1973. The Troodos ophiolite complex was probably formed in an island arc. *Earth Planet. Sci. Lett.*, 19: 218—224.

Noble, J.A., 1970. Metal provinces of the western United States. *Geol. Soc. Am. Bull.*, 81: 1607—1624.

Noble, J.A., 1974. Metal provinces and metal finding in the western United States. *Miner. Deposits*, 9: 1—25.

Ovchinnikov, L.N., 1971. Estimate of world reserves of metals in terrestrial deposits. *Dokl. Akad. Nauk S.S.S.R.*, 196: 200—203.

Parak, T., 1975. The origin of the Kiruna iron ores. *Sver. Geol. Unders., Ser. C.*, No. 709, 209 pp.

Park, C.F. Jr. and MacDiarmid, R.A., 1970. *Ore Deposits*. W.H. Freeman, San Francisco, Calif., 2nd ed., 522 pp.

Pavillon, M.J., 1969. Les minéralisations plombo-zincifères de Carthagène (Cordilléres bétiques, Espagne), Un exemple d'héritages successifs en métallogénie. *Miner. Deposita*, 4: 368—385.

Pearce, J.A., 1975. Basalt geochemistry used to investigate past tectonic environments on Cyprus. *Tectonophysics*, 25: 41—67.

Pereira, J. and Dixon, C.J., 1971. Mineralization and plate tectonics. *Miner. Deposita*, 6: 404—405.

Petraschek, W.E., 1965. Typical features of metallogenic provinces. *Econ. Geol.*, 60: 1620—1634.

Poldervaart, A., 1955. Chemistry of the earth's crust. In: A. Poldervaart (Editor), *Crust of the Earth. Geol. Soc. Am. Spec. Paper* 62, pp. 119—144.

Pratt, W.P. and Brobst, D.A., 1974. Mineral resources: potentials and problems. *U.S. Geol. Surv., Circ.* 698, 20 pp.

Radkevich, Ye.A., 1961. On the types of metallogenic provinces and ore districts. *Int. Geol. Rev.*, 3: 759—783.

Routhier, P., Brouder, P., Fleischer, R., Macquar, J.C., Pavillon, M.J., Roger, G. and Rouvier, H., 1973. Some major concepts of metallogeny (Consanguinity, heritage, province). *Miner. Deposita*, 8: 237—258.

Scott, R.B., Rona, P.A., McGregor, B.A. and Scott, M.R., 1974. The TAG Hydrothermal Field. *Nature (London)*, 251: 301—302.

Sederholm, J.J., 1925. The average composition of the earth's crust in Finland. *Bull. Comm. Geol. Finlande*, 12: 1—20.

Sekine, Y., 1963. On the concept of concentration of ore-forming elements and the relation of their frequency in the earth's crust. *Int. Geol. Rev.*, 5: 505—515.

Shatalov, Ye.T., 1961. Some suggestions concerning the principles of classification of ore-bearing areas. *Uzbek. Geol. Zh.*, 6: 62—84. English transl. in *Int. Geol. Rev.*, 1964, 6: 2110—2126.

Shaw, D.M., Reilly, G.A., Muysson, J.R., Pattenden, G.E. and Campbell, F.E., 1967. Composition of the Canadian Precambrian Shield. *Can. J. Earth Sci.*, 4: 829—853.

Simons, F.S. and Prinz, W.C., 1973. Gold. In: D.A. Brobst and W.P. Pratt (Editors), *United States Mineral Resources. U.S. Geol. Surv., Prof. Paper* 820, pp. 263—275.

Skinner, B.J., 1969. *Earth Resources.* Prentice-Hall, New York, N.Y., 150 pp.

Smirnov, V.I., 1959. Experience in the metallogenic regional zonation of the U.S.S.R. *Izv. Akad. Sci. U.S.S.R. Geol. Ser.*, 4: 1—15.

Spurr, J.E., 1923. *The Ore Magmas. A Series of Essays on Ore Deposition.* McGraw-Hill, New York, N.Y., 2 volumes.

Stanton, R.L., 1972. *Ore Petrology.* McGraw-Hill, New York, N.Y., 713 pp.

Taylor, S.R., 1964a. Abundance of chemical elements in the continental crust; a new table. *Geochim. Cosmochim. Acta*, 28: 1273—1285.

Taylor, S.R., 1964b. Trace element abundance and the chondritic earth model. *Geochim. Cosmochim. Acta*, 28: 1989—1998.

Thayer, T.P., 1973. Chromium. In: D.A. Brobst and W.P. Pratt (Editors), *United States Mineral Resources. U.S. Geol. Surv., Prof. Paper* 820, pp. 111—121.

Turekian, K.K. and Wedepohl, K.H., 1961. Distribution of the elements in some major units of the earth's crust. *Geol. Soc. Am. Bull.*, 72: 175—192.

Turneaure, F.S., 1955. Metallogenetic provinces and epochs. *Econ. Geol., 50th Anniv. Vol.*, 1: 38—98.

Tweto, O., 1960. Scheelite in the Precambrian gneisses of Colorado. *Econ. Geol.*, 55: 1406—1428.

Vhay, J.S., Brobst, D.A. and Heyl, A.V., 1973. Cobalt. In: D.A. Brobst and W.P. Pratt, (Editors), *United States Mineral Resources. U.S. Geol. Surv., Prof. Paper* 820, pp. 142—155.

Vinogradov, A.P., 1962. Average contents of chemical elements in the principal types of igneous rocks in the earth's crust. *Geochemistry*, 7: 641—664.

Wedepohl, K.H., 1969. Die Zusammensetzung der Erdkruste. *Fortschr. Miner.*, 46: 145—174.

Chapter 4

GEOGRAPHIC CONCENTRATION OF WORLD MINERAL SUPPLIES, PRODUCTION, AND CONSUMPTION

M.H. GOVETT

> "The single most important fact about mineral resources is that they are not distributed equally over the world."
>
> P.T. Flawn (1966, p. 266)

INTRODUCTION

Few countries in the world can hope to be entirely self-sufficient in mineral resources; the U.S.S.R. probably comes closest to this goal. International trade in minerals has allowed the world to circumvent the consequences of the uneven and erratic geographic distribution of the world's mineral wealth, but the vulnerability of the industrialized mineral-consuming countries, where demand is doubling every 15—20 years, is abundantly clear.

In most of the industrialized countries of the world mineral imports average more than one-quarter of total imports — Canada, Australia, the U.S.S.R., and South Africa are the only notable exceptions. The U.S., long an important mineral producer, is now a net importer of minerals. In Japan minerals account for 40% of total imports, and many of the less developed countries (LDCs) import large quantities of minerals. On the other hand, the economies of a number of LDCs are heavily dependent on the export of one or two minerals, and are therefore extremely vulnerable to changes in demand.

While the concentration of petroleum reserves and production in the Middle East and tin and tungsten supplies in a few countries in Asia are the most obvious cases of geographic concentration of mineral supplies, reserves of many of the world's industrially important minerals are similarly limited to one or at most a few countries. Even where reserves are more widely distributed, present-day mineral production is dominated by a few countries.

In the coming decades the pattern of mineral production, consumption, and trade may alter substantially. Since World War II there have been major shifts in world mineral production away from the traditional producing countries. New reserves may be found in the less explored parts of the globe. Advances in technology and changes in world prices may allow hitherto uneconomic resources to be mined. Nevertheless, the picture presented in this chapter will probably provide the framework for the mineral-consuming countries for at least the next decade and possibly longer. Even if new sources of minerals are found and ways are developed to exploit presently unconventional sources of minerals such as the ocean-floor nodules, clays, shales, and sands, the time span between the decision to undertake exploration and development and the actual time when a deposit can be exploited averages ten years. This lead-time might be improved, but it seems likely that most of the major mineral consuming nations of the world will remain extremely dependent on a few mineral producers and on increasing world trade for some time to come.

DISTRIBUTION OF WORLD MINERAL RESERVES AND RESOURCES

Estimates of mineral reserves and resources on a global basis are extremely difficult to evaluate (see Chapter 1). While the reserve data published by the U.S. Bureau of Mines (1970) are admittedly conservative, they have the advantage of allowing international comparison in most cases; these data, together with recent estimates of resources made for the U.S. Geological Survey (Brobst and Pratt, 1973) are used here. It has been pointed out in Chapters 1 and 2 that most resource estimates have little numerical validity, and their use in this chapter is for comparative purposes only and should not be regarded as endorsement of the actual figures.

The twenty minerals covered in this section are those which are important in industry and agriculture for which data are available; together they account for well over 90% of the total value of all minerals consumed in the world (copper, lead, zinc, iron ore, and aluminium alone account for 88% of the value of all metals consumed in the U.S., which is probably typical for most industrialized countries, Petrick, 1967).

Geographic concentration

Fig. 4-1 gives the percentage share of world reserves held by the five most important countries — the U.S., the U.S.S.R., Canada, Australia, and South Africa — for each of twenty minerals. The data dramatically illustrate the concentration of world reserves in these countries; they account for more than one-quarter of the world's reserves for sixteen of the minerals considered (tungsten, cobalt, tin, and oil are exceptions). For ten of the twenty minerals, the five countries are responsible for more than 75% of total world reserves.

Three countries — the U.S., the U.S.S.R., and Canada — together hold more than one-half of the world's total reserves of molybdenum, silver, lead, zinc, vanadium, ilmenite, and the platinum group of metals; they account for more than 40% of the world's copper, phosphate, and uranium reserves (data for Soviet uranium reserves are unavailable, but large deposits are known to exist). The predominant position of the U.S.S.R. in ownership of world re-

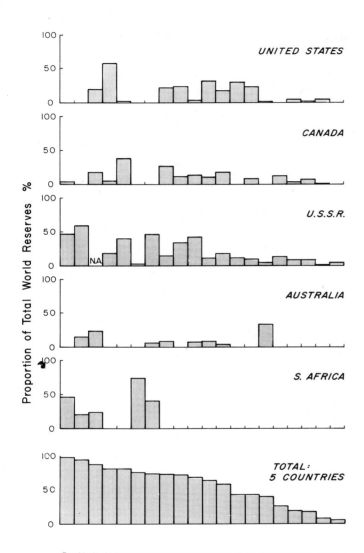

Fig. 4-1. Percentage share of world reserves of 20 minerals in five main mineral producing countries. (Data from U.S. Bureau of Mines, 1970; U.S. National Commission on Materials Policy, 1973b; and *Mining Journal (London)*, 1973.) Pt = platinum group of metals; V = vanadium; U = uranium oxide; Mo = molybdenum; K = potash; Cr = chromium; Mn = manganese; Zn = zinc; Ag = silver; Fe = iron ore; Pb = lead; Ti = ilmenite; P = phosphate; Cu = copper; Al = bauxite; Ni = nickel; Co = cobalt; W = tungsten; Sn = tin.

102

serves is illustrated by the fact that more than 5% of total world reserves of all twenty minerals except chromium and tungsten are in the Soviet Union; the U.S.S.R. probably has significant supplies of virtually every important mineral (Sutulov, 1973; Shimkin, 1964).

TABLE 4-I

Percentage share of total world reserves in countries other than the five main countries (Fig. 4-1) with greater than 5% of total world reserves (data from U.S. Bureau of Mines, 1970; Brobst and Pratt, 1973)

Mineral	Country	Percent of total world reserves
Platinum group Vanadium Potash Zinc Lead	none	—
Uranium	France	9
Molybdenum	Chile	16
Chromium	S. Rhodesia	23
Manganese	Gabon	13
	Brazil	6
Silver	Mexico	13
	Peru	10
Iron ore	European countries	8
Ilmenite	Norway	21
	India	10
	U.A.R.	7
Phosphorus	Morocco	42
Copper	Chile	17
	Zambia	8
	Zaire	6
	Peru	6
Bauxite	Guinea	21
	Jamaica	10
Nickel	Cuba	24
	New Caledonia	22
	Indonesia	11
	Philippines	6
Petroleum	Middle Eastern countries	68
Cobalt	Zaire	31
	New Caledonia	18
	Cuba	15
	Zambia	16
Tungsten	China	75
Tin	Thailand	32
	Malaysia	14
	Indonesia	13
	China	12
	Bolivia	11
	Brazil	6

Fig. 4-1 somewhat, as would the addition of resources contained in the tar sands in Canada and the oil shales in the U.S., but, on balance, it seems likely that the U.S.S.R. and the Middle Eastern and North African countries will continue to dominate world petroleum reserves for some time to come.

Molybdenum

Geographically the world's supply of molybdenum, outside of the U.S.S.R. (which has nearly one-fifth of the world total reserves), is concentrated in the Cordilleran range extending from Canada, through the U.S. and into the Andes. There may be additional large quantities of molybdenum resources in these districts. Hypothetical resources, primarily in porphyries (Brooks, 1966) appear to be divided between the U.S. and the U.S.S.R., although South America also has significant amounts.

Vanadium

Only a few countries have significant reserves of vanadium, dominated by the U.S.S.R. Vanadium resources are very large, but only a few known deposits are rich enough in the metal to be worked for it alone, and most vanadium resources are in deposits from which it is recovered as a co-product or by-product (especially phosphate rock, uranium ore, and titaniferous iron ores). Oil shales and tar sands are a potential source of vanadium in the U.S. and Canada, respectively; beach sands in New Zealand, Australia, India, and Japan contain vanadium in the ilmenite concentrate.

Platinum group of metals

In spite of recent discoveries of significant deposits of the platinum group of metals in southern Rhodesia, and the existence of potential resources in Colombia, Canada, and possibly Ethiopia, world reserves are dominated by the U.S.S.R. and South Africa; if South African deposits were mined to greater depths it is estimated that reserves could be increased three-fold (Page et al., 1973).

There are large hypothetical resources of platinum in Alaska. The occurrence of platinum metals in copper porphyries indicates potentially large speculative resources.

Silver

The geographic concentration of silver reserves is not as extreme as that of molybdenum, vanadium, or the platinum group of metals. However, apart from the U.S., the U.S.S.R., Canada, Mexico, and Peru, no single country has silver reserves in excess of 3% of the world total. An increase in the price of silver could considerably improve the position of U.S., Canadian, Mexican, and Peruvian resources by stimulating exploration for large, low-grade by-product deposits (copper, lead, and zinc) and by the development of more efficient mining methods and environmentally acceptable recovery methods.

Identified resources of by-product silver are fifteen times greater than estimated resources in deposits in which silver is the main product (Heyl et al.,
1973).

Lead and zinc

Although reserves of lead and zinc are highly concentrated in the U.S.S.R.,
the U.S., and Canada, deposits are generally much more widespread geographically than in the case of the minerals considered so far. Lead is found
on all continents except Antarctica; although most of the deposits are small,
lead and zinc are mined in some fifty countries. Morris et al. (1973) believe
that there are large lead resources in Europe, the U.S.S.R., North and South
America, and in the ocean-floor nodules and the thermal deeps of the Red
Sea Basin. Wedow et al. (1973) conclude that there are also large zinc resources; they argue that zinc may be concentrated in the geochemical environment of tropical and semitropical laterites. Brooks (1967) also argues
that there must be large lead and zinc resources in low-grade deposits in
many parts of the world.

Copper

Approximately 70% of the world's known copper resources are concentrated in the porphyry deposits of Chile, Peru, and the U.S., the porphyry
and sedimentary deposits of the U.S.S.R., and the sedimentary deposits of
Zaire and Zambia. Cox et al. (1973) have estimated that some 290 million
tonnes of speculative resources (compared to 280 million tonnes of reserves)
could be found in known copper belts, but they warn that the figures are
at best a guess and could be in error by as much as 100%. They further
speculate that another 346 million tonnes may be contained in very low-grade
deposits in the U.S. and the Andes countries.

On the basis of recent discoveries, it is also possible to speculate on additional resources in the porphyries in the Canadian Cordillera, Iran, and the
Southwest Pacific. Lowell (1970) predicts that in the next three decades the
world's large copper provinces will include the Oceania belt, New Guinea, the
Solomon Islands, and the Philippines.

Titanium

Rutile, which is concentrated in Australia and Sierra Leone (46% and 34%
of the world total, respectively) is the preferred form of titanium for industrial use, but ilmenite is much more abundant and widely distributed.
Norway, the U.S., the U.S.S.R., and Canada each hold approximately one-
fifth of the world's ilmenite reserves. According to Klemic et al. (1973)
identified titanium world *resources* are large, but only about 10% can be
counted as reserves (this explains the very different distribution of reserves
and resources in Table 4-II). If the hypothetical resources in Africa and the
U.S. were converted into reserves, the current pattern of geographic distribu-

tion of reserves would be considerably altered. Exploitation of speculative titanium resources in bauxite and lateritic deposits, sandstone deposits, offshore placer deposits, and ocean-floor nodules would also have a significant impact on the world's titanium supplies.

Nickel

The reserve figures for nickel in Fig. 4-1 and Table 4-I are based on fragmentary data and are generally considered to be low. Lateritic ores, abundant in tropical and subtropical areas, are increasingly important sources of nickel; there are also billions of tons of low-grade material in peridotites and serpentinites throughout the world which could become part of world reserves in the future. The substantial quantity of nickel in ocean-floor manganese nodules is also a potentially important resource, as is nickel in disseminated deposits throughout the world. The potential impact of the development of these sources of nickel on traditional producers — especially Canada — could have serious economic consequences.

Tin

Reserve estimates of tin are also conservative. In addition to Southeast Asia (including China) and Bolivia which hold the bulk of the world's tin reserves, Zaire and Brazil are possible sources of undiscovered tin resources. The recent discoveries of tin in Brazil are the basis for the large resources in Brazil shown in Table 4-II.

Aside from the possibility of future discoveries in South America, it seems unlikely that there will be significant discoveries outside of known tin districts, nor is there much basis for optimism about the exploitation of non-conventional sources of tin (Sainsbury, 1969; Sainsbury and Reed, 1973). Unless there are large price increases, beneficiation problems are solved, and offshore sources can be tapped, it seems reasonable to conclude that Southeast Asia will continue to dominate world tin supplies for some decades. Flawn's (1966) description of tin as an "Asiatic metal" seems well founded.

Cobalt

The data for cobalt reserves in Fig. 4-1 are based on 1960 estimates; if cobalt in low-grade sulphides and laterites, as well as more recent information on known deposits were reflected in the reserve estimates, the size and geographic distribution of cobalt reserves would be significantly different. The distribution of cobalt resources (Table 4-II) puts Cuba as the leading holder of resources (23%); Zaire has an estimated 15% of resources, compared with the 31% of reserves; the U.S. is credited with 17% of resources, while the U.S.S.R., Canadian, Zambian, and New Caledonian shares are lower than the reserve figures.

None of the data presented here include cobalt resources from non-conventional sources, although the ocean-floor manganese nodules are considered to be an important source of cobalt (Vhay et al., 1973). Mero (1965) estimated cobalt in the nodules in the Pacific Ocean alone at 11.6×10^{12} tonnes; this compares to world resources from conventional sources of 9.9×10^9 tonnes.

Manganese, aluminium, and iron ore

The more abundant minerals show remarkable geographic concentration of conventional reserves, considering their abundance. While South Africa and the U.S.S.R. dominate conventional manganese reserves and resources, potential resources on the ocean floor are enormous.

Bauxite reserves are highly concentrated in Australia and Latin America; Australian reserves have risen from practically nothing in 1950 to almost 20% of the world's total; Surinam's reserves have doubled since 1965. Details on bauxite resources by country are unavailable, but the essential pattern of reserve holdings seems to be reinforced. The potential resources of aluminium in clays and shales in many parts of the world could increase aluminium resources enormously in the future and significantly alter the geographic distribution of aluminium production if the problems of high-energy require-ments for aluminium-reduction processes could be solved (Patterson and Dyni, 1973; see also Chapter 3).

Significant iron ore reserves — iron ore is found on all continents — are remarkably concentrated; the U.S.S.R. holds 43% of the world's reserves and almost 40% of the estimated world resources, in spite of the large increase in both reserves and known resources since 1954 which resulted from explora-tion in Canada, Australia, and South America (especially Brazil). While Australian and South American reserves will undoubtedly continue to in-crease in the next decade, it is unlikely that the Soviet Union's predominance in iron ore reserves will be seriously challenged. In the long-run the exploita-tion of ocean-floor nodules and sea-floor sediments may cause a significant change in the present pattern of geographic concentration (United Nations, 1970).

Phosphate and potash

The two agricultural raw materials included in Fig. 4-1 are both abundant. Phosphate rock is widely distributed; there are large, undeveloped deposits in Peru, Australia, and the Spanish Sahara. Most of the world's resources of phosphate are in marine phosphorite deposits in known sedimentary basins.

Present supplies of potash are dominated by the U.S.S.R. and Canada; Canada alone, on the basis of present reserves, could supply world demand through the year 2000. Theoretically potash is available from brines and seawater.

WORLD MINERAL PRODUCTION

In the past seventy years world mineral production increased twelve-fold (Van Rensburg, 1974); in the past ten years alone the production of bauxite increased 137%, petroleum production increased 107%, and the production of iron ore, nickel, vanadium, zinc, and chromium all increased by more than 50% (see Table 4-III). Nevertheless, world mineral production remains highly concentrated. Almost 70% of total world production of metals comes from about 170 mines (Beckerdite, 1972); some 1,000 of the world's largest mines (greater than 150,000 tons of ore produced per year) account for 90% of the value of all minerals produced outside of the centrally planned economies (United Nations, 1972). The concentration of petroleum production is especially marked; 85% of total world production of hydrocarbons (plus reserves) occurs in 238 fields — these represent less than 5% of all producing fields in the world (Pratt and Brobst, 1974).

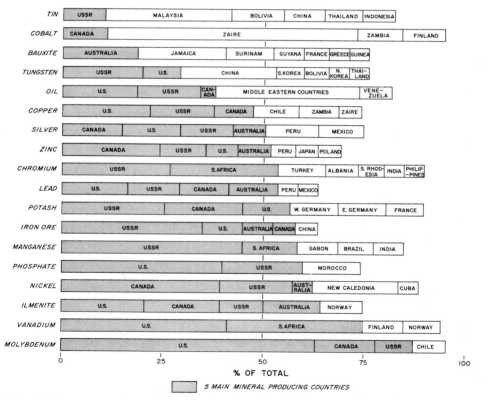

Fig. 4-2. Percentage share of world mineral production by countries producing more than 5% of total world output. (Data from United Nations, 1974c.) Shaded area indicates production in five main producing countries.

TABLE 4-III

Percentage share of mineral production in countries producing more than 2% of total world production in 1972 (data from United Nations 1973b, 1974c)

	1962	1972
Petroleum (world production, mil. tonnes)	(1,220.6)	(2,521.8)
U.S.	29.6	18.5
U.S.S.R.	15.6	16.7
Saudi Arabia	6.2	11.3
Iran	5.4	9.8
Venezuela	13.7	6.7
Kuwait	7.6	6.0
% in countries producing more than 5% of total:	78.1	69.0
Libya	—	4.2
Canada	2.6	3.0
Iraq	4.0	2.8
United Arab Emirates	—	2.3
Algeria	1.7	2.0
% in countries producing more than 2% of total:	88.3	85.4
Iron ore (world production, mil. tonnes)	(288.6)	(434.3)
U.S.S.R.	35.4	41.3
U.S.	13.8	10.5
Brazil	2.5	6.6
China	5.7	5.8
Canada	4.9	5.6
Liberia	0.9	5.2
India	4.2	5.1
Sweden	4.7	5.0
% in countries producing more than 5% of total:	72.1	85.1
France	7.5	3.8
Venezuela	2.9	2.6
% in countries producing more than 2% of total:	82.5	91.5
Bauxite (world production, mil. tonnes)	(28.0)	(66.3)
Australia	—	20.7
Jamaica	27.5	19.5
Surinam	11.8	10.3
U.S.S.R.	—	7.1
Guyana	12.9	5.6
France	7.9	5.0
% in countries producing more than 5% of total:	60.1	68.2

113

TABLE 4-III (continued)

	1962	1972
Guinea	5.4	3.9
Greece	4.6	3.6
Hungary	5.4	3.6
U.S.	6.1	3.3
Yugoslavia	4.6	3.3
India	2.0	2.6
% in countries producing more than 2% of total:	84.5	90.5
Tin (world production, thous. tonnes)	(195.1)	(246.6)
Malaysia	30.5	31.1
Bolivia	11.4	13.1
U.S.S.R.	10.2	10.9
China	15.4	9.3
Indonesia	9.0	8.6
Thailand	7.7	8.2
% in countries producing more than 5% of total:	84.2	81.2
Australia	1.4	4.5
Nigeria	4.2	2.7
Zaire	3.7	2.6
% in countries producing more than 2% of total:	93.5	91.0
Copper (world production, thous. tonnes)	(4,577.9)	(6,824.9)
U.S.	24.3	22.1
U.S.S.R.	13.1	15.4
Chile	12.9	10.6
Zambia	12.3	10.5
Canada	9.1	10.4
Zaire	6.5	6.0
% in countries producing more than 5% of total:	78.2	75.0
Philippines	1.2	3.1
Australia	2.4	2.5
Poland	0.3	2.2
% in countries producing more than 2% of total:	82.1	82.8
Silver (world production, thous. tonnes)	(7.7)	(9.0)
Canada	11.7	16.7
Mexico	16.9	13.3
U.S.	14.3	13.3
Peru	13.0	13.3
U.S.S.R.	13.0	13.3
Australia	6.5	7.8
% in countries producing more than 5% of total:	75.4	77.7

114

TABLE 4-III (continued)

	1962	1972
Japan	3.9	3.3
East Germany	2.6	2.2
% in countries producing more than 2% of total:	81.9	83.2
Nickel (world production, thous. tonnes)	(366.9)	(643.4)
Canada	57.4	36.2
U.S.S.R.	21.8	19.7
New Caledonia	9.3	15.9
Cuba	4.5	5.7
Australia	—	5.5
% in countries producing more than 5% of total:	93.0	83.0
Indonesia	0.1	3.5
Dominican Republic	—	2.7
U.S.	3.4	2.7
% in countries producing more than 2% of total:	96.5	91.9
Lead (world production, thous. tonnes)	(2,541.8)	(3,462.0)
U.S.	8.5	16.2
U.S.S.R.	13.8	13.3
Australia	14.8	12.2
Canada	7.5	10.9
Peru	6.6	5.2
% in countries producing more than 5% of total:	51.2	57.8
Mexico	7.6	4.7
Yugoslavia	4.0	3.5
China	3.5	3.1
Bulgaria	4.2	2.9
Morocco	3.5	2.7
North Korea	2.0	2.3
Sweden	2.7	2.2
Poland	1.5	2.0
% in countries producing more than 2% of total:	80.2	81.2
Zinc (world production, thous. tonnes)	(3,689.0)	(5,621.3)
Canada	12.3	22.8
U.S.S.R.	11.1	11.6
Australia	9.3	8.8
U.S.	12.4	7.7
Peru	5.0	6.2
Japan	5.2	5.0
% in countries producing more than 5% of total:	55.3	62.1

TABLE 4-III (continued)

	1962	1972
Mexico	6.8	4.8
Poland	3.9	4.2
North Korea	2.4	2.5
West Germany	3.0	2.2
Italy	3.6	2.1
Sweden	2.2	2.0
% in countries producing more than 2% of total:	77.2	79.9
Vanadium (world production, thous. tonnes)	(8.0)	(14.4)
South Africa	16.2	30.6
U.S.	58.8	30.6
U.S.S.R.	—	18.0
Finland	7.5	8.3
% in countries producing more than 5% of total:	82.5	87.5
Namibia	11.2	3.5
% in countries producing more than 2% of total:	93.7	91.0
Cobalt (world production, thous. tonnes)	(15.6)	(21.4)
Zaire	62.2	49.5
Finland	13.5	10.7
Zambia	4.5	9.3
Canada	10.3	8.9
U.S.S.R.	—	7.5
Cuba	—	7.0
Morocco	8.9	5.1
% in countries producing more than 2% of total:	94.9	98.0
Chromium (world production, thous. tonnes)	(1,853.3)	(2,818.2)
U.S.S.R.	26.4	27.7
South Africa	21.7	23.5
Turkey	11.8	13.2
Albania	5.8	9.2
Southern Rhodesia	12.4	6.4
India	—	5.0
% in countries producing more than 5% of total:	78.1	85.0
Philippines	9.7	4.4
Iran	2.7	3.1
Finland	—	2.5
% in countries producing more than 2% of total:	90.5	95.0

TABLE 4-III (continued)

	1962	1972
Tungsten (world production, thous. tonnes)	(40.0)	(47.5)
U.S.S.R.	15.8	19.2
China (1972 figure revised)	34.0	18.5
U.S.	11.9	8.4
Thailand	0.5	6.9
North Korea	6.0	5.7
Bolivia	3.8	5.5
South Korea	11.2	5.3
% in countries producing more than 5% of total:	83.1	69.5
Canada	—	4.6
Australia	2.8	4.2
Portugal	3.8	3.6
France	1.0	3.4
Brazil	1.5	3.4
Peru	0.5	2.3
Japan	1.5	2.3
% in countries producing more than 2% of total:	94.2	93.3
Manganese (world production, thous. tonnes)	(6,581.5)	(10,064.2)
U.S.S.R.	49.3	43.8
South Africa	8.8	13.6
Brazil	7.8	11.4
Gabon	1.5	9.5
India	9.5	6.7
% in countries producing more than 5% of total:	76.9	85.0
Australia	0.5	4.0
China	3.6	3.0
Ghana	2.8	2.2
% in countries producing more than 2% of total:	83.8	94.2
Molybdenum (world production, thous. tonnes)	(34.1)	(77.6)
U.S.	68.0	64.0
Canada	1.2	13.3
U.S.S.R.	16.7	10.3
Chile	7.0	8.1
% in countries producing more than 5% of total:	92.9	95.7
China	4.4	2.0
% in countries producing more than 2% of total:	97.3	97.7

Fig. 4-2 illustrates the degree to which the same five countries which dominate world mineral reserves — the U.S., the U.S.S.R., Canada, Australia, and South Africa — also dominate world mineral production; together they account for more than 50% of the production of two-thirds of the industrially important minerals. The U.S., although a net mineral importer, produces about one-quarter of the world's supply of minerals and is the leading producer of those minerals which enter into world trade (U.S. National Commission on Materials Policy, 1972). The U.S. produces a major share of the world's molybdenum, vanadium, phosphate, copper, uranium, and the platinum group of metals (the latter two are not included in Fig. 4-2 since data on world production are incomplete); the U.S. also produces sizeable shares of ilmenite, lead, silver, and petroleum.

The U.S.S.R. is the world's second largest mineral producer, at least 20% of the world's output of minerals is produced in the Soviet Union. The U.S.S.R. leads in world production of iron ore, manganese, potash, and chromium, and produces a large share of the world's nickel, phosphate, tungsten, copper, lead, zinc, silver and petroleum; production of gold, platinum and uranium is large, although data on actual output are unavailable. The Soviet leaders claim that the U.S.S.R. has reached the point of mineral self-sufficiency (Sutulov, 1973); while this may be an exaggerated claim, there is little question that Soviet resources are enormous and that production is expanding rapidly.

Canada is the world's leading mineral producer on a value per capita basis — only the U.S. and the U.S.S.R. produce more in total dollar value (Govett and Govett, 1973). Canada ranks first in world production of nickel, zinc, and silver, and is second in production of uranium outside of the U.S.S.R. Canada also produces significant amounts of ilmenite, lead, copper, the platinum group of metals, potash, iron ore, molybdenum, silver and gold.

Australia is a relative newcomer to the top ranks of world mineral producers. Only ten years ago Australia produced almost no bauxite; today Australian production accounts for about 20% of the total world output. Australia produces significant shares of ilmenite, lead, zinc, silver, and iron ore and could produce considerable uranium in the future (Mining Journal (London), 1974). The future of Australian mineral production will depend, to a very large extent, on government policy toward foreign investment and exports; the increase in Australian iron ore production in the last decade was primarily the result of the lifting in 1960 of the embargo on iron ore exports (imposed in 1939) and the resulting upsurge in exploration (Livingstone, 1967; King, 1967).

South Africa dominates world chrome and manganese production and is the world's largest gold producer, accounting for more than 75% of total gold production in the non-communist world (Canada is the second largest; Mining Journal (London), 1974). South Africa also produces a significant share of the world's uranium.

Chinese mineral production may be more important than is indicated in Fig. 4-2. Information on Chinese production is incomplete, but China ranks first in world production of tungsten and antimony, and ranks just behind the U.S. and the U.S.S.R. in coal production; China ranks fifth in world production of tin and mercury and is a medium-sized producer of aluminium, iron ore, copper, lead, and zinc (*Mining Journal (London)*, 1974).

Some forty other countries produce sizeable shares of the minerals included in Fig. 4-2; New Caledonia (nickel), Morocco (phosphate), the Middle Eastern countries (petroleum), and the Caribbean countries (bauxite) are important producers. However, only eleven countries (Chile, Peru, Bolivia, Mexico, Zambia, Zaire, Thailand, India, Finland, Norway, and France) produce two or more of the minerals in Fig. 4-2; only New Caledonia, the Middle Eastern countries, Jamaica, Malaysia, and Zaire produce very large shares of any one mineral.

Although the countries with large share of reserves of a given mineral are, in most cases, large producers of that mineral, world mineral production reflects the geographic distribution of reserves to a lesser degree than might be expected (see Fig. 4-3). The U.S. share of world reserves of a large number of minerals is less than its share of production of these minerals (the exceptions are silver, lead, and zinc), reflecting in part the declining importance of the minerals industry in the U.S. (see below). The most extreme cases of countries with large reserves relative to production are petroleum in the Middle Eastern countries (where reserves account for nearly 70% of the world's total, while production has been averaging only about one-third of total world production); tungsten (Chinese reserves account for three-quarters of the world total but production is only about one-third); and chromium (South Africa has huge reserves relative to production, while the U.S.S.R. is a large producer relative to its reserves). Some of the discrepancy between the share of reserves and production in the U.S.S.R. may reflect inadequate or incorrect information; this may also be the case for China and Cuba.

The data in Fig. 4-3 indicate something about future patterns of mineral production. Quite clearly production of a number of minerals could be increased in the U.S.S.R., while American reserves of most minerals do not support large increases in future production. Tin production in Thailand could expand considerably, as could bauxite production in Australia, chromium production in South Africa, and petroleum production in the Middle East and North Africa. Chinese tungsten reserves are sufficient for China to produce most of the world's tungsten for many decades; Morocco, in addition to Canada, could supply large shares of the world's potash.

Such conclusions, however, can be misleading. Reserves can change remarkably in a short period of time — witness the increase in Australian bauxite and iron ore reserves in the past two decades. The development of the ocean's resources will have a potentially great impact on the pattern of production and particularly on some of the world's leading producers of

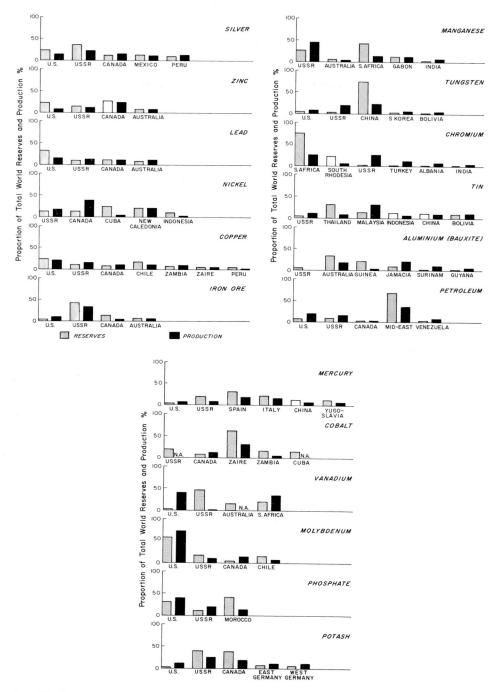

Fig. 4-3. Percentage share of world mineral reserves and production of 18 minerals in main mineral producing countries. (Data from U.S. Bureau of Mines, 1970; U.S. National Commission on Materials Policy, 1973a; *Mining Journal (London)*, 1973; United Nations, 1974c.)

minerals like nickel and manganese; one ship can dredge 10,000 tonnes of manganese nodules a year; this would be sufficient to supply more than the current annual consumption of manganese and nickel in the U.S. (Warren, 1973). The drive for self-sufficiency can lead a country to produce rather than import minerals, even though its reserves may be limited. On the other hand, the existence of large reserves does not, in itself, guarantee an economic minerals industry.

While the existence of reserves is usually a basic requirement for mineral production — although Japan has built up a minerals industry based on importing raw materials for re-export — the existence of mineral deposits alone is not a sufficient basis for mineral production. Low mineral prices and high mining costs (or environmental restrictions) may make a mine economically unworkable; the scale of operations and mining efficiency may make low-grade ores competitive with higher-grade ores. New transport facilities may reduce the advantages of nearness to markets. Only very large-scale operations may justify the expenditure on infrastructure needed to develop a new mine; finance for such ventures may not be available because of political factors.

The essential characteristics of the mining industry — the need to undertake large capital investment programmes with a high degree of risk for exploration and development before a product can be marketed, and the significant economies of large-scale operations — tend to confine mining operations to areas where deposits are large and to companies which have access to finance and technical skills and control of marketing outlets. For these reasons the multinational corporations have become increasingly important in the mining industry; out of a total of fifty-two new mining operations established outside of the centrally planned countries in 1968, twenty-seven had a 1 million ton or greater annual capacity, and most were financed by the multinationals (Beckerdite, 1972, United Nations, 1973a; Warren 1973). A single or a few companies today have a virtual monopoly in the production and marketing of molybdenum, platinum, diamonds, and zirconium, and there are a limited number of companies which substantially control world production of aluminium, nickel, lead, copper, tin, and zinc (United Nations, 1973a). While this pattern was established some time ago, the degree of concentration of ownership is becoming increasingly pronounced.

The main change in the world mining industry this century has been this increasing dominance of the multinational corporation coupled with rising demand for a whole range of new minerals and the declining importance of the proximity of coal and iron ore resources which underlay the industrialization of the 19th and early 20th century. Before World War II it was the mineral wealth of Europe and North America and the colonies of the U.K. and a few European countries which dominated world mineral demand and supply. With the decline of the over-riding importance of coal, the economic growth of the centrally planned countries, and the independence of the former colonies, the whole pattern of mineral production changed. The replacement of coal by relatively cheap imported petroleum combined with rising mining costs, depletion of high-grade deposits, difficulties of foreign operations in the less developed countries, and environmental constraints which limited the development of lower-grade deposits in North America and Western Europe, led the main mineral-consuming countries to increasingly rely on imports. High labour costs made it uneconomic to develop the tin resources of Cornwall in the U.K.; labour disputes drastically reduced Canadian nickel production; high transport costs reduced bauxite production in some of the Caribbean countries; cheap petroleum imports discouraged new exploration and technological developments in the developed countries. Exploitation of the Canadian tar sands and the American oil shales will depend, to a very large extent, on the future price of a barrel of oil; uranium mining in the U.S. and Canada is a direct function of the world price of uranium. Environmental protection measures have directly and indirectly increased mining costs in North America and Western Europe; exploration has been curtailed, and production has been reduced or actually stopped in a number of mines in the U.S. (U.S. Secretary of the Interior, 1972; Flawn, 1973); output from strip-mining operations has been seriously affected in the U.S. and in some European countries (see Chapter 9); government fiscal policies have reduced investment in exploration and development in Australia and Canada (Govett and Govett, 1976); political difficulties have reduced exploration in some of the Latin American countries and reduced production in some cases (such as in Chile).

Petrick (1967) analyzed the changes in world production patterns between 1950 and 1965 for iron ore, copper, lead, zinc, aluminium, petroleum, and the nitrogen fertilizers by a simple statistical technique called "shift analysis" which divides countries into two groups; those in which output of a given mineral increased at a rate faster than the world average (net upwards shift), and those in which output increased at a rate slower than the world average (net downward shift). For each of the minerals considered in the study there was a net downward shift in production in the U.S. and a net upwards shift in the U.S.S.R.; these shifts were attributed primarily to the much more rapid rate of economic growth in the U.S.S.R. than in the U.S. over the period considered and to the very major Russian efforts in mineral exploration. Although U.S. production of the industrially important minerals doubled in value since 1945, world production doubled once every seven or eight years since 1945; the U.S. share of world production declined from 46% of the total in 1945 to 13% by 1971 (Cameron, 1973). (Table 4-IV gives data on this decline for a number of minerals.) Such a decline would not be un-expected in the Hewett-Lovering model of national history (Lovering, 1943) which states that the importance of mineral resources in national develop-ment reaches a peak after industrial development is well underway and thereafter declines. A comparison of U.S. mineral production and consumption as a percentage of Gross National Product (GNP) since 1900 shows that both production and consumption have been a decreasing share since the late 1950s (Cheney, 1974).

A net downward shift in production in Europe in favour of Australia, South America, Asia, and Africa is probably due to the same factors. Also, increased exploration activities (particularly in Australia, Venezuela, and Liberia) and rising costs and labour problems in the European countries must have been contributing factors.

For specific minerals considered in Petrick's (1967) study, a net downward shift in aluminium production in the U.S., Guyana, and Surinam in favour of Jamaica, Australia, Guinea, Greece, Yugoslavia, and the U.S.S.R. is attributed, in part, to the low-reserve, low-grade status of U.S. bauxite, high transport costs in Guyana and Surinam, and favourable transport and distribution networks developed for Australian ores. A net downward shift in copper production in the U.S. and Chile is a result of increasing use of substitutes for copper in the U.S. and increased recycling of scrap copper, lower mining costs (in some cases for higher-grade ores) in other countries, and the political difficulties in Chile. The most extreme case of a net downward shift — petroleum production in North and South America — was the result of the dramatic increase in Middle Eastern and Soviet petroleum production.

Table 4-III gives data on changes in the geographic distribution of mineral production in the ten years between 1962 and 1972. The shifts observed by Petrick (1967) for the minerals considered in his study continue, and the same patterns are noticeable for other minerals. Of the fifteen minerals included in Table 4-III, the U.S. share in world production declined for all eleven minerals produced in the U.S. except lead; capital expenditure in the mining industry in 1971 was unchanged from the 1967—1968 level (Landsberg, 1974). The share of Western European producers also declined between

TABLE 4-IV

Percentage share of U.S. production in total world production of selected minerals 1900—1970 (data from Schroeder and Mote, 1959; U.S. National Commission on Materials Policy, 1972, 1973a, b)

Mineral	1900	1957	1970
Aluminium	—	~47	< 5
Iron ore	31	25	13
Copper	~35	27	26
Lead	30	13	15
Silver	33	17	14
Zinc	62*	16	9
Petroleum	43	41	14

* Data for 1920.

1962 and 1972; only Sweden, France, and Finland now account for more than 5% of world production of any of the minerals included in Table 4-III.

In contrast, the percentage share of only four of the fifteen minerals produced in the U.S.S.R. declined (nickel, molybdenum, manganese, and lead); the Soviet share of world production of petroleum, iron ore, copper, bauxite, vanadium, cobalt, tungsten, and tin increased and shares of the other minerals remained virtually unchanged. Canadian and Australian shares also generally increased, especially Australian nickel, bauxite, and tin and Canadian lead, silver, zinc, tungsten, and molybdenum, although the Canadian share of world nickel production fell from 57% of the total in 1962 to 36% in 1972.

The LDCs as a whole accounted for an increasing share of world iron ore, nickel, cobalt, manganese, and tungsten production, while their shares of copper, bauxite, and lead production generally decreased between 1962 and 1972. The most severe decline was for bauxite, where the Caribbean countries' share in world production fell from 52% in 1962 to 35% in 1972. The share of copper production in the traditional copper producing countries — Chile, Zaire, and Zambia — also declined. Zaire's domination of world cobalt

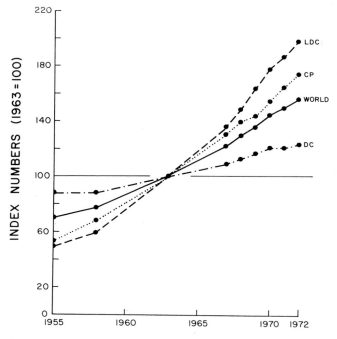

Fig. 4-4. Total world mineral production (value added) index numbers, 1963 = 100. (Data from United Nations, 1974c.) LDC = less developed countries; CP = centrally planned economies; DC = developed countries.

production was reduced as Cuban and Soviet production increased. On the other hand, nickel production in New Caledonia and Cuba rose, as did iron ore production in Liberia and Brazil. Asia continued to dominate world tin and tungsten production, and the Middle East and North Africa dominate petroleum production.

On balance, the LDCs share in total world mineral production has increased (Fig. 4-4), but much of this increase is due to petroleum production (see Fig. 4-5). If metallic minerals alone are considered (Fig. 4-6) the picture is considerably altered. Since 1963 the LDCs production of metallic minerals has risen at a much slower rate than production in the centrally planned economies and has grown only slightly faster than production in the non-communist developed countries. This may reflect the fact that in the past decade the level of exploration has been relatively low in the LDCs, while Soviet expenditures have been very large; more than 500,000 people, some 120,000 of them with university or technical institute degrees, were employed

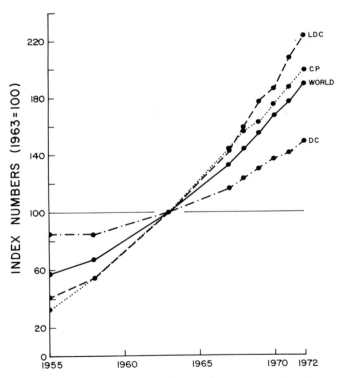

Fig. 4-5. World petroleum and natural gas production (value added) index numbers, 1963 = 100. (Data from United Nations, 1974c.) LDC = less developed countries; CP = centrally planned economies; DC = developed countries.

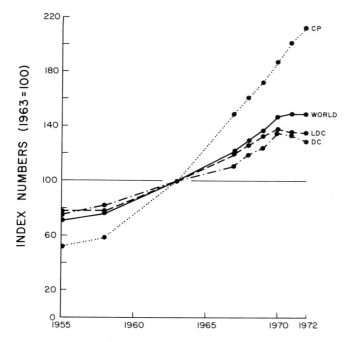

Fig. 4-6. World production of metallic minerals (value added) index numbers, 1963 = 100. (Data from United Nations, 1974c.) LDC = less developed countries; CP = centrally planned economies; DC = developed countries.

in geological prospecting in the U.S.S.R. in 1972 (*Mining Journal (London)*, 1973).

On a regional basis (Fig. 4-7), Asia recorded the largest increase in total mineral production (this includes petroleum production in the Middle Eastern countries), followed by Australia and the U.S.S.R. and Eastern Europe; Latin America and Western Europe fared relatively badly. The low rate of growth in Latin America was largely the result of political events and the effect of nationalization of foreign mines in Peru and Chile (in Chile copper production fell from 600 thousand tonnes in 1967 to 490 thousand tonnes in 1970, *World Mining*, 1974). The high cost of imported energy fuels, rising mining costs, and lack of exploration all contributed to the slow rate of growth in Western Europe.

The most important fact that emerges from the data in this section is that, in almost every case, the number of countries which produce a sizeable share of a given mineral is small, and that the changes in the pattern of production in the last decade have done little to alter the high degree of geographic concentration of mineral production in the U.S., the U.S.S.R., Canada, Australia, South Africa and a dozen or so Latin American, Asian, and African countries.

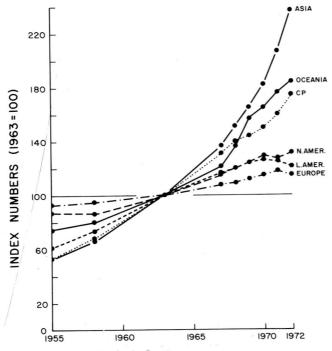

Fig. 4-7. Total world mineral production (value added) by region. Index numbers 1963 = 100. (Data from United Nations, 1974c.) Asia excludes Japan. CP = centrally planned economies; N. Amer. = North America; L. Amer. = Latin America.

The argument that the LDCs have a "strangle-hold" over the industrialized countries by their control of mineral supplies is largely unsubstantiated, although this is not to deny that the Middle Eastern countries presently control world petroleum prices nor that Chile, Zambia, and Zaire control a major share of world copper production. It does, however, indicate that the so-called confrontation between the "haves" and the "have-nots" will not be confined only to the developed versus the less developed countries. Many of the LDCs themselves, especially those who are becoming important mineral consumers, are mineral "have-nots".

This chapter deals only with primary mineral reserves and production; however, it must be noted that the use of recycled minerals, especially in the more developed countries, is becoming an important factor in the total supply of a number of minerals. If secondary production were added to primary production, the geographic distribution of world mineral production would alter somewhat. In the U.S. 35% of the lead consumed, 28% of the iron, and 20—30% of the copper, nickel, antimony, mercury, silver, gold, and platinum metals are recovered from scrap (U.S. Secretary of the Interior, 1972); almost one-half of the steel produced by the British Steel Corporation comes from scrap (Barnes, 1974). Increased recycling in other countries could significantly affect primary metal production.

Also, for most minerals there are a score or more countries which produce small amounts of minerals and which have reserves which could form the basis of a minerals industry; new methods of production could allow considerable expansion of small-scale mining (United Nations, 1972), but to the degree that the general trend is towards increasingly large-scale mining operations on large, low-grade deposits financed by governments and the multinational corporations, small-scale operations will probably remain fairly limited.

It appears likely that the U.S.S.R., Canada, and Australia will become increasingly important mineral producers in the future, given continued high levels of exploration and investment. Countries like Brazil and Mexico, as well as some of the politically stable Asian countries and China will also probably play an increasingly important role in world mineral production. Petroleum production will no doubt continue to be dominated by the Middle Eastern and North African countries, but it would be foolhardy to predict which country will be the world's leading oil producer in the 1980s — who could have foreseen the current position of Libya in 1960?

WORLD MINERAL CONSUMPTION

The pattern of world mineral consumption is largely a function of the stage of economic development in the various countries and regions of the world and the rate of economic growth; therefore, it is not surprising that, except for the U.S.S.R. and the U.S., the main mineral-producing countries are not the main consuming countries, and that the industrialized countries consume the major share of the world's minerals. North America and Western Europe together consume nearly two-thirds of the world's petroleum (Darmstadter, 1971); 20 tons of new mineral supplies per person are needed each year by the American economy (U.S. Secretary of the Interior, 1972).

Since 1950 the world has consumed more minerals than were produced in the previous recorded history of the world; most of this consumption took place in North America and Western Europe. Table 4-V gives available data on the current share of world mineral consumption accounted for by the main mineral-consuming nations and illustrates the degree to which the U.S.S.R. and Japan are now important mineral consumers. The centrally planned economies increased their share of world consumption of aluminium, copper, iron ore, lead, manganese, nickel, phosphorus, tin, and zinc from 16% in 1950 to 26% in 1970; Japan's share increased from 1% in 1950 to 12% in 1970; in the same period Western Europe's share declined from 30% to 27%. (United Nations, 1975.) For the developed countries as a whole, the

TABLE 4-V

Percentage share of world mineral consumption by country or region 1970—1972 (data from *Mining Journal (London)*, 1974; OECD, 1973; U.S. National Commission on Materials Policy, 1973a)

Mineral	U.S.	Japan	W. Europe	U.S.S.R.	Canada	Australia	Asia	Africa
Zinc	23	13	30	20	2	2	4	2
Lead	28	5	31	25	1	1	3	1
Copper	33	15	40	n.a.	4	2	2	n.a.
Aluminium	47	12	29	n.a.	3	1	4	1
Tin	28	17	30	n.a.	3	n.a.	n.a.	n.a.
Petroleum	38	10	30	11	4	n.a.	n.a.	n.a.

n.a. = not available.
Note: data for copper, aluminium, and tin exclude the communist countries. U.S.S.R. data for copper and aluminium are estimated at 13% and 12% of total world consumption, respectively.

share in world mineral consumption of the nine minerals fell from 80% in 1950 to 68% in 1970, while in the LDCs it rose from 4% to 6% during the same period (United Nations, 1975).

The decline in the U.S. share in mineral consumption has been much sharper than that of Western Europe; today the U.S., although still an important mineral consumer, is no longer the world's principle market for many minerals. In 1950 the U.S. consumed 42% of the world's minerals; this share dropped to 27% in 1970. As Table 4-V shows, American consumption of copper, lead, zinc, and tin is smaller than Western Europe's. For eighteen industrially important minerals (including the main non-ferrous metals, the ferro-alloy metals, the main fertilizer minerals, and iron and cement) Cameron (1973) calculated that the U.S. share had dropped to 16% of the world total in 1971. In 1930 the U.S. consumed 90% of the world's copper; it now consumes about one-third; the U.S. share of lead and zinc consumption has fallen from around 50% in 1930 to one-quarter now (U.S. Bureau of Mines, 1970; Landsberg, 1974). Japan consumes more iron and steel per capita than the U.S.; Canada consumes more copper on a per capita basis; Belgium-Luxembourg consumes more zinc per capita (U.S. National Commission on Materials Policy, 1973a).

The decline in U.S. dominance of world mineral consumption seems likely to continue; the growth of mineral consumption in the rest of the world is currently increasing at a faster rate than in the U.S. (where it has been increasing at a rate of approximately 3% annually). Japan, where real Gross National Product (GNP) has been doubling every seven or eight years, is rapidly challenging the American position as the world's most important mineral consuming country; recession in 1974—1975 may reduce the Japanese rate of growth, but the U.S. rate of growth will also be affected. On the basis of available data, the share of the U.S.S.R. and Eastern European countries in world mineral consumption is also increasing at a rate well in excess of the American rate of growth. Currently the LDCs consume only about 10% of the world's mineral output, but consumption in a number of Asian and Latin American countries is increasing at an annual rate exceeding that in both the U.S. and Europe.

The U.S. Bureau of Mines (1970) forecasts for the growth in mineral consumption in the next three decades for the U.S. and the rest of the world are shown in Table 4-VI. For the fifteen minerals covered, only demand for uranium, tungsten, and vanadium are forecast to increase less in the rest of the world than in the U.S. The high projections for petroleum demand in the rest of the world reflect the fact that, while demand for energy fuels in the U.S. is currently increasing at an annual rate of about 3%, it is increasing at a rate nearly double this in the rest of the world; between 1965 and 1970 energy consumption in Japan increased annually by nearly 14% and even in the LDCs it grew by 8% per year (United Nations, 1974a). Even in the poor and heavily populated countries of Asia the rate of growth of per capita energy consumption in the last decade has exceeded that of the U.K. (a 12% increase between 1961 and 1971); Pakistan, with an increase of 35%

TABLE 4-VI

Ratios of forecast mineral demand in 2000 to mineral demand in 1968 for selected minerals in the U.S. and the rest of the world (data from U.S. Bureau of Mines, 1970)

Mineral	U.S.		Rest of the world	
	high projection	low projection	high projection	low projection
Petroleum	3.3	1.5	5.6	3.0
Uranium	29.1	25.0	24.8	22.6
Chromium	2.8	1.9	2.8	1.8
Iron ore	2.1	1.5	2.2	1.6
Nickel	3.5	2.4	n.a.	2.3
Aluminium	9.5	4.8	10.6	4.9
Copper	1.3	0.8	6.1	2.9
Lead	3.1	1.5	2.1	1.7
Mercury	2.5	1.6	2.7	2.0
Silver	3.1	1.6	3.5	1.6
Tin	1.7	1.2	2.1	1.0
Zinc	2.8	1.5	2.8	2.2
Phosphorus	2.8	1.6	7.8	4.0
Tungsten	5.7	3.7	2.4	1.9
Vanadium	6.5	4.4	3.6	2.6

n.a. = not available.

TABLE 4-VII

Per capita consumption (in kg) of selected minerals by region (data from West, 1973)

Metal	World average	U.S.	Japan	U.K.	South America	Africa
Aluminium	13.5	20	10.5	9.6	0.4	0.2
Copper	2.4	10.8	8.8	11.5	1.0	0.2
Lead	1.0	5.7	n.a.	6.9	1.0	1.5
Zinc	1.2	5.9	n.a.	7.0	0.6	2.4
Tin	0.06	0.25	0.26	0.40	0.02	0.01
Nickel	0.02	0.70	n.a.	0.62	0.01	0.02
Steel	250	650	650	450	50	20

n.a. = not available.

between 1961 and 1971, and China, with an increase of 37% in the same period, nearly matched the U.S. rate of growth of 39%, although none of these countries approached the Japanese rate of 149% (United Nations, 1973b).

In the next few decades the present level of 10% of world energy consumed in the LDCs, as well as the share of metallic minerals consumed in a number of Asian and Latin American countries should increase significantly. Table 4-VII, which shows the consumption of a number of minerals on a per capita basis, indicates the potential in the LDCs for increased consumption. Intensity-of-use indices (based on the ratio of the quantity of a mineral consumed within a country to its GNP) for a number of minerals were calculated by Malenbaum et al. (1973) to illustrate the degree to which the demand for minerals increased during the early stages of economic development, reaches a peak as per capita income rises above a certain level (e.g., it seems to peak for energy at about $2,000 per capita), and declines thereafter. Many of the indices reached a peak in 1966—1969 in Western and Eastern Europe, the U.S.S.R., Canada, Australia, and Japan. In contrast, in the rest of the world the indices have not yet reached a peak and are expected to continue to increase in Africa (with the exception of South Africa), Asia, Latin America, and China until the year 2000. The share of non-U.S. demand for iron ore is expected to grow from the present 79% of the world total to 87% by the year 2000; for copper and aluminium, the share of demand in the rest of the world is expected to grow from 71% and 60% of the world total, respectively, to 78% and 67% respectively. Energy consumption outside of the U.S. is forecast to increase from 66% of the world's total to 72% by the end of the century (in Japan per capita energy consumption is expected to grow from 2.4 tonnes (coal equivalent) in 1966—1969 to 11.5 tonnes in the year 2000, and in Asia and Africa it is expected to double in the next 30 years).

Except in the U.S.S.R., most of the increase in mineral consumption in

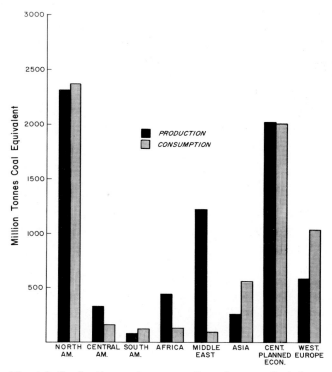

Fig. 4-8. Production and consumption of energy fuels by region, 1972. (Data from United Nations, 1974a.)

both the developed and less developed countries will have to be met by increased imports. The disparity between consumption and production of most minerals is large for most of the industrialized countries; it is especially acute for energy fuels in Western Europe (see Fig. 4-8). The ratio of energy production to energy consumption fell from 1.03 in 1925 to 0.46 in 1967 in Europe; in the U.S.S.R. it rose during the same period (Darmstadter, 1971). The deliberate policy of reducing the share of coal in electricity generation in the U.K. in the 1960s and the policy of controlled cut-backs in coal production in a number of European countries has meant increased petroleum imports, as has the increasing reliance on petroleum for energy in Asia and Latin America.

Whether the increase in the price of oil will result in a reversal of the trend toward an increasing share of energy requirements being met by imported petroleum — particularly in Western Europe — is uncertain. OECD (1974) forecasts a reduced rate of growth in petroleum demand which may result in a considerably lower level of oil imports into OECD countries, especially after 1980; the share of oil in total energy consumption in OECD member countries is forecast to fall from the current level of about 55% to approximately 45% by 1985. This projection is based on the assumption that the annual rate of growth of total energy consumption in OECD member countries will be about 3.5—4% (instead of 5% as earlier forecast), that the declining role of coal will not continue but that the trend will be reversed, and that nuclear energy will provide more than 10% of primary energy needs (see Chapter 7).

WORLD TRADE IN MINERALS

The role of minerals in world affairs has long been a matter of concern (Lovering, 1943). The recent efforts by OPEC to use oil as a political weapon is yet another example of the degree to which the world is dependent on trade. While world trade as a whole has been growing at a rate twice that of global GNP in the past 15 years, mineral exports have been increasing at a

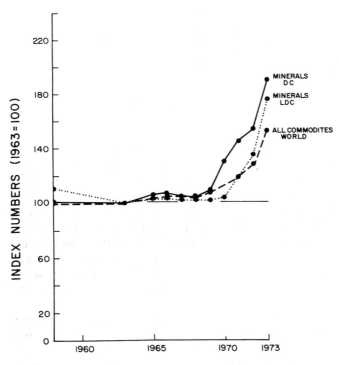

Fig. 4-9. World mineral exports and world exports of all commodities. Index numbers 1963 = 100. (Data from United Nations, 1974b.) Excludes trade by centrally planned economies. DC = developed countries; LDC = less developed countries.

even more rapid rate (see Fig. 4-9); between 1966 and 1972 exports of minerals increased by 83%, while exports of all other commodities increased by 72% (U.S. Bureau of Mines, 1973, 1974). Minerals as a group now account for about one-quarter of total world trade.

Until after World War II most of the world's mineral exports went to the Western European countries, the U.K., and the U.S. Since 1950 Japan has become a major mineral importer, and the centrally planned economies, especially the U.S.S.R., have begun to import and export on a large scale; even China has increased its minerals trade, and in the other countries of Asia, Latin America, and Africa — traditional mineral exporting regions — the level of mineral imports has been growing rapidly.

Trade by the communist countries is not included in Fig. 4-9 (nor in Figs. 4-10 and 4-11) since comparable data for these countries are not available. It is known that trade between the various communist countries is large, and that trade between the U.S.S.R. and the rest of the world is growing. Sutulov (1973) estimates that Soviet mineral exports account for about 12% of total Soviet mineral production; this is an increase from the 9% estimated for 1961 by Shimkin (1964) and may be a conservative figure.

It is estimated that the centrally planned economies together account for about 11—12% of total world mineral exports (U.S. Bureau of Mines, 1973). Approximately 16% of total world aluminium exports, 8% of world copper exports, and 10% of the zinc and tin exports are from the U.S.S.R.; exports of gold and platinum metals are very large. Approximately 65% of Soviet exports are to other communist countries, and about one-quarter are to Western industrialized countries. Soviet imports of minerals are more limited, although imports have been increasing recently as a result of barter agreements with some LDCs; the Eastern European countries are larger importers, although a large share of their imports are from the U.S.S.R. Data on Chinese trade are limited; there was a dramatic increase between 1972 and 1973 when trade increased by 74% (United Nations, 1974a).

The main mineral-importing nations, outside of the communist countries, are shown in Fig. 4-10. Only in the case of petroleum do the LDCs constitute a major factor in world demand. The concentration of imports of non-ferrous metals (the United Nations series includes copper, lead, zinc, tin, aluminium, and nickel), iron ore, and petroleum in the U.S., Japan, the U.K., and Western Germany is clear, as is the domination of world mineral imports by the developed countries generally.

The U.S. imports about 15% of the minerals it consumes; American mineral imports average one-fifth of total commodity imports into the U.S. (U.S. National Commission on Materials Policy, 1973b). While the U.S. has never been entirely self-sufficient in minerals, the increasing importance of mineral imports in the U.S. economy has been an important factor in world mineral trade in the last few decades. In the early part of this century U.S. mineral imports were limited to the few minerals not found in the continental

134

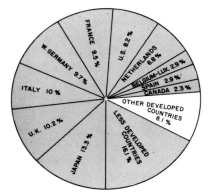

CRUDE PETROLEUM IMPORTS

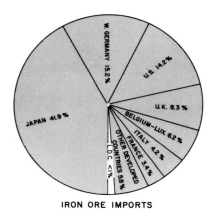

IRON ORE IMPORTS

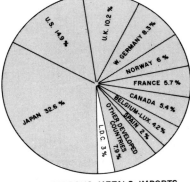

NON-FERROUS METALS IMPORTS

Fig. 4-10. Mineral imports by countries importing more than 2% of total imports. (Data from United Nations, 1974b.) Excludes trade by centrally planned economies. Shaded area indicates developed countries.

U.S.; the U.S. had a net export surplus of about 3% of consumption in 1900. This surplus began to decline after 1910 and, since 1920, mineral imports have exceeded exports in all but five years. Table 4-VIII gives the percentage share of imports to total U.S. demand for a number of minerals; it is this data which has caused concern in the U.S. recently, in spite of a generally optimistic view about future sources of supply taken by writers like Landsberg (1964, 1974) and Fischman and Landsberg (1972). The National Commission on Materials Policy, set up by the U.S. Congress in 1970 to develop a national materials policy concluded: ". . . as the Nation's needs continue to grow and as per capita consumption of materials in other countries increases at an even faster rate than ours, it becomes increasingly difficult for

TABLE 4-VIII

U.S. imports as a percentage of total U.S. demand for selected minerals 1971 (data from U.S. Secretary of the Interior, 1972)

Mineral	Imports as a percentage of total demand
Titanium (rutile)	100
Manganese	96
Aluminium	92
Chromium	89
Cobalt	75
Platinum group of metals	75
Nickel	66
Tin	64
Antimony	54
Gold	48
Zinc	45
Mercury	42
Silver	34
Iron ore	30
Vanadium	28
Petroleum	24
Lead	18
Titanium (ilmenite)	16
Copper	6

the United States to fill its ever-growing deficit by imports, even at increasing prices." (U.S. National Commission on Materials Policy, 1972, p. 4).

The Western European countries, the U.K., and Japan are even more dependent on mineral imports than the U.S., as shown on Tables 4-IX and 4-X. Japan's imports of minerals account for 40% of her total imports; 100% of the bauxite, phosphate, nickel, and cobalt, more than 95% of the iron ore and petroleum, and more than one-half of the zinc, copper, manganese, and coal consumed in Japan are imported. As Japan's import requirements have grown, an increasing share have come from the LDCs (Hruz, 1967); the share from the LDCs rose from 8% in 1960 to more than 13% in 1970, and only the U.S. now imports more from the LDCs than Japan (United Nations, 1971a).

Japan's need to import minerals is primarily a function of the lack of domestic sources of supply; American and European reliance on mineral imports is a result of a combination of factors, most notably the enormous increase in mineral consumption in these countries since World War II, rising domestic mining costs, lack of new exploration, environmental protection measures, and the conversion from coal to petroleum, especially in Europe. In 1973 OECD (1973) expected the demand for oil in member countries to increase

TABLE 4-IX

Ratio of net imports to consumption 1954—1956 and 1967—1969 (data from UNCTAD, 1973)

Mineral	EEC countries		Japan		U.K.		U.S.	
	1954—56	1967—69	1954—56	1967—69	1954—56	1967—69	1954—56	1967—69
Iron ore	32	66	84	97	62	74	18	32
Copper	92	95	19	68	n.a.	n.a.	10	3
Bauxite	36	48	100	100	100	100	75	87
Lead	n.a.	72	n.a.	59	100	93	36	22
Zinc	56	67	15	58	97	100	49	51
Tin	99	98	11	39	97	94	100	100
Petroleum	93	96	99	99	100	100	10	14

n.a. = not available

by a further 75% this decade; this has been revised downward (OECD, 1974) and oil imports after 1980 are expected to decline.

The increased use of petroleum in place of coal in many Asian and Latin American countries has meant that the LDCs share of world petroleum imports has risen recently. The rise in oil prices in 1973—1974 has caused serious balance of payments problems in a number of developing countries; in India the total import bill increased by one-third between 1973 and 1974. The World Bank has estimated that the over-all deficit of the non-OPEC countries in 1974 was $2.6 billion and could be as high as $6.8 billion in 1975 (Tuke, 1975).

Aside from the importance of petroleum imports, minerals are not a major factor in LDCs imports; in Latin America, Africa, and Asia mineral imports account for only 12—15% of total imports (see Table 4-X). On the other hand, mineral exports are extremely important; if Fig. 4-10 is compared with Fig. 4-11 the relative importance of mineral exports and imports in the LDCs is clear. The distribution of petroleum exports is not shown on Fig. 4-11 since the Middle Eastern countries account for approximately two-thirds of total exports; Venezuela, Indonesia, Nigeria, Canada, and the U.S.S.R. are the only other countries which export sizeable quantities of petroleum.

While the developed countries are still the main mineral exporters, the trend has been toward an increasing share of exports from the LDCs. The United Nations index of non-ferrous minerals trade based on 1963 = 100 was

TABLE 4-X

Mineral exports and imports as a percentage of total trade (data from U.S. Bureau of Mines, 1973)

Country or region	Mineral exports as a percentage of total exports	Mineral imports as a percentage of total imports
Developed countries		
U.S.	9	24
Canada	29	14
Japan	15	40
Australia	11	14
EEC countries	16	25
Other W. Europe	13	21
E. Europe (incl. U.S.S.R.)	23	19
Less developed countries		
Latin America	39	13
Africa	49	12
Middle East	84	14
Asia	16	15

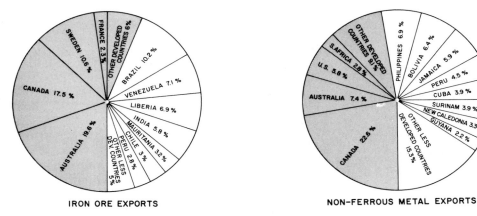

IRON ORE EXPORTS   NON-FERROUS METAL EXPORTS

Fig. 4-11. Mineral exports by countries exporting more than 2% of total exports. (Data from United Nations, 1974b). Excludes trade by centrally planned economies. Shaded area indicates developed countries.

150 for the developed countries in 1972, while it was 161 for the LDCs; preliminary estimates put it at 235 for the LDCs in 1973, compared with 198 in the developed countries (United Nations, 1974b). Copper exports from Zaire, Zambia, Chile, and Peru account for about one-half of the world's total copper exports; lead and zinc from these same four countries average between 10 and 15% of total world exports. Iron ore exports from Brazil, Liberia, Venezuela, and India account for almost one-third of total world iron ore exports.

The importance of mineral exports to the economies of the exporting nations is enormous. Copper exports account for 95% of Zambia's total exports and 67% of Zaire's; exports of iron ore make up 72% of Liberia's total exports; aluminium accounts for 87% of Surinam's total exports (UNCTAD, 1973). Among the Latin American countries mineral and fuel exports are 94% of Bolivia's total exports, 97% in Venezuela, 85% in Chile, and more than 50% in Peru and Guyana (United Nations, 1971a).

Fig. 4-11 also illustrates the degree to which world mineral exports are dominated by a relatively few countries; about a dozen countries are responsible for more than 90% of the world's exports of aluminium, copper, lead, zinc, tin and nickel, and 75% of the iron ore. The number of countries which account for more than 5% of total world exports of these minerals are limited to six in the case of the non-ferrous metals and seven in the case of iron ore.

In view of the relative concentration of mineral exports and the vulnerability of many of the mineral-consuming nations in the face of rising world demand, the conflict between mineral producers and mineral consumers could become acute. The success of the OPEC countries in increasing the price of oil four-fold — and severely upsetting the economies of a number of developed and less developed countries — is proving to be a model for other mineral producers (e.g., the creation of the Council of Copper Exporting Countries (CIPEC), the formation of an Intergovernmental Bauxite Association by Australia, Guinea, Guyana, Jamaica, Sierra Leone, Surinam, and Yugoslavia (United Nations, 1975), and the setting up of a study group by thirteen of the iron-exporting countries). In a report on a seminar at the University of Sussex on the problems of the mineral-exporting LDCs (Page, 1974) attended by senior officials of the main petroleum-, copper-, and bauxite-producing countries, the consensus was that the mineral-producing countries had been "exploited" by the consuming countries in the past, and that the inequality of power and standards of living must be reduced. The more optimistic hoped that the change would come through international cooperation; the pessimists foresaw increasing nationalization and OPEC-style action.

Producer associations may become an increasingly important factor in attempting to redress the terms of trade in favour of primary producing countries. To the extent that these associations are able to maintain the purchasing power of their exports and organize producers in such a way that all countries benefit from trade, such associations could encourage mining development. On the other hand, action to cut back production in order to maintain or raise prices (cartel-type action) could have serious effects on both producing and consuming nations. Fortunately cartels for most minerals, except petroleum, would be difficult to operate, since for many minerals substitutes are available, recycling can be increased, and domestic production by countries which currently choose to import lower-cost minerals from abroad could be increased if import prices rise; also, new sources of supply, particularly non-conventional sources such as the tar sands and oil shales for oil, may be developed. There are two criteria for successful OPEC-style action: exporters must control a very large share of total production of a mineral, and price increases must not result in a decrease in demand nor call forth increased supplies in other countries. Even in the case of copper, where a few countries dominate world production, the attempt by CIPEC to cut back production by 10% in 1974 in order to stabilize declining copper prices was not successful; production cut-backs in an over-supply situation, especially during a period of general economic recession, poses difficulties unless the share of production controlled by the cartel is very much larger than it is in the case of copper.

As an alternative to cartels, a number of countries have tried to set up producer-consumer associations and have created new agencies for regional cooperation. Bilateral arrangements have been particularly successful; Japan, following an established policy of combining investment with trade and guarantees of raw material supplies, signed an agreement with an Indonesian oil company in 1973 for a twenty-year period and has also negotiated an agreement with China for petroleum exports (*Mining Journal (London)*, 1974). Unfortunately attempts at broader international action have largely failed; even the more limited attempts to set up consultation or negotiations for specific commodities have not been very successful (UNCTAD, 1973).

The importance of world trade and cooperation has never been more apparent than it is today. The conclusion of the U.S. National Commission on Materials Policy (1973b, p. 9-6) that "The trade-offs between sovereignty and cooperation plus integration will be debated by all nations in the late 20th century" and that "Peoples' rising aspirations for a better life will weigh heavily in the calculus of benefits and costs from participating in and encouraging the growth of interdependence" echos the conclusions of the Third Session of UNCTAD in Santiago in 1973 that new means of expanding world trade and achieving international cooperation are more than ever necessary in this decade.

CONCLUSIONS

The concentration of mineral reserves in a few countries is likely to continue for most *conventional* sources of minerals in the next decade. An economist has perhaps best summed up the reason for the present geographic distribution of reserves: ". . . known reserves represent the reserves that have been worth finding, given the price and prospects of demand and the costs of exploration" (Beckerman, 1974, p. 118). The view expressed in Chapter 10, that the LDCs will become increasingly important mineral producers, reflects optimism about future exploration in Asia, Africa, and Latin America.

Any significant changes in the pattern of reserve distribution will depend on the rate at which resources can be converted into reserves — either by the development of new mining and processing techniques, new methods of exploiting low-grade deposits, price changes, or new exploration efforts. A major increase in exploration in the less explored parts of the world could alter the present pattern — two-thirds of Burma have not been effectively surveyed, only 5% of Indonesia has been geologically mapped, and much of Latin America is unexplored; exploration and development in these areas will depend primarily on political factors, the level of bilateral and multilateral assistance, and the development of new methods of financing that are acceptable to both the developing country and the investor (see Ridge, 1973).

If the current pattern of exploration continues — in the last decade three-quarters of the non-communist world's exploration activity was concentrated in Canada, Australia, the U.S., and South Africa, while a major exploration programme was undertaken in the U.S.S.R. — there is little likelihood of any significant change in the geographic concentration of reserves in the next decade. This conclusion does not preclude the possibility of finding another

North Sea oilfield or of new discoveries of base metals in Africa, nor does it negate the other main conclusion about reserves and resources — that non-conventional sources of minerals may be the most critical factor in the future pattern of world mineral supply.

While the geographic concentration of mineral reserves is a result of geologic events in the distant past and the rate at which resources have been converted into reserves, the domination of mineral production, consumption, and trade by a relatively few countries is largely the result of the pattern of economic development in the world in this century, particularly since World War II. The traditional position of the U.S. and Western Europe as the main markets for the world's minerals has been undermined by the growth of demand in Japan, the centrally planned economies, and a number of Asian and Latin American countries since 1950. The increase in exploration for and exploitation of minerals in Australia, Canada, the U.S.S.R., and Asia has altered the pattern of mineral production considerably. The phenomenal increase in world trade in the past two decades has been both a cause of and a result of the very rapid rise in world mineral demand and the changing pattern of consumption, particularly the shift from coal to petroleum as the main energy fuel.

The future growth of world mineral consumption will depend, to a very large extent, on the rate of economic growth in the developed and the less developed countries; most forecasts see a continuation of the rapid rate of increase in the past 20 years, particularly in the centrally planned economies, Japan, and some of the LDCs. The future of world mineral production will be determined primarily by the success of exploration efforts, the rate of technical progress in developing new extraction and processing techniques, and the structure of mineral prices and costs.

Most of the conventional deposits which have a surface expression in the intensively prospected parts of the world have been found; rising exploration costs, political and environmental barriers, the need to invest in remote and inhospitable parts of the globe all make the future of exploration for conventional ore deposits uncertain. Assuming the technology to exploit unconventional sources of minerals can be developed in the next few decades and that the problem of the enormous energy requirements to mine and process these ores can be overcome, there will still be a serious problem in determining who will control the ocean's resources and in the potentially disruptive effect of the exploitation of non-conventional resources on present mineral producers.

Soaring energy costs will also affect the mining of conventional ores; the most immediate effect may be price increases for some minerals, but high energy costs can also raise the cut-off grade for some ores, thereby reducing reserves. The effect of energy costs on mineral production will clearly vary; the effect may be minimal for minerals like copper, but may be very large for minerals like aluminium. Increased petroleum cost may also result in a change in the energy-mix, particularly by increasing the use of coal and accelerating the development of nuclear energy.

The future pattern of world mineral production will depend on whether the development of high-cost mines in the developed countries will be economic, whether environmental constraints will allow the mining of low-grade deposits, and on which countries are willing to spend the money needed for new exploration and development. The experience of the last decade indicates that exploration and development will probably continue at a high level in the centrally planned economies; the future in Asia, Latin America, and Africa will depend, to a very large extent, on political factors and the willingness of both the developed and the less developed countries to find new methods of resource finance and ownership (see Chapter 10). Continued growth of the minerals industry in Australia will depend primarily on the attitude of the Australian government; this is also the case in Canada. There is a real possibility of an economic-political axis incorporating a number of Southeastern Asian countries with Japan as the link between China and Southeast Asia (*Manchester Guardian Weekly*, 1974). Environmental constraints and the recent low level of exploration and investment in the mining industry in the U.S. will make it difficult for the U.S. to improve its position in the immediate future.

While it is beyond the scope of this chapter to forecast mineral demand and supply, the following trends may be noted:

(1) Increasing mineral trade and less national self-sufficiency in most of the mineral-consuming nations.

(2) Changes in traditional trading patterns, with the LDCs increasing their demand for minerals at a rate faster than that experienced in the developed countries, and with Japan and the centrally planned economies continuing their rapid rate of growth.

(3) Increasing conflict between those countries which are rich in resources and those which have to import to survive.

(4) A continuation and increase in the importance of large-scale mining and the development of new methods for financing exploration and development.

(5) Increasing exploitation of non-conventional sources of minerals, and a possible conflict over ownership of the wealth of the oceans and the seas.

To conclude on a hopeful note, it is possible that the world will recognize the need to reduce the rate of population growth and thereby reduce the strain on mineral reserves and that exploration for new ore deposits and research and development for new extracting and processing techniques will allow low-grade deposits and non-conventional deposits to be mined. Given the geological pattern of resource distribution, world trade in minerals must expand and ways must be found for the mineral "have-nots" to develop within a system of international cooperation.

REFERENCES CITED

Adelman, M.A., 1974. Oil demand, supply, cost and price in the world market. In: United Nations, *Petroleum in the 1970s*. New York, N.Y., pp. 163—170.

Banquis, P.R., Brasseur, R.B. and Masseron, J., 1974. Trends and prospects in crude petroleum production with particular reference to the period 1971—1980. In: United Nations, *Petroleum in the 1970s*. New York, N.Y., pp. 139—162.

Barnes, R.S., 1974. The material resources for the iron and steel industries. *Resour. Policy*, 1: 66—74.

Beckerdite, A.D., 1972. A presentation on world minerals. Stanford Research Institute— Geneva Roundtable on World Business, 20 pp.

Beckerman, W., 1974. *In Defense of Economic Growth*. Jonathan Cape, London, 287 pp.

Brobst, D.A. and Pratt, W.P. (Editors), 1973. *United States Mineral Resources. U.S. Geol. Surv. Prof. Paper* 820, 722 pp.

Brooks, D.B., 1966. *Low-grade and Non-conventional Sources of Manganese*. Resources for the Future, Washington, D.C., 123 pp.

Brooks, D.B., 1967. The lead-zinc anomaly. In: R.B. Toombs (Editor), *Proceedings of the AIME Annual Meeting 1967*. AIME, New York, N.Y., pp. 144—159.

Cameron, E.N., 1973. The contribution of the United States to national and world mineral supplies. In: E.N. Cameron (Editor), *The Mineral Position of the United States, 1975—2000*. University of Wisconsin Press, Madison, Wisc., pp. 9—28.

Cheney, E.S., 1974. U.S. energy resources: limits and future outlook. *Am. Sci.*, 62: 14—22.

Cox, D.P., Schmidt, R.G., Vine, J.D., Kirkemo, H., Tourtelot, E.B. and Fleischer, M., 1973. Copper. In: D.A. Brobst and W.P. Pratt (Editors), *United States Mineral Resources. U.S. Geol. Surv., Prof. Paper* 820, pp. 163—190.

Darmstadter, J., 1971. *Energy in the World Economy*. Resources for the Future, Baltimore, Md., 876 pp.

Fischman, L.L. and Landsberg, H.H., 1972. Adequacy of nonfuel minerals and forest resources. In: R.G. Ridker (Editor), *Population, Resources, and the Environment. The Commission on Population Growth and the American Future Research Report*. U.S. Government Printing Office, Washington, D.C., 3: 79—101.

Flawn, P.T., 1966. *Mineral Resources*. Rand McNally, Chicago, Ill., 406 pp.

Flawn, P.T., 1973. Impact of environmental concern on the mineral industry 1975—2000. In: E.N. Cameron (Editor), *The Mineral Position of the United States, 1975—2000*. University of Wisconsin Press, Madison, Wisc., pp. 95—108.

Govett, G.J.S. and Govett, M.H., 1973. Mineral resources and Canadian-American trade — double-edged vulnerability. *Can. Inst. Min. Metall. Bull.*, 66: 66—71.

Govett, M.H. and Govett, G.J.S., 1976. The problems of energy and mineral resources. In: F.R. Siegel (Editor), *Review of Research on Modern Problems in Geochemistry*. UNESCO, Paris, in press.

Heyl, A.V., Hall, W.E., Weissenborn, A.E., Stager, H.K., Puffett, W.P. and Reed, B.L., 1973. Silver. In: D.A. Brobst and W.P. Pratt (Editors), *United States Mineral Resources. U.S. Geol. Surv., Prof. Paper* 820, pp. 581—604.

Hobbs, S.W. and Elliot, J.E., 1973. Tungsten. In: D.A. Brobst and W.P. Pratt (Editors), *United States Mineral Resources. U.S. Geol. Surv., Prof. Paper* 820, pp. 667—678.

Hruz, F.M., 1967. Japan's mineral trade and balance of payments problem. In: R.B. Toombs (Editor), *Proceedings of the AIME Annual Meeting 1967*. AIME, New York, N.Y., pp. 250—277.

King, H.F., 1967. Base metal resources and prospects in Australia. In: R.B. Toombs (Editor), *Proceedings of the AIME Annual Meeting 1967*. AIME, New York, N.Y., pp. 226—239.

Klemic, H., Marsh, S.P. and Cooper, M., 1973. Titanium. In: D.A. Brobst and W.P. Pratt

144

(Editors), *United States Mineral Resources. U.S. Geol. Surv., Prof. Paper* 820, pp. 653—665.

Landsberg, H.H., 1964. *Natural Resources for U.S. Growth: A Look Ahead to the Year 2000.* Johns Hopkins, Baltimore, Md., 257 pp.

Landsberg, H.H., 1974. Policy elements of U.S. resources supply problems. *Resour. Policy,* 1: 104—114.

Livingstone, D.F., 1967. Australia looks to the future — prospects and problems in mineral development. In: R.B. Toombs (Editor), *Proceedings of the AIME Annual Meeting 1967.* AIME, New York, N.Y., pp. 160—191.

Lovering, T.S., 1943. *Minerals in World Affairs.* Prentice-Hall, New York, N.Y., 394 pp.

Lowell, J.D., 1970. Copper resources in 1970. *Min. Eng. (N.Y.),* 22: 67—73.

Malenbaum, W., Cichowski, C. and Mirzabagheri, F., 1973. *Material Requirements in the United States and Abroad in the Year 2000. A Research Project Prepared for the National Commission on Materials Policy.* University of Pennsylvania, University Park, Pa., 30 pp.

Manchester Guardian Weekly, 1974. 111: 26.

Mero, J.L., 1965. *The Mineral Resources of the Sea.* Elsevier, Amsterdam, 312 pp.

Mining Journal (London), 1973. *Mining Annual Review, 1973.* 497 pp.

Mining Journal (London), 1974. *Mining Annual Review, 1974.* 505 pp.

Morris, H.T., Heyl, A.V. and Hall, R.B., 1973. Lead. In: D.A. Brobst and W.P. Pratt (Editors), *United States Mineral Resources. U.S. Geol. Surv., Prof. Paper* 820, pp. 313—323.

OECD, 1973. Oil. *The Present Situation and Future Prospects.* Paris, 293 pp.

OECD, 1974. *Energy Prospects to 1985.* Paris, Vol. 1, 224 pp.

Page, N.J., Clark, A.C., Desborough, G.A. and Parker, R.L., 1973. Platinum-group metals. In: D.A. Brobst and W.P. Pratt (Editors), *United States Mineral Resources. U.S. Geol. Surv., Prof. Paper* 820, pp. 537—546.

Page, W., 1974. Conference report. Development strategies for mineral exporters. *Resour. Policy,* 1: 120—121.

Patterson, S.H. and Dyni, J.R., 1973. Aluminium and bauxite. In: D.A. Brobst and W.P. Pratt (Editors), United States Mineral Resources. *U.S. Geol. Surv., Prof. Paper* 820, pp. 35—44.

Petrick, A., Jr., 1967. World demand for mineral products and the shifting supply of mineral raw materials. In: R.B. Toombs (Editor), *Proceedings of the AIME Annual Meeting 1967.* AIME, New York, N.Y., pp. 69—117.

Pratt, W.P. and Brobst, D.A., 1974. Mineral resources: potentials and problems. *U.S. Geol. Surv., Circ.* 698, 20 pp.

Ridge, J.D., 1973. Minerals from abroad; the changing scene. In: E.N. Cameron (Editor), *The Minerals Position of the United States, 1975—2000.* University of Wisconsin Press, Madison, Wisc., pp. 127—152.

Sainsbury, C.L., 1969. Tin Resources of the World. *U.S. Geol. Surv. Bull.,* 1301: 55 pp.

Sainsbury, C.L. and Reed, B.L., 1973. Tin. In: D.A. Brobst and W.P. Pratt (Editors), *United States Mineral Resources. U.S. Geol. Surv., Prof. Paper* 820, pp. 637—652.

Schroeder, W.C. and Mote, R.H., 1959. Dimensions and changing patterns of supply and demand. In: E.H. Robie (Editor), *Economics of the Mineral Industries.* AIME, New York, N.Y., pp. 351—392.

Shimkin, D.B., 1964. Resource development and utilization in the Soviet economy. In: M. Clawson (Editor), *Natural Resources and International Development.* Johns Hopkins, Baltimore, Md., pp. 155—205.

Sutulov, A., 1973. *Mineral Resources and the Economy of the U.S.S.R.* McGraw-Hill, New York, N.Y., 192 pp.

Thayer, T.P., 1973. Chromium. In: D.A. Brobst and W.P. Pratt (Editors), *United States Mineral Resources. U.S. Geol. Surv., Prof. Paper* 820, pp. 111—122.

145

Tuke, A.F., 1975. Statement by the Chairman of Barclays' Bank International, 49th Annual General Meeting. *Economist*, January 4, pp. 40—41.

UNCTAD, 1973. *Third Session. Vol. 2: Merchandize Trade*. United Nations, New York, N.Y., 267 pp.

United Nations, 1970. *Survey of World Iron Ore Resources*. New York, N.Y., 479 pp.

United Nations, 1971a. *Yearbook of International Trade Statistics 1969*. New York, N.Y., 936 pp.

United Nations, 1971b. *World Economic Survey 1969—1970*. New York, N.Y., 248 pp.

United Nations, 1972. *Small-scale Mining in the Developing Countries*. New York, N.Y., 171 pp.

United Nations, 1973a. *Multinational Corporations in World Development*. New York, N.Y., 195 pp.

United Nations, 1973b. *The Growth of World Industry, 1971 Edition. Vol. 2: Commodity Production Data 1962—1971*. New York, N.Y., 542 pp.

United Nations, 1974a. *World Economic Survey, 1973. Part 2: Current Economic Developments*. New York, N.Y., paged by chapter.

United Nations, 1974b. *Yearbook of International Trade Statistics 1972—1973*. New York, N.Y., 1117 pp.

United Nations, 1974c. *The Growth of World Industry, 1972 Edition. Vol. 2: Commodity Production Data 1963—1972*. New York, N.Y., 589 pp.

United Nations, 1975. *Problems of Availability and Supply of Natural Resources*. Report by the Secretary-General for Committee on Natural Resources, 4th Sess., Tokyo, 24 March—4 April 1975. Mimeogr. draft.

U.S. Bureau of Mines, 1970. *Mineral Facts and Problems*. U.S. Government Printing Office, Washington, D.C., 1291 pp.

U.S. Bureau of Mines, 1973. *Minerals Yearbook, 1971*. U.S. Government Printing Office, Washington, D.C., Vol. 1, 1303 pp.

U.S. Bureau of Mines, 1974. *Minerals Yearbook, 1972*. U.S. Government Printing Office, Washington, D.C., Vol. 1, 1370 pp.

U.S. National Commission on Materials Policy, 1972. *Towards a National Materials Policy. Basic Data and Issues. Interim Report*. U.S. Government Printing Office, Washington, D.C., 63 pp.

U.S. National Commission on Materials Policy, 1973a. *Towards a National Minerals Policy. World Perspective. Second Interim Report*. U.S. Government Printing Office, Washington, D.C., 87 pp.

U.S. National Commission on Materials Policy, 1973b. *Material Needs and the Environment Today and Tomorrow. Final Report*. U.S. Government Printing Office, Washington, D.C., paged by chapters.

U.S. Secretary of the Interior, 1972. *First Annual Report of the Secretary of the Interior under the Mining and Minerals Policy Act of 1970*. U.S. Government Printing Office, Washington, D.C., 142 pp.

Van Rensburg, W.C., 1974. Changing patterns in the world's supply of minerals. *Miner. Sci. Eng.* (Johannesburg), 6: 142—153.

Vhay, J.S., Brobst, D.A. and Heyl, A.V., 1973. Cobalt. In: D.A. Brobst and W.P. Pratt (Editors), *United States Mineral Resources*. U.S. Geol. Surv., Prof. Paper 820, pp. 143—156.

Warren, K., 1973. *Mineral Resources*. John Wiley and Sons, New York, N.Y., 272 pp.

Wedow, H., Kiilsgaard, T.H., Heyl, A.V. and Hall, R.B., 1973. Zinc. In: D.A. Brobst and W.P. Pratt (Editors), *United States Mineral Resources*. U.S. Geol. Surv., Prof. Paper 820, pp. 697—712.

West, E.G., 1973. Factors in the future demand for metals, with special reference to usage in the United Kingdom. *Trans. Inst. Min. Metall.*, 82: A45—A51.

World Mining, 1974. Catalog, Survey, and Directory Number, 27: 7.

Chapter 5

RECENT TRENDS IN ENERGY CONSUMPTION AND SUPPLY

T.M. THOMAS*

INTRODUCTION

In the last few years a great deal has been written on the demand and supply of energy fuels, both from a national and an international point of view. The most obvious factors in the current energy situation are the enormous increases in demand for energy in most countries of the world, the shift in the relative importance of the various fuels in energy consumption, and the change in the relative importance of the main energy producing countries.

The world's consumption of energy is doubling every decade. Hubbert (1973) has noted that one-half of the coal mined in the world up to the year 1969 was burned *after* 1940, and one-half of the petroleum produced was burned *after* 1959. Since 1900 energy consumption in the U.S., where per capita consumption is six times the world average, increased thirteen-fold; it is now increasing at a rate in excess of 3% annually. In the rest of the world demand for energy is presently increasing at a rate nearly twice that of the U.S. The highest rates of growth have been in Europe and Japan. Japanese energy consumption increased nearly four-fold in the fourteen years between 1950 and 1964 alone, and by the beginning of this decade Japan's per capita energy consumption was only slightly less than that of the six original EEC countries; total Japanese energy consumption per capita today surpasses that of the U.K. The rate of growth of energy consumption in some of the less developed countries has outstripped even that of Japan; between 1961 and 1971 the less developed countries as a whole almost doubled their energy consumption (demand has been growing at an annual rate of approximately 7%). Table 5-I illustrates the phenomenal rise in world energy consumption in the ten years between 1961 and 1971 (see also Chapter 4).

The change in the composition of energy fuels in energy consumption has been equally dramatic. The decline in the role of coal in most of the world and the rising importance of petroleum has been one of the most notable features of the past few decades. In the past ten years alone the shift to liquid fuels has been especially pronounced in both the developed and the less

* This chapter was compiled and edited by G.E. Thomas and G.J.S. and M.H. Govett from a draft prepared by the late T.M. Thomas in 1973–74. The opinions expressed are those of the author.

148

TABLE 5-I

World energy consumption, 1961—1971 (data from United Nations, 1973)

Region	Total energy consumption 1961—1971		
	million tonnes coal equivalent 1961	1971	percentage change 1961—1971
World	4,192	7,096	69.3
Developed countries	2,654	4,416	66.4
Less developed countries	313	612	95.5
Centrally planned countries	1,224	2,068	69.0

developed countries; even in the communist countries there has been a noticeable change toward liquid fuels (Table 5-II). While the change has been least noticeable in the less developed countries, there seems little doubt that in the future they will try to increase the proportion of petroleum and natural gas in their energy "mix".

If a longer term view is taken (Darmstadter, 1971), the shifts in energy fuel composition are even more remarkable. In 1925 coal accounted for 83% of energy consumption (down from the 95% level at the beginning of the century); today it accounts for less than one-third of world energy consumption. In 1925 liquid fuels produced about 17% of total world energy; 25 years ago they accounted for one-quarter of total world energy consumption; today they account for nearly 45%. Only hydroelectric power

TABLE 5-II

Percentage share of energy consumption by type of fuel, 1961—1971 (data from United Nations, 1973)

Region	Solid fuels		Liquid fuels		Natural and imported gas		Hydro and nuclear power	
	1961	1971	1961	1971	1961	1971	1961	1971
World	48.5	33.6	33.4	44.7	15.9	21.3	2.2	2.3
Developed countries	39.0	24.0	38.0	48.9	20.2	24.4	2.7	2.8
Less developed countries	27.2	18.8	60.1	62.6	10.5	15.4	2.6	3.3
Centrally planned countries	74.7	58.4	16.7	23.5	7.7	16.6	0.9	1.2

TABLE 5-III

Percentage share of principal producing countries in world output of liquid fuels, 1925–1973 (data from Darmstadter, 1971; British Petroleum, 1973)

Country	1925	1950	1960	1968	1973
U.S.	70.9	52.7	34.9	25.5	16.1
U.S.S.R.	4.7	7.2	13.8	15.6	14.8
Venezuela	1.9	14.5	13.7	9.6	6.3
Iran	3.1	6.0	4.8	7.2	10.3
Saudi Arabia	—	4.9	5.7	7.1	12.9
Libya	—	—	—	6.3	3.7
Kuwait	—	3.2	7.5	6.1	4.9
Iraq	—	1.2	4.3	3.7	3.4
Canada	—	0.7	2.4	3.0	3.6
Algeria	—	—	0.8	2.2	n.a.
Indonesia	2.1	1.2	1.9	1.4	2.3
Abu Dhabi	—	—	—	1.2	n.a.
Kuwait neutral zone	—	—	0.7	1.1	n.a.
Mexico	11.6	1.9	1.4	1.1	n.a.
All other countries	5.7	6.5	8.1	8.9	n.a.

n.a. = not available.

has shown no major change; at the beginning of the century it accounted for less than 1% of total energy consumption, and today it only produces about 2% of the total.

As these changes have taken place in the composition of energy fuels, the sources of supply have also changed radically. Until very recently the U.S. was the leading petroleum producing country (see Table 5-III); in 1950 the U.S. was still responsible for over 50% of the world's petroleum production. Six of the world's ten leading oil-producing countries today had virtually no known output in 1925; most of the big oil discoveries in the Middle East have taken place in the last three decades.

Together with the energy crisis of 1973–1974 — brought about largely by political and economic factors — the main feature of the world energy picture today is the growth in geographical imbalance between the location of energy supplies and the main areas of energy demand. Most of the less developed countries of the world are energy deficient; nearly 10% of the merchandize imports of the less developed countries are energy fuels, and, as prices rise, this share will increase. Japan, which imports all of its energy fuels, is the most extreme case, but many European countries (most notably the U.K.) and the U.S. are presently experiencing serious problems in paying for oil imports. World energy trade has grown much faster than the aggregate dollar value of world trade, and it may well grow even faster in the next few years than it has in the past decade.

This chapter surveys the present sources of supply and demand. It is, in some sense, an introduction to Chapter 7 which assesses the future potential of the energy-producing and -consuming world.

SOLID FUELS

Solid fuels now account for about one-third of the world's energy consumption (Table 5-II). Only in the centrally planned economies are solid fuels a major source of energy; in the developed countries they account for 24% of energy consumption and in the less developed countries they account for 19%. In the last four decades coal's share in energy production dropped throughout the world, but in absolute terms the use of coal increased at about the same rate as population growth; therefore, world-wide per capita coal consumption has remained relatively stable (although in North America and Europe it declined). Coal has, in effect, held its own, but it has not been responsible for the enormous world-wide growth in energy consumption (Darmstadter, 1971).

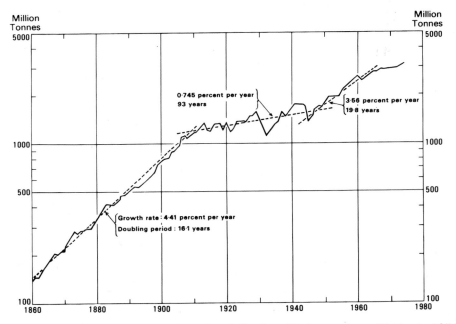

Fig. 5-1. World production of coal and lignite. (Redrawn from Hubbert, 1971, with permission from International Atomic Energy Agency, *Environmental Aspects of Nuclear Power Stations*, Fig. 4, p. 18.)

TABLE 5-IV

Coal production in selected countries, 1972 (compiled by T.M. Thomas from various sources)

Country	Production (million tonnes)
U.S.S.R.	652 (including 152 mil. tonnes lignite)
U.S.	547
China	400
East Germany	249 (including 248 mil. tonnes brown coal)
West Germany	213 (including 110 mil. tonnes brown coal)
Great Britain	120
Czechoslovakia	113 (including 82 mil. tonnes brown coal)
India	78 (including 3 mil. tonnes lignite)
Australia	83 (including 24 mil. tonnes brown coal)
South Africa	58

Between the beginning of World War I and the end of World War II the annual growth rate in world coal production averaged only about 0.75% (equivalent to a doubling period of 95 years). After World War II a more rapid growth rate of 3.5% per year was established (Fig. 5-1). Production exceeded 3 billion tonnes in 1972 when the major producing countries were the U.S.S.R., the U.S., China, East Germany, West Germany, U.K., Czechoslovakia, India, Australia, and South Africa (Table 5-IV).

Main producing countries

U.S.S.R.
Within the U.S.S.R. about 30% of the coal output is from open-cast mining. More than one-third of the coal mined is used in electricity generation and about one-quarter is used for coking and in the metallurgical industries. Since 1967 only a small degree of increase in coal production has been recorded. Ninety percent of the Soviet Union's total energy resources are coal; claimed resources (8.7×10^{12} tonnes) are enormous (Sutulov, 1973). Coal currently contributes more than 40% of the Soviet Union's total energy production.

U.S.
In contrast, coal provides only about 23% of the domestically produced energy in the U.S. (OECD, 1974). Risser (1973) estimates recoverable coal in the U.S. at 200×10^9 tonnes. Coal accounts for about 88% of the total fuel reserves of the U.S. and nearly three-quarters of the estimated ultimate resources (OECD, 1974).

In the last decade there has been little increase in coal production; between 1965 and 1972 some 398 electricity-generating companies changed their furnaces from coal to oil, first for economic reasons and, subsequently, as a consequence of environmental protection measures. Strip mining accounts for about one-half of total production, and its future is severely affected by environmental constraints. The sulphur content of the individual coal seams will also have an important bearing on their future utilization; it ranges from 0.2 to 7.0%, with an average value of 1.0—2.0% (Averitt, 1973).

China

Some 90% of China's energy comes from coal. There are about 70 significant coal production centres in the country; seven of these produce more than 10 million tonnes of coal annually. The coalfields are generally smaller in the south and west of the country, and there is considerable transportation of supplies by rail from the north. In recent years the most significant development work has been concentrated on opening up new small- to medium-sized mines located south of the Yangtze river to ease the burden on the country's inadequate railway network.

Western Europe

In the first third of this century Western Europe was a major coal exporter; this is no longer the case. When all of Britain's major coal mines were nationalized in 1947 the National Coal Board inherited 958 productive collieries; by mid-1974 these had been reduced to 260. During the 1960s uneconomic mines with limited reserves or those with complex geological structures were closed at a rate of 40 per year; output over the decade was reduced by one-quarter. From a 90% share in the U.K. energy market in 1950, coal has slipped back to a point where it accounts for not much more than one-third of total energy production (strip-coal production reached a peak of 14.5 million tonnes in 1958; in 1973 it was 10.7 million tonnes). While the introduction of mechanization has raised productivity, in some of the older coalfields adverse mining conditions have made it difficult to attain satisfactory levels of production.

Some of the same problems were experienced in West Germany, where the period 1958—1973 saw the closure of 110 mines. In the main Ruhr coal basin many of the workable seams plunge steeply to depths in excess of 2,100 m and average thicknesses of 1.7 m or less are common.

East Germany and Poland

Coal is the base for more than 70% of the energy consumed in East Germany and Poland. Poland has several well-equipped modern pits; since 1967 production has been increasing, and Poland now ranks as Europe's leading producer of bituminous coal. Polish exports are expected to be of the order of 43 million tonnes in 1974 (about one-half of this is destined for Western Europe), and Poland is now second only to the U.S. as a coal exporter.

Australia

There has been a spectacular rise in black coal production in Australia recently; output rose from 46 million tonnes in 1971 to 56 million tonnes in 1972. Much of this increase was from open-cast mines. In 1972—1973 black coal provided 30% of the primary energy used in Australia. The current annual production of 24 million tonnes of brown coal comes from the State of Victoria where the thick Oligocene deposits of the Latrobe Valley, lying about 160 km east of Melbourne, show individual seams ranging from 70 to 150 m in vertical thickness; more than 30% of Victoria's electricity is derived from modern thermal power stations using this indigenous brown coal.

South Africa

South Africa has increased production of bituminous coal during every year since 1967 (the 6,800 MW of installed electrical generating capacity in the country is produced mainly by coal-fired power stations). Pithead costs of coal in South Africa are claimed to be the lowest in the world, amounting to about one-third those currently prevailing in the U.S.

Canada

The Canadian coal mining industry continues to show a slow upward growth trend. Estimated gross production in 1973 of all types of coal was 19.8 million tonnes; this compared with 18.8 million tonnes in 1972. Nevertheless, Canada is not a major coal producer and, given the present environmental constraints, it is unlikely that the coal industry will expand to any large degree.

Conclusions on solid fuels

Coal — the one fossil fuel which is in abundant supply in the world — is no longer the major energy fuel for the world, although it remains important in the U.S.S.R., Eastern Europe, and China. Asia has by far the largest share of the world's resources, concentrated in the U.S.S.R. and China, followed by the U.S. and Europe. In spite of its abundance and relatively low cost, especially as the cost of petroleum increases, coal is not likely to regain its former pre-eminent position in the industrial world. Environmental concern has resulted in the virtual shut-down of large parts of the industry in North America and Western Europe; even in the U.S.S.R. and Eastern Europe, where environmental pressures are virtually non-existent, there is a noticeable trend away from coal toward petroleum and natural gas as more sophisticated technology develops. The future of coal depends, to a very great extent, on technical breakthroughs in environmental control, relative prices of other energy fuels, and the development of new methods of mining, distributing, and utilizing coal (see Chapter 13).

154

LIQUID FUELS

In the last decade petroleum (and natural gas) have replaced coal in much of
the world as the leading source of energy (Table 5-II). World petroleum production increased at an annual average rate of 7% from 1890 to 1970 (Fig.
5-2), resulting in a doubling period of 10 years (Hubbert, 1973). Cumulative
world production of crude oil amounted to 233 billion barrels by 1970. The
first 50% of this was produced during the 103-year period from 1857 to
1960 — a doubling of yield occurred in the short span of the ten years from
1960 to 1970.

World crude oil production reached 21.1 billion barrels in 1973, with the
Middle Eastern countries producing 37% of the total. The two largest oilfields
in the world are in Saudi Arabia and Kuwait (in 1973 they produced 13%
and 5%, respectively, of the world's total crude oil). Although the U.S.
reached its peak output of petroleum in 1970, it is still the world's single

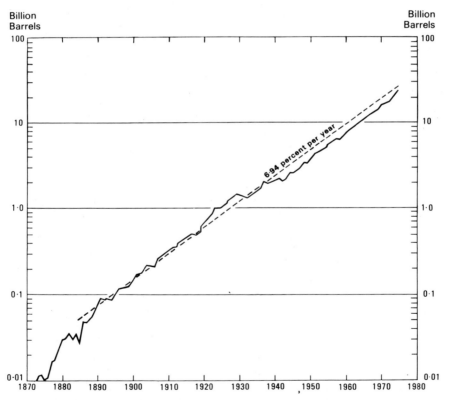

Fig. 5-2. World production of crude oil. (Redrawn from Hubbert, 1971, with permission
from International Atomic Energy Agency, *Environmental Aspects of Nuclear Power
Stations,* Fig. 6, p. 19.)

largest producing country (producing 16% of the world's total in 1973), followed by the U.S.S.R. with 15% of the 1973 total. Other substantial contributors outside of the Middle Eastern countries are Venezuela, Canada, Indonesia, and Mexico (see Table 5-III and Chapter 4).

In 1973 offshore petroleum production reached about 3.8 billion barrels (18% of total production). A gain of 2.5 billion barrels per annum from offshore oil was registered in the last decade; the most significant gain (more than 1.0 billion barrels) was recorded in the Middle East, followed by 0.7 billion barrels in the Western Hemisphere. There has been only limited progress in offshore production in the U.S.S.R.

The growth in the exploitation of petroleum resources is clearly related to the build-up of refining capacity in the major petroleum-consuming areas (Western Europe and Japan) and to an enormous expansion in world trade in crude oil. This trade, largely undertaken by fleets of supertankers, has been in large part responsible for the growing geographical imbalance between the location of energy supplies and the areas of energy demand. The largest importers of crude oil in 1973 were the Western European countries, the U.S., and Japan (see Table 5-V for data on imports). Japanese industrial growth in the past few decades has been built primarily on imported fuel oil for power generation; in 1971 oil provided three-quarters of Japan's energy requirements (compared with 40% in 1961). Oil-fired power stations are the source of 90% of Japan's electricity, compared with less than 30% in the U.K. and less than 20% in West Germany.

TABLE 5-V

Petroleum imports by main importing countries, 1973 (compiled by T.M. Thomas from various sources)

Country	Import (million barrels)
Western Europe	5,621
from:	
Middle East	3,823
North Africa	900
West Africa	375
U.S.S.R. and East Europe	362
U.S.	2,443
from:	
Caribbean	977
Canada	491
Middle East	304
West Africa	188
Japan	2,114
from:	
Middle East	1,608
Southeast Asia	362

Main petroleum-producing areas

U.S.

In the U.S. during 1972 there were 28,755 exploratory, development, stratigraphic, and service wells drilled; these wells accounted for nearly 80% of the total wells drilled for oil or gas in the non-communist world (Inglehart, 1973). Although significant discoveries were made, new domestic reserves continue to be increasingly difficult to find. The average depth of new field wildcat wells has increased from 1,220 m in 1946 to 1,840 m in 1972; of the 566 fields discovered in 1972, 453 had recoverable reserves of less than 1 million barrels and only two had reserves greater than 50 million barrels. Of the more than 10,000 oilfields which have recorded production within the U.S. and its offshore area, only 36 can be classified as "giant" fields, i.e., those in which cumulative production and estimated reserves exceed 500 million barrels. Only two fields have been found in the U.S. since 1965 with reserves estimated to be greater than those of Brent, the largest oilfield outlined to date in the British sector of the North Sea; these two fields are Prudhoe Bay in Alaska (the largest oilfield discovered to date in North America) and Santa Yuez in California. The Prudhoe Bay field has recoverable reserves of at least 9.6 billion barrels; the structure comprises a west-plunging anticlinal nose faulted to the north and truncated by an unconformity on the east; Mississippian through Jurassic rocks contain the field's important reservoirs which have a combined closure covering at least 505 million m^2 (Morgridge, 1970).

Whether other such discoveries will be made is a matter of conjecture. Examination of reservoir age, lithology, trap and depositional environment leads to the conclusion that there are no diagnostic characteristics common to the known giant oilfields of the world, although Palaeozoic age, shallow marine environments, sandstone lithology, and structural traps which have surface expression are the preferred modes (Mooding et al., 1970). Nearly one-half of the "giants" are related directly to evaporite occurrences, mostly as components of facies systems containing carbonate reservoirs.

Unless major new discoveries are made, the U.S. will experience a large shortfall of conventional oil and gas supplies from domestic sources; in 1970 the U.S. was already importing 23% of its requirements, and this figure is rising (U.S. National Commission on Materials Policy, 1972); it has been estimated that the recent level of imports of 3.4 million barrels daily will increase by 1980 to at least 8 million barrels daily. There is little reason for optimism given the present trend in domestic exploration and development; for example, in California between the peak year of 1955 (when 651 wells were drilled) and 1972, when the number of exploratory wells decreased by 71%, one-half of the decline was concentrated in the 1967—1972 period (Higgins, 1972). The causes of this trend include deteriorating economic climate and a post-maturity state of exploration in several offshore basins, as well as the fact that promising offshore areas have remained largely unexplored because of the activities of environmentalists.

U.S.S.R.

The area underlain by marine sedimentary strata in the U.S.S.R. is estimated at about 14 million km^2 (compared to 4.6—6.3 million km^2 estimated for the U.S.). The West Siberian Lowland basin alone covers some 2 million km^2; Sutulov (1973) cites the Russian claim that when the large oil deposits in Siberia are exploited, the Soviet Union will become the largest single holder of world petroleum reserves. Major discoveries clearly remain to be made; the fact that only 1,200 oilfields in the U.S.S.R. are in production, compared to more than 23,000 gas- and oilfields in the U.S., is an index of the relative immaturity of the petroleum industry in the U.S.S.R. and its future potential (Halbouty et al., 1970).

In 1972 some twenty-six of the known U.S.S.R. oilfields were of "giant" proportion. Between the years 1963 and 1973 Soviet oil production doubled, with much of the increase in the early part of the period due to the progressive development of the older Tatar and Bashkirian fields lying west of the Urals; these fields provided 70% of the national output in 1970. They are now on the decline, and their significance is diminishing as production rises in the newly developed fields of the West Siberian Lowland (production increased from nothing in 1963 to 220 million barrels in 1970 and 645 million barrels in 1973). Further increases are expected since more than 120 new fields with recoverable reserves exceeding 30 billion barrels have been discovered in the area since 1959. These fields are in a major topographic and structural basin lying immediately east of the Urals; the reservoir rocks are sandstones of late Jurassic or Cretaceous age disposed in domed structures; the depth to the igneous and metamorphic Palaeozoic basement ranges from 3,050 m in the south to 3,450 m in the north (Dickey, 1972). The productive wells have a very high yield of light gravity oil which is low in sulphur. However, a practically virgin environment, with major physical impediments of extensive boglands and widespread permafrost, has presented complex engineering, technological, organizational, and transportation problems; nevertheless, a 2,000-km pipeline from Samoltor, the largest oilfield discovered to date in the area, to the Volga-Ural Basin has been built, and a further 5,000 km of high-capacity oil and gas pipeline was expected to be completed in 1974. With the experience gained in West Siberia, attention is now being directed to the possible future development of the resources of East Siberia, notably in Yakutia Buryat, Irkutsk, and Krasnoyarsk where recoverable oil is considered to lie at depths of 2,420—3,450 m.

Several offshore fields have already been brought into production in the Caspian Sea off Azerbaydzhan where drilling platforms have been installed in water depths ranging up to 30 m. Subsea drilling operations have also been initiated in the Ukraine offshore area and to the south of Okha on Sakhalin Island in the Pacific in the 1971—1972 period. It is, nevertheless, evident that Soviet offshore technology has not yet advanced to the stage where it can adequately cope with the deeper-water stormy environments now being encountered by international oil companies in some sectors of the northern North Sea or the outer shelf of the Gulf of Mexico.

The Middle East

With something like 160 major oilfields already outlined and with no country in the region likely to run out of reserves within the next 25 years, it is not surprising that exploration activity has not been undertaken in the Middle East recently with any air of urgency, although new finds have been made in Iran, Saudi Arabia, Sharjah, and Qatar. There was a five-fold increase in production in both Saudi Arabia and Iran between 1963 and 1973, although increases in Iraq and Kuwait were smaller.

Saudi Arabia, where exploration only began in 1936, is the largest of the Middle Eastern oil producers, and the third largest oil producer in the world (see Table 5-III). The Saudi Arabian oilfields are in the nature of large anticlinal oil accumulations lying within a broad north-south zone of major uplift; the reservoir rocks are Upper Jurassic limestones lying at depths of 1,560–3,020 m. Each productive oilfield has estimated recoverable reserves plus cumulative production exceeding 500 million barrels and thus can be classified as "giants"; individual wells are characterized by exceptionally high rates of productivity. The Ghawar field is one of the largest in the world; it is 217 km long and has six culminations, each containing a major oil pool. This one field alone is responsible for about one-half of total Saudi production; cumulative production and remaining reserves are reckoned to be of the order of 75 billion barrels.

Reserve estimates for petroleum are notoriously difficult to evaluate (see Chapter 1). However, the Middle East countries (including Iran and North African states) were estimated to have approximately 58% of the world's proved reserves (Thomas, 1973).

Central America, the Caribbean, and South America

Ten producing countries in this area provided 1,824 million barrels of crude oil in 1973, with Venezuela contributing nearly three-quarters. Peak production in Venezuela, which was a major oil-producer in the 1950s and 1960s, was reached in 1970; one of the major disappointments in recent years has been the failure to discover new oil reserves on the South Lake Maracaibo prospective acreage after numerous probes, some of which attained depths of 5,900 m.

An ambitious continental shelf effort has been carried out in Brazil since 1968, but without proving the presence of major oil accumulations. The vast trans-Andean Basin, extending from Colombia southwards through Ecuador and Peru into northeast Bolivia has also been explored. In adjoining northeast Ecuador production began during 1972 from the first three of several evidently important oil finds (Jacobsen and Neff, 1973). In 1974 there were reports of an enormous oil potential in Mexico; preliminary reserve estimates range from 3 to 20 billion barrels (*Oil Industry Scouting Reports*, 1974).

Canada

Crude oil production in Canada almost trebled between 1963 and 1973, when oil exports, almost exclusively to the U.S., exceeded 1 million barrels per day. Early in 1973, as a result of criticism about continued exports at a time when some observers argued that Canadian reserves were rapidly running out, controls were placed on exports (Govett and Govett, 1973). Exploratory activity has been increasing, but the results have been largely disappointing. All five of Canada's "giant" oilfields are located in the Alberta sector of the major West Canada Sedimentary Basin. Development of the Athabasca tar sands in Alberta could well provide enormous new oil reserves in Canada (see Chapter 7).

Africa

Crude-oil production in Algeria doubled between 1963 and 1973, while that from Libya, where oil was only discovered in the 1960s, increased seven-fold during the same period. Political unrest plus expropriation of the international oil company holdings have brought most exploration activity to a halt. From lowly beginnings (only 75,000 barrels a day in 1963), Nigerian oil output has surged ahead to 2 million barrels a day in 1973. Oil exports in 1973 provided 83% of Nigeria's revenue from overseas trade; with an exceptionally strong demand for its virtually sulphur-free oil, Nigeria has now reached the position of being seventh amongst the world's oil-exporting nations. By the end of 1973 there were 84 oilfields in production in Nigeria. In the southern half of Africa several oilfields have been discovered in offshore areas of northern Angola, Zaire, and Congo (Brazzaville); whether further exploration will result in large finds remains to be seen.

Southeast Asia

In 1973 crude-oil production in Southeast Asia, including China, amounted to some 2.3 million barrels per day, with Indonesia contributing 1.3 million barrels per day. A total of nineteen oilfields were discovered in Indonesia in 1972 (Kennett, 1973). According to press reports, China's oil production in 1972 exceeded 600,000 barrels per day; major increases in output have been recorded in the prolific Ta-Ch'ing field, and it is alleged that an important discovery was made in the Yellow Sea in 1970. It is impossible to speculate on future Chinese exploration or production, although it seems likely that exploration is under way.

Other countries

In Australia a six-year programme of exploration has established reserves of at least 1.6 billion barrels of oil off the coast of Victoria; these fields have provided more than 50% of Australia's production in recent years. Interest in European offshore projects has spread to the entire Atlantic shelf and large parts of the Mediterranean (Hubbert, 1973). Spectacular discoveries have been made in the northern waters of the North Sea, particularly in the East Shetlands Trough (or Viking Graben) lying east and northeast of the Shetlands, where several "giant" oilfields have already been outlined (Fig. 5-3). Ultimate production levels remain a matter of speculation, partly because information is not released by the oil companies and partly because of the rate of new discoveries (see Chapter 7).

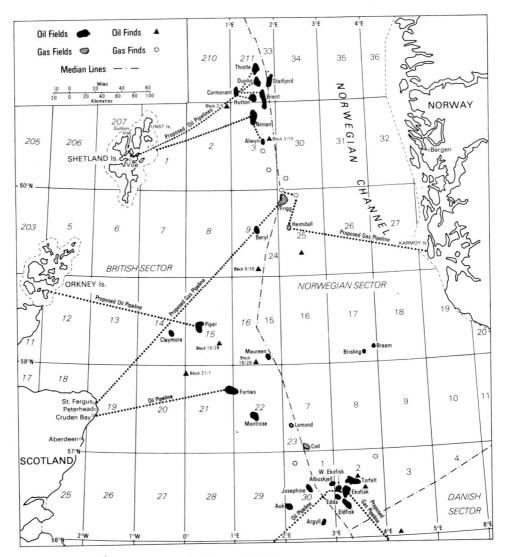

Fig. 5-3. North Sea oil- and gas fields (mid-1974).

Conclusions on liquid fuels

World production of crude oil grew at an annual rate in excess of 10% in the period 1950–1970; petroleum is now the world's leading energy fuel. The most spectacular increase in petroleum consumption occurred in Japan where the oil consumed in 1970 was nearly one hundred times the amount used in 1950.

This growth rate was not spread evenly throughout the world; in the centrally planned countries coal still predominates as an energy fuel, although the share of petroleum and natural gas is now increasing rapidly. In 1970 over three-quarters of the world's oil consumption was concentrated in twelve countries (see Table 5-VI), with the U.S. dominating consumption (32% of the world total), followed by the U.S.S.R. and Japan (with 10% and 8%, respectively).

The rapid increase in the demand for oil was met largely by exploiting the large known deposits in the Persian Gulf area and Iran and by new exploration, chiefly in Libya, Algeria, and Nigeria. These sources together produced more than 40% of the world's oil in 1970. The construction of supertanker fleets, pipelines, and new port facilities was vital to the enormous growth in trade, as were new refineries (such as those in Iran and the Caribbean). Crude oil is now the largest commodity (in terms of tonnage) in world trade; more than

TABLE 5-VI

World consumption of refined oil products, 1950–1970 (data from U.S. National Commission on Materials Policy, 1973)

Country	Consumption (million barrels)		
	1950	1960	1970
World total	3,688	7,725	16,534
U.S.	2,352	3,537	5,374
U.S.S.R.	267	729	1,607
Japan	15	242	1,373
West Germany	25	211	941
U.K.	124	366	783
Italy	32	199	690
France	73	209	684
Canada	118	314	531
Netherlands	26	102	299
Sweden	26	97	266
Spain	10	45	209
Belgium-Luxembourg	17	57	208
Others	583	1,617	3,569

one-half of the oil produced in the world today is exported. Only the U.S.S.R. and Canada, among the world's industrialized countries, are self-sufficient in oil reserves.

Approximately 90% of world oil production is accounted for by fourteen countries; of these, the U.S., the U.S.S.R., and the Middle Eastern countries (including Iran and Libya) supply well over one-half of the world's oil (see Table 5-V). Given the present political division in the world and the rising wave of nationalism, the implications of the current petroleum situation are clear — unless new sources of energy can be found to augment or replace oil, or major new discoveries are made, many countries in the world will face a continuing oil "crisis". Except in the U.S.S.R. (and possibly China) the prospects for major new "giant" discoveries in continental areas seem limited; the prospect of finds in offshore areas and the oceans are considerably better. As discussed in Chapter 7, one of the most likely sources of vast new petroleum resources will have to be the development of presently unconventional sources of oil.

NATURAL GAS

The importance of natural gas as an energy fuel has been increasing rapidly in the last decade; today it accounts for more than 20% of the world's energy consumption (Table 5-II). Natural gas is often found in association with crude oil, but recent finds of non-associated gas in several parts of the world suggest that it has a considerably wider distribution than oil. This applies both to its greater range in depth and to its occurrence towards the peripheries of basins bordering on tectonically disturbed belts. Porosities of most clastic rocks approach limiting values close to 5% at depths of 3,050—6,100 m (McCulloh, 1967).

Until quite recently difficulties of storage and transportation resulted in large-scale wastage of natural gas by flaring or venting of most of the associated natural gas within the oil wells of the Middle East and North Africa and even within some of the remoter North American fields. However, the laying down of large-diameter pipelines and the construction of special tankers for the transport of liquified natural gas has transformed the situation. Thus, world output of natural gas increased from 48 million tons of coal equivalent in 1925 to 1,307 million tons in 1969, a rate of increase far greater than that of any other source of energy (Darmstadter, 1971). In 1973 the consumption of natural gas reached 1,595 million tons of coal equivalent, of which 53% was consumed in the U.S., 22% in the U.S.S.R., China, and Eastern Europe, and 12% in Western Europe.

As was the case with oil, the known reserves of natural gas are largely concentrated in giant fields containing original recoverable reserves in excess of 3.5 trillion (10^{12}) cu. ft. At the beginning of 1972 there were 83 such "giants" in the world, each with a cumulative production and estimated remaining reserves exceeding this level. No less than 42 of these fields are located in the U.S.S.R.; the U.S. has seventeen, the British sector of the North Sea has four, Algeria has three, and Mexico has three.

If a "super-giant" field is considered to hold recoverable reserves of more than 10 trillion cu. ft, then the world total of such fields is now 32, with the U.S.S.R. holding 21 of the total and the U.S. only three. Altogether the giant gas fields known at this time contain an estimated ultimate minimum recoverable reserve of 1,200 trillion cu. ft. Three-quarters of these reserves are in sandstones and the other one-quarter is in carbonate reservoirs. A total of 10% of the reserves occur in beds of Tertiary age, 65% are in Mesozoic strata, and 25% are in Palaeozoic reservoirs (Halbouty et al., 1970). These giant fields make up nearly 70% of the total world reserves which were estimated at 1,725 trillion cu. ft in 1973.

Main areas

U.S.

Reserves in the U.S. are estimated at 279 trillion cu. ft, accounting for 16% of the world total. In 1973 the marketed production of natural gas amounted to 23 trillion cu. ft and represented one-third of the primary energy consumed. A total of 270 gas discoveries were made in 1972; of these 61 could be classified as significant. The Prudhoe Bay oilfield in Alaska is expected to contain recoverable reserves of 26 trillion cu. ft (Morgridge, 1970). It is not, however, fully appreciated that for a number of years the successive reserve additions in the U.S. have consistently fallen short of depletion rates (Inglehart, 1973). The majority of giant gas fields were discovered prior to 1940.

The marketing of natural gas within the U.S. quadrupled between 1940 and 1960. The commercial exploitation of many of the fields was made possible by large-diameter pipelines; the country is now covered by a 363,000-km network of trunk transmission lines. A shortage was nevertheless evident by the winter of 1972—1973, due to a large extent to the policy of the Federal Power Commission which had discouraged gas exploration in the preceding years by keeping well-head prices too low, and wastage at power stations (in Texas and Luisiana, which supply more than 70% of the national output, gas had been so cheap and plentiful that little thought was given to conservation). The Federal Power Commission decision to raise prices recently has led to a sharp increase in gas drilling activities in 1974. It has been suggested that a new price structure could more than double the payments for gas from new wells and, accordingly, marginal reserves could be brought into production.

U.S.S.R.

In recent years the production of natural gas within the U.S.S.R. has averaged 7—8 trillion cu. ft, with annual increases in output of 6—7% being recorded. More than three-quarters of the Soviet production is from the European part of the U.S.S.R.; the Ukraine is responsible for nearly one-third of the national output. Completion of several pipelines recently has allowed Soviet gas to be exported in increasing quantities to both Eastern and Western Europe.

Although no official figures have been published, it is considered that proved recoverable reserves are of the order of 1,300 trillion cu. ft. More than 80% lie within Siberia where a "super-giant" field in the Karkan desert in Turkmenia came into production in 1974; it holds at least 50 trillion cu. ft of recoverable reserves. Gas from this field is supplied to the rest of the country by a pipeline passing through the gas centre at Khiva in Uzbekistan. In August of 1974 another major gas field was announced lying beneath the Yamal peninsula and the adjacent waters of the Arctic Ocean; it contains well over 35 trillion cu. ft of recoverable reserves and represents the largest of five recent discoveries in the northern half of the West Siberian basin.

The world's largest gas field, the Urengot field, lies south-southeast of the recent Yamal peninsula discovery; it has proved plus probable reserves of 175 trillion cu. ft (and proved, probable, and potential reserves of 210 trillion cu. ft). Two other giant fields and four smaller units have been discovered in the Vilyuy intracratonic basin centred on the confluence of the Lena and Vilyuy rivers; development of these fields, and others likely to found in this area, awaits transportation and the development of markets (Meyerhoff and Meyerhoff, 1973).

The Middle East

Production of natural gas in the Middle Eastern countries has been on a comparatively limited scale; much gas associated with the production of oil has been wantonly wasted. In Iran the "super-giant" fields of Khangiran and Pazana, which were discovered in 1969, have combined reserves exceeding 70 trillion cu. ft.

Canada

Natural gas production in Canada has recently been increasing at an annual rate of 15%; in 1974 the level of output is expected to exceed 3 trillion cu. ft, with exports to the U.S. accounting for about one-third of the production. The Canadian Petroleum Association has estimated that at the end of 1972 the proved recoverable reserves within Canada amounted to 71.5 trillion cu. ft. In view of these limited reserves, the Canadian National Energy Board is not issuing any more export permits for gas; export levels will therefore decline in the future.

Australia

Australian reserves of natural gas are considered more than sufficient for domestic use, and Australia may well become a significant exporter of liquid natural gas (LNG) by the early 1980s (OECD, 1974). The largest fields and the greatest reserves are found in the Cooper basin of South Australia. There are also large reserves in the extensive shelf area off the northern coast of Western Australia, where the North Rankin, Goodwyn, Angel, and Rankin fields total some 20 trillion cu. ft (Durkee, 1973); exploration work has proved the existence of several geologic provinces and sub-basins in which post-Permian sedimentary successions exceeding 9,100 m in overall thickness are present (Martison et al., 1973).

Europe

In the past ten years gas discoveries have been recorded in Europe, most notably in the Netherlands, West Germany, and Italy; none of these have approached the size of the major Groningen field in the northern Netherlands where a Lower Permian sandstone reservoir shows a thickness of 91—213 m (Stauble and Milius, 1970). Successful offshore gas wells have been reported from the Adriatic, Aegean, and Celtic Seas; these discoveries, however, bear no comparison with the scale of discoveries in the North Sea, where proved recoverable reserves within southern British waters now exceed 35 trillion cu. ft. In the northern area of the North Sea the major discovery to date is the Frigg field (which straddles the British-Norwegian median line); it holds estimated recoverable reserves of 10—12 trillion cu. ft. In addition, a number of lesser fields have been outlined in these waters.

Other countries

Algeria has been involved in much of the pioneering work on the development of liquified natural gas schemes; it contains the major Hassi er R'Mel field with 70 trillion cu. ft of reserves (Magloire, 1970). While Algerian gas reserves have remained largely neglected recently, accelerated programmes of exploration in Indonesia and Malaysia have resulted in the discovery of several new onshore and offshore gas fields; while a clustering of successful wells is a feature in these countries, most of the finds have been relatively small. Half a dozen gas fields, each with recoverable reserves exceeding 1 trillion cu. ft have been discovered at relatively shallow depths in reservoirs of Triassic age in China; a current trade agreement between China and Rumania calls for an annual delivery of thirty drilling rigs capable of operating to depths of 5,000 m (Kennett, 1973), but few details have been released on recent exploration results. A number of gas fields have been outlined in the central part of the Indus basin in Pakistan; a recent evaluation suggests that the largest of these fields has recoverable reserves aggregating 8.62 trillion cu. ft (Kennett, 1973).

South America, to date, has only limited natural gas production. At the

166

end of 1972 the proved recoverable reserves of gas in Brazil were placed no higher than 1 trillion cu. ft (Jacobsen and Neff, 1973). Since 1970 annual production in Brazil has averaged around 650 million cu. ft; no other South American country has reported major production to date.

HYDROELECTRICITY

Water power has been utilized since Roman times; prior to the adoption of other sources of energy, it was extensively employed in operating textile mills, saw mills, and other manufacturing establishments in the older in-

TABLE 5-VII

Countries with installed hydroelectric generating plants with capacity exceeding 1,000 MW, 1961—1970 (data from United Nations, 1972)

Country	Installed capacity (1,000 MW)		Percentage increase
	1961	1970	1961—1970
U.S.	36.3	55.8	54
U.S.S.R.	16.4	31.4	92
Canada	19.0	28.3	44
Japan	13.5	20.1	49
Italy	13.1	15.4	17
France	11.0	15.2	38
Norway	7.0	12.8	83
Spain	4.7	10.9	132
Sweden	7.5	10.9	45
Switzerland	6.0	9.6	60
Brazil	3.8	8.8	132
India	2.4	6.5	171
Rumania	1.6	6.1	281
Austria	3.0	5.5	83
West Germany	3.5	4.8	37
Australia	1.5	3.6	140
Mexico	1.4	3.4	143
Yugoslavia	1.6	3.3	106
New Zealand	1.6	3.2	100
U.K.	1.3	2.2	69
Finland	1.7	2.1	24
Egypt	0.4	1.9	375
Portugal	1.2	1.6	33
Czechoslovakia	1.3	1.5	15
Chile	0.6	1.1	83
Greece	0.2	1.0	400
Peru	0.5	1.0	100

dustrial areas of Western Europe and the U.S. Large-scale generation and transmission of electrical power did not commence until the beginning of the present century, and even now hydroelectric power contributes only about 2% of total world energy production, although in some European countries, notably Norway and Switzerland, hydro power is the main source of electricity.

In 1970 a total of 27 countries, headed by the U.S., the U.S.S.R., Canada, and Japan had a combined installed capacity exceeding 267,000 MW. Table 5-VII shows the limited reliance on hydroelectric power in most countries of the world, outside of the four noted above and a few European countries. Even in the U.S., which has the world's highest installed capacity, hydroelectric power provides only 4% of the total energy produced. The U.S.S.R., with the second highest installed capacity in the world plans to substantially increase the number of hydro plants. While notable degrees of expansion in hydroelectric production were recorded in Japan, Italy, France, and Norway between 1961 and 1971, they were overshadowed by advances in Brazil, Spain, Rumania, and India and, on a percentage basis, significant increases were made in Egypt, Greece, Mexico, and Australia. Brazil is almost as dependent as Norway and Switzerland on hydroelectric power since only about 10% of its electricity is based on thermal stations.

NUCLEAR ENERGY

In 1970 the production of electricity from the fission of uranium (^{235}U) in nuclear reactors amounted to 77 billion kWh; this was 1.5% of total production from all sources, and compares to 0.2% (4 billion kWh) in 1960 (United Nations, 1972). In 1960 electricity production from a nuclear source was limited to the U.K., the U.S., France, West Germany, and Canada; by 1970 Japan, India, Belgium, Italy, the Netherlands, Sweden, Switzerland, Spain, East Germany, and the U.S.S.R. had been added to the list of countries operating nuclear power plants. As of June, 1973, the installed world capacity of operative nuclear plants was 40,802 MW, with the U.S. accounting for approximately one-half of the total (see Table 5-VIII for details).

Of the installed capacity, no less than 71% was based on light-water reactors which consisted of two main types: pressurized light-water reactors, and boiling light-water reactors. Apart from reactors operated on a prototype basis, other reactors in use included light-water gas reactors, gas-cooled graphite units, and pressurized heavy-water reactors (Candu); prototype fast breeders were also producing limited quantities of electricity in the U.S., U.S.S.R., and U.K.

The U.K. was the first country to embark on a nuclear power programme, largely in response to the Suez crisis of 1956 which threatened to disrupt its traditional source of oil supplies. A number of gas-cooled graphite reactors

168

TABLE 5-VIII

Installed world capacity of operating nuclear plants and nuclear plants under construction, 1973 (data from Australian Atomic Energy Commission, 1973)

Country	Operating nuclear plants, installed capacity (MW)	Nuclear plants under construction (MW)
Main countries		
U.S.	20,462	47,746
U.K.	4,128	6,463
France	2,711	n.a.
West Germany	2,225	7,526
Japan	2,179	12,083
Canada	2,022	3,522
Sweden	—	5,429
Spain	—	3,570
World	40,802	106,851

n.a. = not available.

(Magnox), utilizing natural uranium (with a total capacity of 5,000 MW) were constructed at coastal sites. The oldest of these reactors has been fully operative since 1961. In 1965 the U.K. began its second nuclear programme, based on the advanced gas-cooled reactor (A.G.R.), a hotter, more pressurized version of Magnox which burns an enriched uranium fuel. This has proved to be a complex system to engineer in large sizes; five stations, each of 1,250—1,300-MW capacity were begun between 1966 and 1971, but problems of safety, carbon dioxide corrosion, boiler reliability, and other factors have dictated major design changes so that construction costs have more than doubled from the original estimates. Nine years since the start of this programme no power had been produced; the project is likely to be at least £750 million above its original budget, and delays of up to six years have occurred. The programme is perhaps the greatest unforeseen drain on resources in British industrial history; Dungeness B station will cost £200 million instead of the original estimate of £85 million (estimated in 1974).

On a world-wide basis nuclear reactors are turning out to be more expensive to build and to operate than was originally estimated. This is due in part to world-wide inflation, but is also due to unexpected problems in construction and operation and the demands for modifications required by environmentalists. Ideally a reactor should have a commercial life of at least 25 years preceded by 10 years of research-development prototype operation; thus the nuclear industry must plan on a timescale of 35 years or more, and increasing size and complexity and lengthier licensing procedures have increased the time required to construct a nuclear plant from 5 years to as much as 5—8 years.

At the end of 1973 there were 39 nuclear reactors producing electricity in the U.S.; these had a capacity of 27,408 MW. Delays in the commercial start-up of 27 nuclear-fueled facilities reported in 1974 were due to design changes, late deliveries of equipment, labour problems, and the regulatory procedures of government agencies. To date nuclear reactors provide only about 6% of U.S. electrical generating capacity.

Canada has the largest operative nuclear power station in the world at Pickering, Ontario, on the lakeside east of Toronto where the first of four 514-MW Candu heavy-water reactors (which burn natural rather than enriched uranium) was commissioned in 1971. The more modest progress in the establishment of nuclear stations in the U.S.S.R. can be related to the country's tremendous oil, natural gas, and coal resources. The total installed capacity of all four Soviet atomic power plants was 1,370 MW, which accounted for only 0.6% of installed electrical capacity in 1971. The U.S.S.R. is the only country which can supply enriched uranium to Eastern Europe, although Czechoslovakia has her own uranium resources; in Eastern Europe two plants of limited capacity were in operation in East Germany and Czechoslovakia in 1973.

In West Germany six light-water reactors, imported from the U.S., were in operation in 1973; the first substantial quantity of electricity from a nuclear source was recorded in 1966. French production started a little earlier, largely from gas-cooled graphite reactors; plans for the construction of a number of gas-cooled reactors where abandoned in 1969, and six pressurized light-water reactors were ordered. In 1974 orders were placed for a further sixteen light-water reactors, eleven of which were expected for delivery in 1979 or 1980.

Since the beginning of nuclear power in the 1950s its adherents have been predicting that cheap, reliable systems were around the corner and that, within a relatively few years, a significant proportion of world energy would be produced in nuclear power plants. These predictions have fallen by the wayside as nuclear power stations have encountered construction and technical difficulties, and, especially in the U.S., opposition from environmentalists. The teething problems have gone on far longer than anyone had anticipated; it would appear that nuclear power will always involve construction and operational problems of some complexity (see Chapter 7).

The world's reserves of exploitable uranium oxide ore were estimated at about 3 million tonnes in 1974; 75% of these reserves are in Canada, South Africa, the U.S., and Australia (data on U.S.S.R. reserves are not available); the only European country with sizeable reserves is France. At the end of 1973 one-half of the uranium production in the non-communist world was in the U.S., 20% was produced in Canada. The crash programme to open up the Blind River uranium fields in northern Ontario in the late 1950s was based on Western nuclear defence needs; once initial commitments had been met, commercial nuclear power demand lagged seriously behind uranium production capacity and several mines were closed with severe consequences to the economy in the Elliott Lake area. It has only been in this decade that new exploration for uranium has been again undertaken in Canada.

CONCLUSIONS

The last half century has seen the world's energy consumption increase by more than five times. At the same time, the pattern of both consumption and production has changed, much of the world has shifted away from solid fuels to petroleum, and the trade in energy fuels has grown enormously.

The brief summary presented in this chapter leads to the conclusion that, at present, there are only two abundant sources of energy — coal and nuclear power. Coal is most abundant in the U.S.S.R. and China, countries which still remain the main users of coal, but even in the U.S.S.R. there is a noticeable trend toward an increasing use of petroleum and natural gas. Outside the communist countries, the future of coal in most of the world depends primarily on the development of new methods of mining and using coal to make it environmentally acceptable (see Hubbert, 1969).

Hydroelectric power, the most environmentally acceptable source of energy, is not likely to increase its share in the energy "mix" and, at present, has also come under attack by environmentalists. Petroleum is now the world's leading energy fuel — although consumption is concentrated in a few countries, as is production — but conventional sources are limited outside of the Middle East and the U.S.S.R. The present energy crisis may lead to new efforts to discover additional "giant" oil- and gas fields, but it seems unlikely that they will be found in the main oil-consuming regions (except, possibly, the U.S.S.R.); the exploitation of unconventional sources of petroleum, including offshore and marine sources, seems more promising, although no one can rule out new discoveries in some of the world's remote, unexplored areas (see Chapter 7).

Nuclear power, which now accounts for less than 1% of the world's energy consumption, is considered by many to be the energy fuel of the future. Yet few observers feel that large supplies will be "on line" in the next decade, and the problems in the development and use of nuclear power are proving to be even greater than anticipated.

The current energy "crisis" is actually a crisis in the pattern and distribution of energy supplies, compounded by the long lead-times necessary to develop new technology to exploit new energy sources and the lack of active new exploration efforts in unexplored areas of the world. In the final analysis, the present situation is just one part of the whole crisis of exponential growth and the drive for a continuation and spread of world affluence. The physical means by which the world may be able to supply its seemingly insatiable appetite for energy are discussed in Chapter 7; however, the conclusion reached by Hubbert (1973, p. 3), that "The real crisis confronting us is . . . not an energy crisis but a cultural crisis" is beyond the scope of the geologist.

REFERENCES CITED

Australian Atomic Energy Commission, 1973. *21st Annual Report, 1972—1973*.
Averitt, P., 1973. Coal. In: D.A. Brobst and W.P. Pratt (Editors), *United States Mineral Resources. U.S. Geol. Surv., Prof. Paper* 820, pp. 134—142.
British Petroleum, 1973. *Statistical Review of the World Oil Industry 1973*. London, pp. 4—16.
Darmstadter, J., 1971. *Energy in the World Economy*. Johns Hopkins, Baltimore, Md., 876 pp.
Dickey, P.A., 1972. Geology of oil fields of West Siberian Lowlands. *Am. Assoc. Pet. Geol. Bull.*, 56: 454—471.
Durkee, E.F., 1973. Petroleum developments in Australia and Oceania in 1972. *Am. Assoc. Pet. Geol. Bull.*, 57: 2116—2125.
Govett, G.J.S. and Govett, M.H., 1973. Mineral resources and Canadian-American trade — double-edged vulnerability. *Can. Inst. Min. Metall. Bull.*, 66: 66—71.
Halbouty, M.T., Meyerhoff, A., Dott, A., King, R.E. and Klemme, H.D., 1970. World's giant oil and gas fields; geological factors affecting their formation, and basin classification. In: M.T. Halbouty (Editor), *Geology of Giant Petroleum Fields. Am. Assoc. Pet. Geol., Mem.* 14, pp. 502—555.
Higgins, J.W., 1972. Developments in West Coast area in 1972. *Am. Assoc. Pet. Geol. Bull.*, 57: 1437—1447.
Hubbert, M.K., 1969. Energy resources. In: National Academy of Science and National Research Council (Editors), *Resources and Man*. W.H. Freeman, San Francisco, Calif., pp. 157—242.
Hubbert, M.K., 1971. Energy resources for power production. In: International Atomic Energy Agency, *Environmental Aspects of Nuclear Power Stations*. Vienna, pp. 13—43.
Hubbert, M.K., 1973. Survey of world energy resources. *Can. Inst. Min. Metall. Bull.*, 66: 37—53.
Inglehart, C.E., 1973. North American drilling activity in 1972. *Am. Assoc. Pet. Geol. Bull.*, 57: 1375—1405.
Jacobsen, P. and Neff, C.H., 1973. Petroleum development in South America, Central America and the Caribbean in 1972. *Am. Assoc. Pet. Geol. Bull.*, 57: 1868—1933.
Kennett, W.E., 1973. Petroleum development in the Far East, 1972. *Am. Assoc. Pet. Geol. Bull.*, 57: 2085—2108.
McCulloh, T.H., 1967. Mass properties of sedimentary rocks and gravimetric effects of petroleum and natural-gas reservoirs. *U.S. Geol. Surv., Prof. Paper* 528-A, pp. A1—A50.
Magloire, P.R., 1970. Triassic gasfield of Hassi er R'mel, Algeria. In: M.T. Halbouty (Editor), *Geology of Giant Petroleum Fields. Am. Assoc. Pet. Geol., Mem.* 14, pp. 489—501.
Martison, N.W., McDonald, D.R. and Kaye, P., 1973. Exploration on the continental shelf off northwest Australia. *Am. Assoc. Pet. Geol. Bull.*, 57: 972—989.
Meyerhoff, H.A. and Meyerhoff, A.A., 1973. Arctic geopolitics and geology. In: *Proceedings, 2nd International Symposium on Arctic Geology. Am. Assoc. Pet. Geol., Mem.* 19, pp. 646—670.
Mooding, J.D., Mooney, J.W. and Spivak, J., 1970. Introduction. In: M.T. Halbouty (Editor), *Geology of Giant Petroleum Fields. Am. Assoc. Pet. Geol., Mem.* 14, p. 8.
Morgridge, D.L., 1970. Geology and discovery of Prudhoe Bay field, eastern Arctic Slope, Alaska. *Am. Assoc. Pet. Geol. Bull.*, 54: 1969.

OECD, 1974. *Energy Prospects to 1985*. Paris, 2 volumes.

Oil Industry Scouting Reports, 1974. Private reports, Calgary, Alta., October.

Risser, H.E., 1973. The U.S. energy dilemma. *Ill. State Geol. Surv., Environ. Geol. Notes*, 64, 64 pp.

Stauble, A.J. and Milius, G., 1970. Geology of the Groningen gas field, Netherlands. In: M.T. Halbouty (Editor), *Geology of Giant Petroleum Fields. Am. Assoc. Pet. Geol., Mem.* 14, pp. 359—369.

Sutulov, A., 1973. *Mineral Resources and the Economy of the U.S.S.R.* McGraw-Hill, New York, N.Y., 192 pp.

Thomas, T.M., 1973. World energy resources; survey and review. *Geogr. Rev.*, 63: 246—258.

United Nations, 1972. *World Energy Supplies, 1961—1971*. New York, N.Y., 373 pp.

United Nations, 1973. *World Energy Supplies, 1968—1971*. New York, N.Y., 187 pp.

U.S. National Commission on Materials Policy, 1972. *Towards a National Materials Policy. Basic Data and Issues — An Interim Report*. U.S. Government Printing Office, Washington, D.C., 63 pp.

U.S. National Commission on Materials Policy, 1973. *Towards a National Minerals Policy. World Perspective — Second Interim Report*. U.S. Government Printing Office, Washington, D.C., 87 pp.

PART II

A PERSPECTIVE ON WORLD MINERAL SUPPLIES

Problems of future demand and supply

Chapter 6

WORLD ECONOMIC DEVELOPMENT AND FUTURE MINERAL CONSUMPTION

EVAN JUST

INTRODUCTION

Statistical indicators of affluence and mineral consumption for all segments of the world community have moved steadily upwards in recent decades. Statistical projections of consumption and gross national product data suggest that the world will nearly quadruple its 1971 rate of mineral consumption by the year 2000 and that its capacity to pay will grow somewhat more rapidly. However, because of ominous aspects of the world scene — belligerence, modern weaponry, and impending depression and mass starvation — the writer considers that a 2.5-fold increase is most probable, and that a level as low as 1.5 times that of 1971 is reasonably possible.

RECENT TRENDS IN WORLD ECONOMIC DEVELOPMENT

The trends of population, economic development and mineral consumption in the past 40 years have been so startlingly upward, that what appear to be simple, conservative projections to the end of the century challenge the imagination. The humanist, perceiving the extent to which the world already seems crowded and the current threat of mass starvation, finds it difficult to imagine reasonable nutrition and comfort for a doubled world population, which is a virtual certainty by the year 2010 unless catastrophe intervenes in the way of war, pestilence or starvation. The environmentalist, already struggling to prevent the existing populace from wrecking the natural environment and habitability of "Spaceship Earth", despairs at coping with a population explosion. All of us, witnessing implacable ideological differences, the centres of conflict, and the spread of instruments of mass destruction, wonder how a peaceful world can be maintained.

In like manner, the mineral specialist is staggered by the growth of mineral requirements. He is aware that the "cream" of the world's mineral deposits has already been skimmed off, and that we are now subsisting in large measure on deposits so low in grade that they were not viewed as potentially usable two generations ago. We are reaching to the most remote and inhospitable wildernesses, and to greater and greater depths — 3,000 m in mines,

10,000 m in wells, and to the deepest ocean floors — to supply present needs.

Technology has been strained very hard to glean minerals from nearly barren rock. Many of the great mineral deposits — the European and Appalachian coalfields, the European and Lake Superior iron deposits, the bonanza silver camps, many great oil- and gas fields, and the famous gold and base metal areas — are all severely depleted. Where will the minerals be found to fuel, feed, clothe, and house a doubled, tripled, or quadrupled world population and a rising per capita demand?

Despite the great demands of this century, the mineral industries have been able to supply these ever-increasing requirements with few shortages and very cheaply. No economic mineral can ever be fully depleted if the consumer can pay the price; at a price, mineral requirements can always be supplied from lower-grade sources, or, except for fuels, can be more fully reclaimed. Our problem in the remainder of this chapter is to look at future requirements and to examine the capacity of the world to pay.

Population

The population explosion, which shows no signs of abating on a worldwide basis, has been truly fantastic in this century (Putnam, 1953; United Nations, 1958, 1974). In the year 1000 the world population was somewhat less than 300 million. It took 700 years for this population to double, but only 160 years for it to double again. It redoubled in the next 85 years, and that figure will double again, 35 years later, in the early 1980s. The earth's present human population is about 4 billion, and only a major catastrophe or an almost unbelievable change in attitudes can prevent it from becoming 7 billion by the end of this century. Any slowdown by non-catastrophic means, whether by education or by compulsion, can hardly be expected to take effect before the 7-billion mark is passed. If no slowdown occurs, there will be 14 billion people in the year 2040, and 30 billion one hundred years hence. Is it possible that even 10 billion people can enjoy a good living standard by any reasonable definition, or that 15 billion people can have anything beyond starvation and misery? This is not a remote problem. Many people now living can expect to be part of the 15 billion if growth continues unchecked.

The population problem can be solved very readily by simply reducing the birth rate, but the obstacles preventing accomplishment of this solution are difficult to overcome. Beyond the deeply embedded instincts leading to procreation are the religious and social convictions of large numbers of people and the ignorance of birth control measures by even greater numbers. Certain ethnic groups loudly insist that attempts at birth control reflect conspiracies designed to keep them forever in minority status, and refuse to cooperate.

Despite the forbidding statistics summarized above, at the United Nations conference on population at Bucharest in late 1974 the majority adopted the incredible position that population control is not necessary and that advocates of birth control are simply conspiring to retain their present large share of the world's food and raw materials! (*Wall Street Journal*, 1974). Quoting Dr. Han Suyin of China, "Family planning is not a cure for poverty and never will be. Population growth is not the world's basic problem, but social injustice and imperialism are. The reality that all should face is that the rich must now divest themselves of their property for the benefit of the poor".

Many factors beyond mineral scarcity or those mentioned above can interfere with population growth; none of us can predict how they will interplay. Therefore, instead of speculating at long range about the unpredictable, let us consider the mineral requirements and the capacity to pay of a world that is barely over the horizon — a world of seven billion people in the year 2000.

Mineral requirements

Fig. 6-1 is a chart of world population expressed in index numbers (1963 = 100), and includes, on the same basis, a curve of the value of world mineral production-consumption; for this purpose, annual production is assumed to equal annual consumption (stockpiling is therefore ignored). (The source material for Fig. 6-1 is given in Table 6-I.) The values of mineral production for the years 1950 and 1955 are estimated by comparison with similar index numbers available for value added in the mineral industries, although the value added indices increase somewhat more rapidly than the indices for the value of world mineral production. The per capita curve in Fig. 6-1 is derived directly from the other two curves.

The value of production indices increased by an average of 25% every five

TABLE 6-I

United Nations world indices for mineral production and gross product (1963 = 100) in constant dollars (data from U.N. Statistical Yearbooks)

Year	Value of mineral production	Value added by mineral production	Gross product per capita (excl. services)
1950	58*	56	67
1955	71*	70	82
1960	88	86	93
1965	111	111	109
1970	142	145	129
1971	145	150	132**
1972	148	156	136

* Estimated from value-added indices.
** World Bank gives $911 for this year.

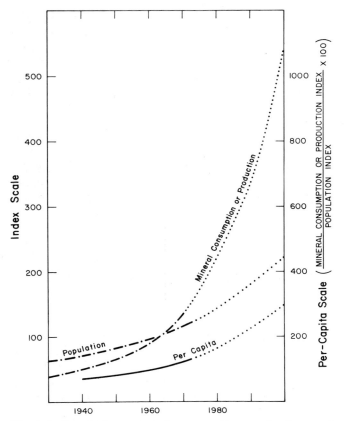

Fig. 6-1. Population, mineral consumption-production, and per capita mineral consumption-production indices (1963 = 100) for the world projected to the year 2000. (Data from U.N. Statistical Yearbooks.)

years from 1950 to 1970. Projecting production to the year 2000 at the same rate gives an index of 540; such a projection is considered to be conservative, because the rate of increase has tended to rise as the years have passed.

The mineral index for 1971 was 145; thus, the projection to the year 2000 indicates a mineral consumption value in that year of 3.75 times the 1971 level (in constant dollars). Also, dividing the mineral index for the year 2000 by the population index of 220 (reflecting a population of 7 billion) gives a per capita level of 245. Compared to the per capita level of 125 in 1971, this indicates a per capita increase by the year 2000 of 96% over 1971.

Unfortunately, at the time of writing, there was no reliable estimate of the total value of world mineral production or consumption in 1972. The U.S. Bureau of Mines (1973) gave a figure of $102 billion for 1971, based on its

adjustment of data from *Annales des Mines.* This 1971 figure was altered to
$103 billion in the 1972 Minerals Yearbook (U.S. Bureau of Mines, 1974).
Sutulov (1972) gives a figure of $166 billion for world production in 1970
(at 1972 prices), which would rise to $173 billion in 1972, using United
Nations indices.

The writer, using U.S. Bureau of Mines figures, estimates production of
about $386 billion in the year 2000 (expressed in 1971 dollars) on the basis
of a simple projection of statistical data (Fig. 6-1). This agrees with the U.S.
Bureau of Mines (1970) low forecast of $365 billion (1968 dollars) in the
year 2000, which was based on an annual growth rate of 3.6%. The high
range of the U.S. Bureau of Mines forecasts is $645 billion, based on an
average annual growth rate of 5.5%.

The capacity to pay

Turning to the capacity of the mineral-consuming countries to pay, the
best indicators of the state of world-wide affluence are index numbers of
gross national production (excluding services) expressed in constant U.S.
dollars reported in the annual issues of the United Nations Statistical Yearbook. These can be translated to absolute figures by using World Bank data
for 1971 covering population, gross national product, and per capita shares
of gross national production (International Bank for Reconstruction and
Development, 1973).

Data derived from these sources are shown in Fig. 6-2. This chart indicates
startling differences in the affluence of various segments of the world population, ranging from $157 per capita in Asia (excluding Japan but including
the wealthy Middle East) to $5,310 for the U.S. and Canada (expressed in
1971 dollars). On a country basis the range is even wider, from $70 in
Bangladesh to $5,400 in the U.S.

From the point of view of mineral consumption, it is significant that all
of the curves point steadily upward, with the exception of Asia (excluding
Japan) where the curve was flat from 1971 to 1972. Hence, on an historical
basis, we have every reason to expect a continued upward trend of world
affluence. The world gross national product per capita increased rather consistently at a rate of 18% every five years from 1950 to 1970. Projecting this
rate of increase forward to the year 2000, we derive a figure of $2,390 per
capita, a figure 2.6 times the 1971 level.

The projected total mineral consumption of 3.75 times the 1971 level is
2.3% of the projected total gross national product, whereas the 1971 consumption was 3.4%. Thus, on a statistical basis, we can expect purchasing
power to be ample to handle the projected increase in mineral consumption.
This view is supported by the fact that from 1963 to 1972 affluence grew
more rapidly than per capita mineral consumption; the former grew by 87%,
while the latter grew by 69%.

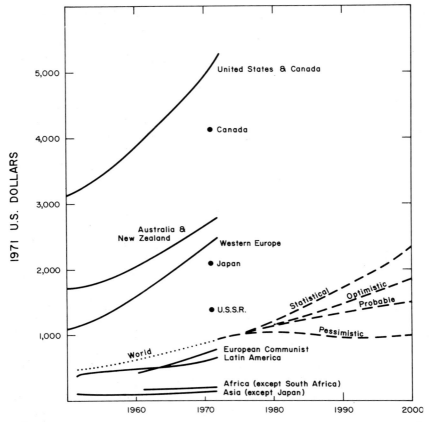

Fig. 6-2. Per capita gross production (excluding services) in 1971 U.S. dollars for some countries and regions and various possible projections to the year 2000 of world per capita gross product. (Data from International Bank for Reconstruction and Development, 1973; U.N. Statistical Yearbooks.)

The conclusion that the world's purchasing power will be sustained, barring catastrophe, is also supported by the data in Table 6-II. This table shows that the growth rate of world affluence is larger than that of the U.S. and Canada. Moreover, it shows that Asia and the Middle East, the European centrally planned economies, Western European countries, and the Latin American countries have all been growing faster than the U.S. and Canada. A recent study by the Wharton School of the University of Pennsylvania (Malenbaum et al., 1973) concluded that the demand for minerals in the "rest of the world" will also continue to grow in the next three decades at a rate faster than that in North America; this agrees with the conclusions reached by the forecasts of the U.S. Bureau of Mines (1970).

TABLE 6-II

Gains in affluence, 1950—1972

	Per capita indices of gross products (excl. services) 1963 = 100, constant dollars			Percent gain ÷ years
	1950	1960	1972	
Asia, East and Southeast, except Japan	79		120	2.35
Asia and Middle East		90	158	6.25
Africa (excluding S. Africa)		97	119	2.00
Latin America	77		132	3.23
European centrally planned economies		89	169	7.50
Western Europe	62		139	5.62
U.S. and Canada	79		132	3.04
Australia and New Zealand	83		128	2.45
World (1971 dollars)	462		938	4.68

A PERSPECTIVE ON WORLD ECONOMIC DEVELOPMENT

Despite the strength and persistency of the trends cited above and the confidence that they may inspire for the continuation of the trends over a mere 25 years, there are several factors in the world scene that tend to undermine such confidence. A pessimistic viewpoint is suggested by the following conditions.

(1) The "cream" of the earth's mineral resources is being rapidly skimmed off. As a result, minerals are likely to become relatively more expensive in the future, even allowing for the present inflationary trend. This argument becomes more credible when one considers how a cartel such as OPEC has quadrupled the cost of crude oil in 1973—1974; crude oil accounts for approximately one-half of the world's mineral bill. If the cost of minerals rises to 15% or 20% of the gross national product of the world, many consumers may be forced to economize, and economic growth may be sadly disrupted.

(2) The threat of continued, spreading mass starvation seems very real. Apart from the civil disorders that may result if starvation becomes widespread and persistent, there is the simple fact that hungry people are unlikely to contribute to world affluence.

182

(3) The World Bank, in its Annual Report for 1974 (International Bank for Reconstruction and Development, 1974) states, ". . . it appears now that the rate of growth in the poorest countries, including the most populous, will be so low that per capita incomes will either stagnate or rise very little between now and 1980."

(4) Population growth simply cannot persist indefinitely in a world of fixed dimensions. Tolerable limits may well be less than seven billion people.

(5) Democracy is showing an inability to generate leadership capable of formulating constructive solutions to national and world problems, and inspiring the people to adopt them. Instead of facing up to problems, electorates are showing an increasing disposition to ignore them. The constructive potential of dictatorship seems even worse from a world viewpoint.

(6) Continuation of the growth of affluence for an increasing world population will require capital-intensive technology. However, the world populace exhibits a growing intolerance towards "savers" and towards policies which nurture private capital formation and utilization. On the other hand, the accomplishments of nations and segments of nations where state capitalism prevails do not encourage the notion that forced saving can propel affluence effectively. The productivity per capita of the European centrally planned economies is below the world average, even though the group includes such advanced countries as East Germany, Poland, and Czechoslovakia;

TABLE 6-III

Probable limits of mineral demand

| | 1971 | 2000 | | |
		highest	most probable	lowest
Population, millions	3,685	7,000	6,000	5,000
Per capita annual product (constant 1971 dollars)	911	1,800	1,500	1,000
Total annual product (billions of constant 1971 dollars)	3,354	12,600	9,000	5,000
Per capita annual mineral demand (constant 1971 dollars)	28	52	45	31
Total annual mineral demand (billions of constant 1971 dollars)	103	360	258	155
Total annual mineral demand ÷ 1971 mineral demand	1	3.5	2.5	1.5

also, Yugoslavia has had considerable help from western countries. Considering the resources of the U.S.S.R., its position is likewise unimpressive. The nationalization of certain industries in the socialistically inclined democracies has demonstrated nothing but the weight and inefficiency of bureaucracy, effective though it may be in perpetuating itself.

(7) A world-wide economic crisis impends. Despite economic theories about curbing inflation through increasing productivity rather than austerity, no nation has even been able to make the first alternative work. Labour unions and liberal politicians have been more concerned over dividing existing economic rewards than over increasing their total. As a result of demands for benefits, regardless of productivity, the world is now confronted with an inflation out of control, but is unwilling to accept the adversities that austerity would impose. Thus, inflation is unlikely to be controlled before it spirals into economic disaster.

(8) The widening availability of "horror weapons", the persistence of ideological conflicts, the growth of terrorist movements, and the increasing belligerence at all levels, make it improbable that the end of the century can be reached without a major eruption of violent destruction.

CONCLUSIONS

Considering all the potentials for interruption of the historical trend, any prophecy of the average status of world affluence from now to the end of the century is necessarily highly subjective. The writer believes that the threats to continued economic growth at the established rate are impressive enough to make him view such continuation as unreasonably optimistic. An increase of 3.5 times the 1971 level for both total economic growth and for total mineral demand is regarded as "reasonably optimistic"; an increase of 2.5 times the 1971 level is regarded as "most probable"; and an increase of only 1.5 times the 1971 level is "reasonably pessimistic". These views are summarized in Table 6-III. Even these scaled-down figures challenge the imagination as to where the new mineral deposits needed will be located and how the necessary capital to discover and exploit them will be obtained.

REFERENCES CITED

International Bank for Reconstruction and Development, 1973. *World Bank Atlas, 1973.* Washington, D.C., 15 pp.
International Bank for Reconstruction and Development, 1974. *Annual Report.* Washington, D.C., 142 pp.
Malenbaum, W., Cichowski, C. and Mirzabagheri, F., 1973. *Materials Requirements in the United States and Abroad in the Year 2000. A Research Project Prepared for the National Commission on Materials Policy.* University of Pennsylvania, University Park, Pa., 30 pp.

Putnam, P.C., 1953. *Energy in the Future*. Van Nostrand, New York, N.Y., pp. 16—17.

Sutulov, A., 1972. *Minerals in World Affairs*. University of Utah, Salt Lake City, Utah, 186 pp.

United Nations, 1958. *The Future Growth of World Population*. New York, N.Y., 75 pp.

United Nations, 1974. *Statistical Yearbook, 1974*. New York, N.Y., 877 pp. See also earlier yearbooks.

U.S. Bureau of Mines, 1970. *Mineral Facts and Problems*. U.S. Government Printing Office, Washington, D.C., 1291 pp.

U.S. Bureau of Mines, 1973. *Minerals Yearbook, 1971*. U.S. Government Printing Office, Washington, D.C., Vol. III, 1103 pp.

U.S. Bureau of Mines, 1974. *Minerals Yearbook, 1972*. U.S. Government Printing Office, Washington, D.C., Vol. III, 1111 pp.

Wall Street Journal, 1974. August 23.

Chapter 7

FUTURE WORLD ENERGY DEMAND AND SUPPLY

T.M. THOMAS*

INTRODUCTION

The five-fold increase in world energy consumption in the past 50 years, the shifting patterns of energy production and consumption, and the growth of trade as the use of solid fuels has been replaced by petroleum have been discussed in Chapter 5. Attempts to forecast future demand and supply are subject to many pitfalls — few could have foreseen the four-fold price increase in petroleum imposed by the OPEC countries in 1973—1974. The price increase has not only had serious repercussions on importing countries' balance of payments, but will certainly affect the level of consumption and trade in many countries. Even more importantly, the price increase could stimulate new exploration, may convert currently uneconomic deposits into economic deposits, may give new impetus to the development of new methods to exploit the tar sands and oil shales, and may accelerate the development of nuclear energy.

Assessment of future energy resources must necessarily be highly speculative, not least because of a lack of adequate coverage of drill holes to requisite depths and the fact that many companies regard their findings as confidential. Nevertheless, this chapter attempts to evaluate potential supplies of energy resources in the light of information on the probable changes in world energy demand.

FUTURE ENERGY DEMAND

The U.S. Bureau of Mines (1970) forecast that world-wide demand for energy could grow at an annual rate between 3.6 and 5.3% in the next three decades. Darmstadter (1972) forecast that the U.S. annual rate of growth would be in the range of 2.3—4.3%; OECD (1966) predicted a 5% annual rate of growth for its member countries, with the exception of Japan where demand is expected to increase by 8% a year. Demand for petroleum is

---

* This chapter was compiled and edited by G.E. Thomas and G.J.S. and M.H. Govett from a draft prepared by the late T.M. Thomas in 1973—74. The opinions expressed are those of the author.

expected to increase at an annual rate between 1.5 and 3.3% in the U.S. and between 3.0 and 5.6% in the rest of the world in the next three decades (see Chapter 4, Table 4-IV).

The high rates of growth in energy demand expected in the future in Japan, some Western European countries, the U.S.S.R. and Eastern Europe, and in a number of less developed countries is a reflection of the fact that during a period of rapid growth and industrialization the demand for energy grows very fast. Between the beginning of the 20th century and 1970, energy consumption in the U.S. increased thirteen-fold; now it is only increasing at a rate of about 3% annually; in contrast, in much of the rest of the world demand is increasing at twice or more this rate. The need for energy in developing countries today is likely to be greater than it was in the U.S. when that country was at the same stage of development — greater amounts of energy are needed as lower-grade ores are mined, pollution controls often mean larger energy inputs for the same product, and urbanization is a rapidly spreading phenomena.

Japan has recently been reproducing the phenomenal rate of increase of energy consumption which the U.S. experienced in the past 50 years; there is good reason to expect that the centrally planned economies and some of the developing countries of Asia and Latin America will also reach and possibly even surpass American per capita energy consumption in the next few decades. The fact that increasing levels of per capita energy consumption are inherent in the early stages of economic growth, and in the more developed countries the rapid growth in road transport, the development of the petrochemical industry, the continued substitution of liquid for solid fuels, and the demand for increased electricity generation capacity as a function of growing industrialization and urbanization will virtually guarantee a high and rising rate of growth for energy demand, especially in the less developed countries.

Price increases for petroleum may result in some slackening of the rate of growth in the short-term, especially in countries which are dependent on imported oil. OECD (1974) forecast an annual rate of growth in its member countries of 3.5—4% between now and 1985, revising its earlier forecast of a 5% annual rate of growth for the same time period. Conservation measures may help to reduce demand; Cheney (1974) estimates that the U.S. could reduce its energy consumption by about one-quarter without seriously affecting the American "life-style", and OECD (1974) states that the potential for conservation in OECD countries is 15—20% lower consumption than forecast levels for 1985. In the short-term there may also be some change in the pattern of consumption — especially a reduction of or a reversal of the trend of the past two decades toward increasing use of petroleum as an energy source. The share of petroleum in energy consumption in the OECD countries is expected to decline from its present level of 55% to about 45% or even less by 1985 (OECD, 1974). This implies a substantial reduction of oil imports,

for 40—50 years; approximately 85% of the current output of brown coal is used for electricity generation.

Canada, Australia, and South Africa

The coal reserves of Canada have been estimated at 110 billion tonnes. Canada's coal resources are obviously very large, but more detailed information on the workability and composition of the component coal seams is needed for the extensive coal-bearing areas of the plains region, outer foothills belt, and inner foothills zone of Alberta, and for several separate basin areas in British Columbia; only with this information would it be possible to determine future export policy and the best use of coal in Canada. It is, nevertheless, confidently anticipated that considerable quantities of western Canadian coal will be marketed before the end of the present decade.

The use of coal in Australia is expected to double by 1985, with the main growth sectors being electricity generation, steel making, and processing of non-metallic minerals and non-ferrous metals. In New South Wales and Queensland black coal reserves (down to a depth of 600 m) which are mineable under current economic conditions and with existing technology have been estimated at 24.7 billion tonnes; second-category reserves are estimated at 195 billion tonnes. Major opencast coal sites, as in the Bowen Basin in Queensland, are now in operation, but access to the great bulk of the coal resources can only be gained by using underground mining methods.

Coal consumption in South Africa is expected to reach 140 million tonnes by the year 2000, with electricity generation accounting for nearly 70% of the total. There is considerable potential for large-scale opencast mining, particularly within eastern Transvaal, while large-scale prospecting is underway in neighbouring Swaziland.

FUTURE PETROLEUM PRODUCTION

Political and economic problems

During the 1960s global energy consumption was increasing at an annual rate of about 5%, while the consumption of petroleum was increasing at a rate in excess of 8% annually (United Nations, 1973). Shortfalls in production in the U.S., and increasing imports by Japan, the Western European countries, as well as by the U.S., made it possible for the OPEC countries to quadruple the price of oil in 1973—74. This action may result in some decline in petroleum consumption in the short-term; however, it seems likely that there will not be any serious decline in the rate of growth of petroleum demand in the medium-term (see earlier section on energy demand).

In the middle of 1974 crude oil production exceeded demand by about 1.5 million barrels per day; some producer countries were of the opinion that the level of demand in 1980 would not unduly exceed that of 1973 if international oil prices remain at, or near, their present levels. The continuity of supply would still be vulnerable to the policies of the main producing countries, some of whom might prefer to keep oil in the ground in view of their surplus revenues. Many international oil companies, who are aware of the uncertainty about the future, are becoming increasingly involved in other aspects of the energy field, notably in coal production and nuclear power construction schemes.

If the consumption of petroleum, which was expected to practically double in the next ten years, only increases in this period by 25% or even less, the peaking of world oil production (determined by the availability of recoverable reserves and demand) which has been predicted for the end of this century (e.g., Hubbert, 1973) could be delayed well into the first half of the next century. One should not, however, be too complacent; exploration effort is increasingly concentrated in difficult or hostile environments and given high costs, the importance of the giant fields may be enhanced; many of the smaller finds may hardly repay the exploration and extraction costs. On the other hand, it is still possible that "giants" may be found in reservoirs already discovered years ago; tremendous research advances have been made recently by the international oil companies on "tertiary" recovery.

The extent to which oil companies will be able to remain independent operators is questionable. Some oil-producing countries have declared their intentions of increasingly tying the sale of their oil to arrangements that will ensure the industrialization and development of their own territories; such policies could perpetuate bilateral agreements which, in themselves, help to maintain the current high price of oil. While the establishment of new industries in the oil-producing countries is a hopeful development, to the extent that the oil companies find it difficult to operate, exploration efforts may well be reduced. Awareness of potential problems of exhaustion of reserves (even if that date is some five decades or more in the future) has led some countries to plan on reductions of production; Kuwait, Libya, and Venezuela have become quite conservation-minded. If prices of crude oil are not to be driven even higher, a proper balance between the needs of consumers and producers and between conservation needs and energy demands in the main consuming countries will have to be struck.

Despite the moderate views reputedly expressed by Saudi Arabia, the world's largest exporter of crude oil, the oil-producing countries are likely to continue to press for further price increases, and production cut-backs are possible. While the U.S.S.R. might take advantage of the high petroleum prices now prevailing, it seems unlikely that the vast oil reserves of the U.S.S.R. will be allowed to enter freely into world trade.

Political problems may also impede the development of offshore oil deposits. The lack of agreement on the ownership of oceanic resources could seriously affect future exploration. With more than 130 oil companies engaged in oil exploration and production off the coasts of some 80 countries, the need for international agreement on the licensing and development of offshore areas is extremely important (see Chapter 8).

Supplies of petroleum

Proved reserves of conventional oil measure the volume remaining in the ground which geological and engineering information indicate with reasonable certainty will be recoverable in the future from known reservoirs under existing economic and operating conditions (see Chapter 1). The world's proved reserves were estimated at approximately 630 billion barrels at the end of 1973 (*Oil and Gas Journal*, 1973); 55% of the total was located in the Middle East, 16% in the U.S.S.R., Eastern Europe, and China, and 6% in the U.S. The several published estimates of world *ultimate* reserves suggest a total within the range of 1.2—2.5 trillion, although a figure as high as 4 trillion barrels might be possible since discoveries made in recent years have not been fully evaluated, there have been significant new and improved recovery techniques, and there is an expectation of producible oil being found at greater depths and in large unknown offshore areas. Among the many estimates, Warman (1972) favours an ultimate reserve of between 1.6 and 1.8 trillion barrels, but McCulloh (1973, p. 487) in the U.S. Geological Survey resource appraisals is unwilling to make an estimate arguing that "Who can say what the undiscovered resources of petroleum and natural gas are in the world". He also points out that not all the hydrocarbon accumulations in the world are worth finding; there are many small accumulations which are unable to repay the cost of discovery, let alone the costs of development.

The significance of "giant" oilfields to world reserves was stressed in Chapter 5. Within the U.S. most of the "giants" were found by exploration directed towards the discovery and exploitation of structural traps; a high density of completed wells has reduced the scope for such discoveries in the future, but it is still anticipated that more sophisticated forms of exploration aimed at the more subtle types of traps — stratigraphic or palaeogeomorphic — will still be rewarded by the discovery of major fields (Halbouty, 1970). Further exploration in other countries containing deep, oil-bearing sedimentary basins may well be rewarded by the discovery of new giant fields.

Optimism about finding new giant oilfields must be tempered by the recognition that much of the most accessible oil has probably already been discovered. In many instances exploration will have to go to greater depths; a high proportion of future exploratory effort will have to be in "frontier" regions or in inhospitable tracts, such as the outer continental shelves and polar latitudes. Exploration to greater depths and in remote and inhospitable areas implies greatly increased exploration and subsequent development and recovery costs; costs will also be high for anti-pollution measures and to counteract potentially detrimental effects on virgin environments.

Even in shallow water the costs of drilling an offshore well can be several times higher than that of an onshore well, and development costs for deep-water oilfields are extremely high. With increasing distances from the shoreline, many of the smaller discoveries may have to be abandoned as uneconomic. Deeper drilling also raises serious technical problems, and the cost of drilling each metre increases markedly at great depths.

A few major discoveries have been made in quite unexpected places, as for example the Aujila field in Libya where part of the reservoir lies in fractured and weathered granite basement of Lower Palaeozoic or Upper Precambrian age. In the future, a greater proportion of exploration will have to be speculative; although most giant field reservoirs are of Mesozoic or Tertiary age, greater attention will need to be focused on Palaeozoic successions. Marine sediments are dominant as source rocks, but this does not preclude non-marine strata from fulfilling the same role, while carbonates as well as shales can provide a source for organic matter.

The greatest potential for the discovery of major oil accumulations lies in the following situations or locations: (1) on the continental shelves and slopes or the ocean rises beyond the shelves and slopes; (2) onshore in Asia, including the U.S.S.R.; (3) in unexplored basins containing mainly continental strata; and (4) in those areas of the world where the deliberate exploration of the more obscure or subtle traps — stratigraphic, unconformity, palaeogeomorphic — has not been carried out on any scale.

Future of the main petroleum-producing countries

U.S.

Weeks (1972) has estimated that the total oil requirements of the U.S. in 1982 will be 8 billion barrels and that in 1992 it will rise to 11.8 billion barrels. Whether this estimate is realistic or not, it is evident to most observers that unless there are major new discoveries in the U.S., domestic supplies will not be adequate to meet future demand. The U.S. Geological Survey (Theobald et al., 1972), in what is admitted to be a very optimistic estimate, based on the largest proportion of favourable ground for exploration, concluded that there could be 2,900 billion barrels of petroleum liquids in the U.S.; of this 52 billion barrels are identified and recoverable. About one-half of the total resources are expected to be found in offshore regions of the continental U.S.; about 20% are located in Alaska.

Fig. 7-1 illustrates the favourable areas for petroleum basins. While the Gulf Basin coastal states and the Pacific Ocean Basin have the highest proportion of known plus unknown resources, Alaska (where only 15% of the land area has good prospects for hydrocarbon discoveries, see Fig. 7-2) has large

Fig. 7-1. Ultimate recovery of potential and known reserves of crude oil in the U.S.

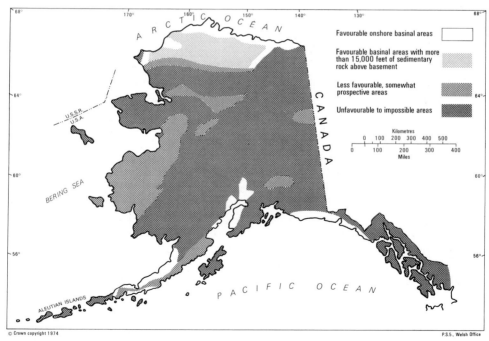

Fig. 7-2. Petroleum basins in Alaska.

potential resources in the North Slope, Prudhoe Bay, and Cook Inlet province. Alaska's continental shelf covers as much area as the land mass itself, although 85% of it is in the ice-bound polar latitudes. The Pacific coastal states, where the greatest potential finds are expected to lie in the offshore areas should pose less problems for exploration.

The Gulf Basin states are considered to have a great potential for petroleum discoveries. After 25 years of active exploratory drilling, the Louisiana-Texas continental shelf remains the region's prime prospective area, with suitable trapping and reservoir conditions within Pleistocene strata likely to continue into deep water.

Estimates of petroleum potential in the Atlantic coastal plain and adjacent continental shelves must await completion of several deep subsea wells. The area south of Cape Hatteras has much in common geologically with the on-shore-offshore region of Saudi Arabia (McCulloh, 1973).

"Project Independence", which was announced in November of 1973, is a plan to reduce America's increasing dependence on foreign energy sources; it proposes a 25% increase in domestic oil and gas production, chiefly from off-shore sources. Companies have been offered rights to drill for oil and gas, and tax changes to encourage new exploration are being made. Conservation programmes are also underway in an attempt to reduce consumption. How successful these programmes will be will depend on whether deep-water terminals can be built, the Alaska pipeline project can be completed (since it was first announced in 1969 the estimated total cost has increased at least four-fold), and the outcome of negotiations on the ownership by the states of the outer continental shelf. Even with the flow of Alaskan oil and intensive exploration for new domestic sources of conventional oil, it seems unlikely that the reliance on imported petroleum can be significantly reduced in the next decade. The possibilities of developing unconventional sources of oil, such as the oil shales, is discussed later.

U.S.S.R.

It is probable that 30—35% of the world's untapped hydrocarbon resources are in Arctic and sub-Arctic U.S.S.R. (Meyerhoff and Meyerhoff, 1973); Fig. 7-3 shows the main petroleum provinces in the U.S.S.R. Exploitation of the vast arctic resources will need a huge investment of both capital and labour in the next few decades; aside from preliminary geological and seismic surveys, no significant oil development seems likely in the arctic continental shelves for some years to come.

The Five-Year Plan for 1971—1975 envisages the development of about 2 billion barrels of new capacity, about one-half of which is designed to cover the decline in production at many of the older oilfields within the Volga-Urals area, the North Caucasus, the republics of Turkmen and Kazakh, and Azerbaydzhan. Although new oil pools could still be discovered in these areas by more intensive or deeper drilling, further expansion of production

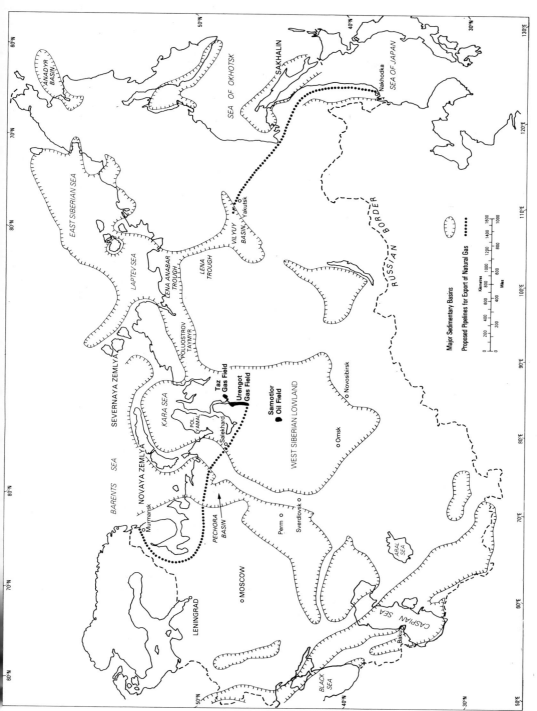

Fig. 7-3. Main petroleum provinces of the U.S.S.R.

at the newer oilfields or recently proved oil-bearing areas in western Siberia, Kazakhstan, Turkmenia, the Komi Republic and similar regions will call for improved communication networks, including the laying of completely new pipeline systems.

Internal demand for petroleum is growing at a rate of 7.5% per annum; doubts have been expressed, given the Soviet level of demand, whether the U.S.S.R. will be able to meet the requirements of Eastern Europe during the next five-year plan period (1976—1980). Up until 1980 it is hoped to increase production by about 250 million barrels per year; a pipeline project linking Tyumen in western Siberia to Nakhodka has been abandoned for the present, and it is therefore quite possible that there will not be sufficient Soviet oil for international consumption and for export to Japan. Large sections of the seemingly most favoured tracts within the western Siberian Basin cover flat marshy lowlands where there are extreme engineering difficulties.

The Middle East and North Africa

There is every prospect of further giant oilfields being found in the Middle East where the bulk of the world's proved reserves of conventional oil are located (see Fig. 7-4). However, many of the productive oilfields of the area are found in clearly discernable anticlinal structures; such features have already been the target of exploratory drilling, and intervening synclinal tracts are probably devoid of sizeable oil accumulations. In the offshore areas of Saudi Arabia lying within the Persian Gulf, the Zuluf and Marharan fields were discovered recently; prospects for further successful drilling are good. In 1974 proved reserves of Aramco were 97 billion barrels and probable reserves were more than 164 billion barrels; at a production rate of 20 million barrels a day (about twice present output) this could last well into the next century. Estimates of possible reserves range between 200 billion and 300 billion barrels.

The major oil reservoirs in Iran lie in fractured carbonates of Oligocene-Miocene age and are concentrated within a zone of intense folding less than 322 km long by 98 km wide; the reservoirs are enormous anticlinal traps, several of which hold recoverable reserves in excess of 5 billion barrels. It is highly unlikely that many reservoirs of this type remain to be discovered.

Libya and Algeria differ from the Persian Gulf states in that reservoir rocks in quite a few of the more productive oil fields comprise sandstones of Lower Palaeozoic age. The use of more sophisticated techniques for the discovery of deeper-lying reservoirs in less obvious structural situations should lead to further discoveries. However, in recent years the scale of exploration work has been reduced because of political difficulties; the future of North Africa, and all of the Middle East, will depend on whether new exploration takes place.

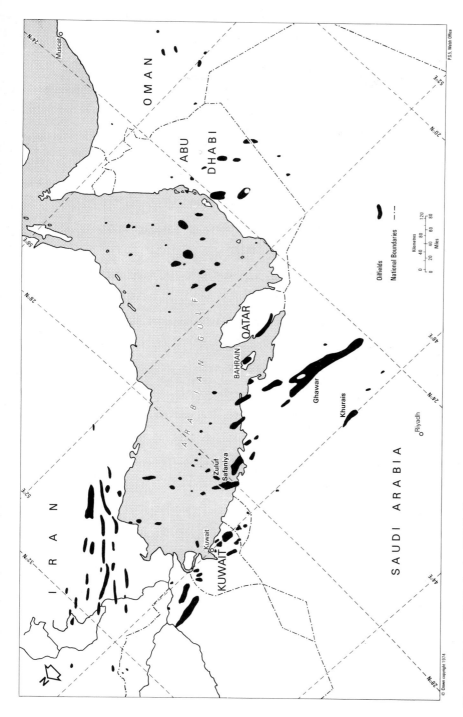

Fig. 7-4. Major oilfields in the Middle East.

200

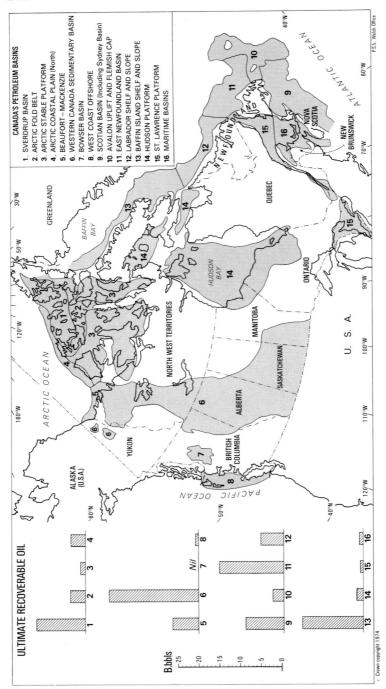

Fig. 7-5. Main petroleum provinces of Canada and estimates of ultimate recoverable oil.

Canada

Fig. 7-5 shows the main petroleum provinces of Canada. Proved oil reserves in Alberta, which provides 90% of Canada's oil output, declined by 446 million barrels to 6,600 million barrels in 1973. Reserves are sufficient for only 12 years at current annual production levels of some 550 million barrels. A lack of fiscal incentives was the main cause for a marked decrease in drilling activity in 1974; adverse taxation policies have also hindered new exploration in other parts of Canada. Faced with the prospect of declining supplies from the prairie fields, and the need to import oil in eastern Canada (there is no transport of oil from Alberta to the eastern provinces), exports to the U.S., which have accounted for something like one-half of the Canadian output, will cease this decade.

On the basis of cumulative probability curves for each of the likely petroliferous sedimentary basins or sub-basins of the continental and off-shore areas, the Geological Survey of Canada (1973) evaluated the country's potential in 1973; ultimate recoverable oil was estimated at 99 billion barrels, of which 12 billion barrels were in the Arctic, 15 billion barrels were in the Baffin Island shelf and slope, and 16 billion barrels were in the shelf and slope of the East Newfoundland Basin. Remoteness and extremely harsh climate in the Arctic archipelago will pose many problems for exploration; average drilling costs could be ten times the average costs encountered in the western provinces.

South America

The ratio of proved recoverable reserves to 1973 production in Venezuela was only about 12/1. Using the threat of complete nationalization (which actually occurred at the end of 1975), the Venezuelan government attempted to bring pressure to bear on the international oil companies operating within its boundaries to invest more on exploration in new areas. These include the tract lying south of Lake Maracaibo, the continental shelf, the Orinoco delta, and the Gulf of Venezuela whose subdivision is being disputed with Colombia. The offshore areas, particularly towards the Guyana border, seemed promising, but no major expansion of output is likely in the near future.

Peru has received a large loan from Japan to build a new pipeline across the Andes to exploit new developing oilfields in the Upper Amazon Basin; extremely difficult terrain will inhibit the rate of progress in exploiting the new oil province. In Brazil, which now imports about 80% of its present oil needs, seismic studies have indicated the presence of a number of large structures in the deeper offshore waters, but the proving and possible exploitation of reserves in such situations is at present beyond the technical capability of the state-controlled oil company; other potentially promising areas in the Upper Amazon Basin close to the Peruvian border would be equally difficult to explore and develop. Geophysical investigations off the east cost of Tierra Del Fuego in Argentina have had highly promising results, but the area is not likely to be fully assessed for a number of years; physical conditions for future drilling and installation of production platforms are likely to be equally as hostile as those in the northern waters of the North Sea. An oil discovery in Mexico was announced at the end of 1974 (see Chapter 5).

South and Southeast Asia

Vast tracts of Southeast Asia and China are virtually unexplored. In China petroleum production in 1974 was estimated at some 400 million barrels — treble the 1970 output. Drilling has started in the northern part of the Yellow Sea area where prospects are sufficiently attractive to warrant the interest of Japanese and international oil companies in the possible future issue of concessions. License blocks have already been allocated for exploration, and drilling rigs are at work in an area nearly 2,410 km long extending northward from the East China Sea to the continental shelf. Oil concessions in the South China Sea under South Vietnam's jurisdiction have been granted.

In Malaysia recent discoveries in Sabah and Sarawak, as well as in peninsular Malaysia, suggest that output could reach at least half a million barrels a day by 1980. In Indonesia the estimate of recoverable reserves is being continually up-graded; daily production could reach 3 million barrels by 1980. In the Bay of Bengal, where the Ganges River offers a highly promising deltaic area with recent thick sediments likely to extend seaward well into the territorial waters of Bangladesh, more than forty international oil companies are interested in exploring for offshore oil. In the Arabian Sea further commercial discoveries are likely to be made in the so-called Bombay High region lying off the west coast of India. In Pakistan the relatively unexplored Indus Basin probably has oil-bearing structures; potential reserves could be as high as 40—50 billion barrels. Other large sedimentary areas, which have not been sufficiently probed, could possibly be oil-bearing; these include the Baluchistan Basin and the shelf zones off Makran and Sind.

Western Europe

More thorough or extended exploration in the continental land mass of Europe is unlikely to result in any major new finds; on the other hand, the discoveries in some of the shelf sea zones have been spectacular, and the possibilities of further major finds have by no means been exhausted. This is particularly true in the North Sea, where proved reserves in the several oilfields in British waters approached 15 billion barrels late in 1974. It is conceivable that, with extended exploration, the reserves within the North Sea as a whole might well exceed 50 billion barrels, and available geophysical information suggests that outside the limits of the North Sea the whole of the shelf sea areas of northwest Europe and the contiguous continental slopes could hold reserves appreciably in excess of 100 billion barrels.

World offshore areas

Repeated reference has been made in this section to the prospects for discovery of offshore oilfields in the continental shelf zones of the countries surveyed. It is generally accepted that convergent plate boundaries, featured by marginal semi-enclosed basins within many of the major shelf areas, constitute some of the most promising areas in the world for petroleum accumulation. There is, however, disagreement as to whether the development of divergent plate boundaries may also create a habitat favourable for the accumulation of oil in some of the existing deep-ocean basins and deeper sectors of the continental slopes and rises. Claims that petroleum accumulation will extend seaward under the continental slope and rises to water depths of about 5,500 m along large portions of both the eastern and western margins of the North Atlantic and South Atlantic are at variance with the views of some who suggest that adequate trapping mechanism and sufficiently thick reservoir rocks for large oilfields could not be expected on the continental slopes except in a few places where great subsidence has occurred. Aside from this doubt, one must question whether new extraction techniques involving the employment of manifold systems, floating platforms, and dynamic positioning will provide sufficient levels of reliability and be financially viable in extreme water depths (exceeding 1,000 m). According to the United Nations (1973), the seabed to a water depth of 200 m covers 21.9×10^6 km^2 entirely within the continental shelf and contains about 68% of the total oil potential for the seabed (1,544 billion barrels). The zones lying beyond the 200-m isobath cover 340.4×10^6 km^2 and include some of the shelf, almost all the slope, and all of the rise, abyssal plain, and other features of the ocean basins. Total estimated ultimate hydrocarbon resources in depths beyond 200 m amount to 728 billion barrels, which is 32% of the total offshore hydrocarbon resources.

Conclusions on petroleum supplies

The above review indicates that, while there are excellent prospects of new discoveries in many parts of the world, particularly in offshore areas, the problems of actually finding new deposits and developing them are severe. There is no doubt that petroleum supplies could be adequate for many decades to come, but whether exploration and development will proceed fast enough to meet rising demands throughout the world will depend on new exploration methods, a willingness to invest in exploration, and the speed with which development can be undertaken. The most serious problems are the difficulty of attracting exploration in politically unstable or unfriendly countries, and the rising costs of exploration in the more remote and physically inhospitable regions of the world, including the offshore areas, which show the greatest potential.

NATURAL GAS SUPPLIES

Problems and prospects

World proved gas reserves increased from 1,144 trillion cu. ft at the end of 1968 to 1,477 trillion cu. ft at the end of 1970 and to 1,730 trillion cu. ft at the beginning of 1972; at 1972 production rates, reserves were sufficient for 39 years. Nearly 30% of the reserves are in the Middle East and North African countries, about one-quarter are in the U.S., and an estimated 13% are in the U.S.S.R. (U.S. Bureau of Mines, 1970). Proved reserves in 1975 are likely to be of the order of 2,000 trillion cu. ft, with much of the increase attributable to new major discoveries in Siberia. (It should be noted that estimates of natural gas reserves are subject to a much greater degree of uncertainty than those for petroleum; abandonment pressures in run-down fields can fluctuate widely, while relatively small increases in prices can provide a major stimulus for bringing into production only partly appraised reserves or those lying within reservoirs of limited porosity and permeability.)

In several major oil-producing countries, especially in the Middle East and North Africa, little interest has been shown in evaluating many of the non-associated gas discoveries. Spacecraft photography also reveals that much wastage takes place at many major oilfields, particularly in the Persian Gulf area and in the U.S.S.R. Much of this wastage could be eliminated by reloca-tion of industries, which are heavily dependent on energy supplies, near the major oilfields or by conservation of gas for later injection when oil pressures begin to fall.

Both the U.S. and the U.S.S.R. have recognized that the extensive pattern of fractures created by underground detonation of nuclear devices can in-crease production of natural gas or oil and can increase ultimate recovery percentages. In the U.S. experimental work has been done on the detonation of devices at moderate depths within natural gas reservoirs consisting of stratigraphic traps in thick, low-permeability clastic rocks. In the U.S.S.R. multiple nuclear devices have been detonated at moderate depths within or below depletion levels in oil or gas reservoirs. However, the high costs of nuclear devices and their installation, taking into account the precautions necessary to combat radiation and seismic hazards, remain a serious deterrent to wider use of the technique.

In many areas of the world, exploration for gas has been limited to those areas where fields would be reasonably accessible to major industrial markets. However, an increase in world trade in liquid natural gas (LNG) is in the offing due to high demand, efforts to curtail the use of petroleum, and the diminishing scale of wastage of formerly unmarketable gas. In 1973, a total of nearly forty specialized LNG carriers were on order throughout the world; these orders were concentrated in French and Norwegian yards, but both the

U.S. and Japan are expected to bid for new orders. By 1980 the level of seaborne natural gas imports in the U.S. could reach 2.2 trillion cu. ft and could rise to 3.6 trillion cu. ft by 1985; in Japan the level could be 1 trillion cu. ft in 1980 and 1.3 trillion cu. ft in 1985.

Main producing areas

Enforced deeper drilling and more speculative ventures in the future should favour the discovery of a greater proportion of gas fields than oilfields. The U.S. and the U.S.S.R. are particularly promising areas. New exploration is particularly important in the U.S. where, at present consumption rates, proved reserves of natural gas are sufficient for only about 12 years. Despite "Project Independence" (referred to above) and policies to promote more efficient use of energy, the shortfall could reach 10 trillion cu. ft per year by 1980. Projects to ease this situation include the importation of LNG from Alaska, Venezuela, Algeria, Nigeria, Trinidad, and the U.S.S.R. Imports from Canada currently account for about 5% of consumption. There is a good possibility that almost one-half of the gas consumed in the U.S. by 1990 will have to be imported or derived from synthesized petroleum liquids or coal.

Estimates of overall potential natural gas supplies in the U.S., exclusive of proved reserves, range from 595 to 1,227 trillion cu. ft of recoverable gas. The proving of potential reserves of this order will involve a far greater exploration effort than is now underway; the increasingly inaccessible habitats expected for most of the new gas reservoirs, and the higher scale of physical, technological, and environmental problems will impose far higher development costs than those now prevailing in existing fields. Nevertheless, recent major discoveries have provided a large measure of optimism that large reserves will eventually be proved.

The U.S.S.R.'s Five-Year Plan for 1971–1975 concentrated on increasing the effectiveness of utilization of associated gas supply from 62% in 1970 to 80–85% in 1975, on tapping more accessible reserves in the European part of the country, and on the completion of 30,000 km of new pipelines. Large trunk gas pipelines have been linked into the nation-wide system of pipelines and will continue to be expanded.

About three-quarters of the known gross reserves of gas within the U.S.S.R. are found in western Siberia, Central Asia, and Kazakhstan. The West Siberian Lowland is still largely unexplored; proved reserves could well be upgraded four-fold to more than 1,200 trillion cu. ft within the next few years (Dickey, 1972). However, the U.S.S.R. has no illusions about the difficulties of developing the important Arctic and sub-Arctic terrain; harsh physical conditions, limited manpower, and restricted financial resources are impeding the rate of progress (Meyerhoff and Meyerhoff, 1973). Nevertheless, the

U.S.S.R. could become the main supplier of natural gas to both Western and Eastern Europe.

While both the U.S. and U.S.S.R. are actively engaged in exploration and development work, only a few of the Middle Eastern countries have shown an interest in natural gas until recently. In Saudi Arabia about 90% of the associated gas from the Aramco Oil Company fields is wasted by flaring, although there has recently been interest in developing facilities to exploit natural gas.

The reserves of natural gas in Iran could be greater than those within the U.S. Algeria is the world's largest LNG exporter. In these, and other Middle Eastern and North African countries, increasing prices and shortages in the U.S. and Europe have made LNG projects increasingly viable.

Other countries

By the late 1980s natural gas production in western Canada could be as much as 15% below Canadian demand. Ultimate recoverable reserves are estimated at 782.9 trillion cu. ft (Geological Survey of Canada, 1973); only about 15% are in the western Canadian sedimentary basins (primarily in Alberta), and about one-quarter are in the Sverdrup Basin in the Arctic Islands. The East Newfoundland Basin, not yet probed by drilling, could contain a few large structures with a high gas potential.

The Indus Basin of Pakistan, where less than 100 exploratory wells have been drilled since 1947, is a highly promising area. In China intensive exploration in Szechwan Province has verified the presence of much larger gas reserves than originally anticipated. In Malaysia prospecting is underway; the greatest interest is being shown off the coast of Sarawak, where huge reserves have already been found. Gas has also been discovered off the eastern coast of peninsular Malaysia; market outlets for this are being sought in the larger towns on the west coast of the peninsula and in Singapore. In Indonesia the gas fields outlined to date have all been relatively small, but the deeper offshore areas which are still relatively unexplored offer good prospects for the eventual delineation of large fields.

Offshore drilling in Australia is currently at its lowest level for many years. On present knowledge, the prospects for discoveries of hydrocarbons in the continental shelf areas are more promising for natural gas than oil. International oil companies have been discouraged by lack of tax incentives, the demand for increased Australian participation, and threats of outright nationalization. There is currently a ban on fuel exports, but future exports to Japan are possible.

There is every prospect of further major gas fields being discovered in the northern half of the North Sea, while commercial finds of smaller scale will probably continue to be made in the southern waters; recoverable reserves within the North Sea province could well be of the order of 80—120 trillion cu. ft. The Celtic Sea could yield a further 30—50 trillion cu. ft. A large increase in consumption of natural gas in the EEC countries is anticipated; natural gas consumption could rise to 25% of total energy consumption by 1985.

HYDROELECTRIC POWER

The total world potential capacity of hydroelectric power has been estimated as 2.8 billion MW; this is 11 or 12 times the estimated 1974 installed capacity and about three times the total installed electric power capacity. Africa has the largest share of the total potential capacity (27%), followed by South America with 20% and Southeast Asia with 16%; North America accounts for 11% and China accounts for about 10% (Hubbert, 1969).

It is virtually impossible to estimate what proportion of the total capacity, either on a world or on a regional basis, will ever be developed because of varying degrees of significance given to land-use conflicts and the extent to which the associated dams, plants, and transmission lines will be allowed to impair the physical landscape in unspoilt or wilderness areas where the scope for new dams is generally the greatest. Most man-made reservoirs have a limited life cycle since impounded streams and rivers are continuously depositing beds of sediments on the floors; a state of practically complete sedimentation of the artificial basins is likely to be reached within two or three centuries. In some areas, notably in semi-arid regions, increasing scales of solution of mineral salts and concentration through evaporation also impose limits on the effective life cycle of dual hydro schemes which supply water for irrigation or domestic use.

Although the world's main hydro resources are in Africa, Asia, and South America, the main development of new capacity will probably take place primarily in North America. In most of the South American countries installed hydroelectric capacity represents only 5—10% of potential output, but there are difficult problems of capital investment and riparian rights where major rivers form international boundaries. Pressures on land and environmental constraints will limit development in Europe, especially in Sweden.

In the U.S. it is forecast that hydro capacity could conceivably double, although most of the suitable sites have already been developed and those that remain are the cause of great environmental concern. In Canada, there is greater scope for development, especially in the Upper Churchill Falls and in James Bay in Quebec, although there are also serious environmental problems.

NUCLEAR ENERGY

Nuclear power provides less than 1% of total world energy capacity today; according to the International Atomic Energy Agency it will have to provide

6% of the world's energy by 1980 and 20% by the year 2000 to meet projected world energy demands (United Nations, 1973). Nuclear power is limited primarily to North America, Western Europe, and the U.S.S.R. The U.S. has the largest share of installed capacity, but only 3% (that is, 10 million kW) of the total U.S. electric utility capacity was nuclear-powered in 1971 (Finch et al., 1973), although an additional 46 million kW capacity was being built and about 52 million kW was on order. It is expected that U.S. nuclear capacity will grow considerably in the next three decades, and it is predicted that by the year 2000 nuclear plants will provide approximately 60% of total American electrical power (U.S. Department of the Interior, 1972). World nuclear electric capacity is expected to increase at least ten-fold by 1980, and total world demand for uranium in 1980 is estimated at 66,000—96,000 tonnes (Finch et al., 1973).

The normal geological problems of assessing the availability of nuclear fuels are compounded by uncertainties regarding the future design of power-generating nuclear reactors. The fission reactor demands fissionable fuel which, in natural uranium, is the isotope ^{235}U. This nuclide constitutes only about 0.7% of natural uranium, the remainder being largely the isotope ^{238}U which is not readily fissionable (a third isotope, ^{234}U, comprises about 0.005% of natural uranium). However, fertile materials such as ^{238}U, when bombarded with neutrons (for example, from ^{235}U) are converted into an artificial fissionable isotope, plutonium ^{239}Pu. It is thus possible to produce more fissionable material than is consumed in the operation of a reactor; this process is called "breeding". Reactors which have a regeneration ratio greater than one are called breeder reactors.

Successful operation of breeder reactors will obviously have an enormous effect on the rate of depletion of uranium reserves. Another potentially significant development of breeder reactor technology is the use of thorium which, upon bombardment with neutrons, is converted into another fissionable uranium isotope ^{233}U. As presently envisaged, the ultimate nuclear power source is a fusion reaction. Its feasibility as a controlled reaction has yet to be demonstrated; it requires tens of millions of degrees of heat to cause fusion of deuterium or a mixture of deuterium and tritium (heavy hydrogen). This may be achieved before 1985 using laser fusion techniques (Roberts, 1974). The basic fuels of deuterium and lithium, which are used for fusion, are so abundant that they can not be considered limiting factors, and the amount of energy released per unit mass of fuel is several times that of nuclear fission (one gram of fused deuterium can produce about 100,000 kW of energy).

Most of the nuclear power reactors now use enriched uranium (i.e., uranium which has its proportion of the fissionable ^{235}U increased through removal of ^{238}U). Enriched uranium is preferred because smaller quantities are required per unit of power output, and a greater proportion is fissioned before the fuel has to be reprocessed.

Even if there was any certainty concerning replacement of fission reactors which use ^{235}U, the general lack of published data on world uranium reserves and resources makes a reasoned assessment of the adequacy of nuclear resources difficult. If it is assumed that fast breeder reactors will provide the major source of power prior to development of fusion reactors, then an order of magnitude of demand can be gauged by the conversion that 1 tonne of uranium burnt produces power equivalent to 25×10^6 tonnes of coal; stated another way, 1 million tonnes of uranium would provide all the world's electricity at the present production level for 1,000 years.

Projections in the field of nuclear energy are difficult under normal circumstances; the impact of the sharp rise in petroleum prices on nuclear power planning compounds the problem, as does the increasing concern for the environment. Identified recoverable resources of uranium for the world (excluding the U.S.S.R., Eastern Europe, and China) are about 1.5×10^6 tonnes, of which Canada, Sweden, and South Africa have about 20% each, the U.S. has 17%, and Australia has 6%. Reserves are clearly not large and depend on world prices; the U.S. has identified reserves which are adequate if consumed in reactors of current design only until the early 1980s (see Chapter 2 for more details on the effect of price on uranium reserves).

The situation with uranium reserves can change dramatically through mining very low-grade deposits and through changes in the type of reactors used. Thus, in the U.S. if the breeder reactor is introduced, the annual consumption of U_3O_8 will peak at about 90,000 tonnes in 1990 and then decline over a 20-year period to an annual consumption of about 9,000 tonnes.

The main problems in developing nuclear power will come in the next decade. According to OECD (1974), presently known reserves of lowest-cost uranium will be committed by about 1981; a major exploration programme should *already* be under way (given the usual lead-times between exploration and exploitation) if shortages are not to occur in the 1980s. Similarly, the present capacity for the production of enriched uranium, at least in the OECD countries (which include the main areas of nuclear power development outside of the U.S.S.R.), is expected to be fully saturated by 1984. OECD concludes that only "very early" introduction of fast breeder reactors (or a massive change to natural uranium-consuming reactors) could change the situation.

TAR SANDS AND OIL SHALES

If the economic fabric of the western world is not totally destroyed by the four-fold price increase for petroleum and Arabian geopolitics, these factors should be enough to bring down the last barriers to commercial exploitation of the tar sands and oil shales in the U.S., Canada, and possibly

210

Venezuela. The main problems for the future are largely technical and financial, and development depends on the future price of conventional oil and the willingness of industry to invest in research and equipment.

Tar sands

Synthetic oil from the Canadian tar sands was trickling out of the bleak muskeg swamps of northern Alberta at the rate of 51,000 barrels a day in

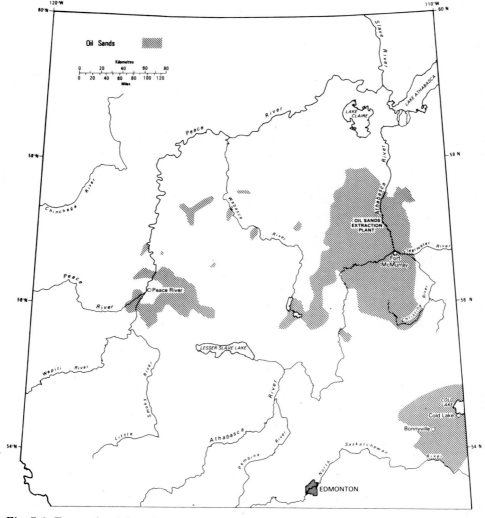

Fig. 7-6. Tar sands of Canada.

1973; location of the Fort McMurray plants and distribution of the 800-km-long Cretaceous heavy oil and tar sand belt (extending from outcrop to drilling depths of 760 m) are shown in Fig. 7-6. There are four deposits of varying size: the Peace River Sands in the northwest, the Wabasca and Athabasca Sands near Fort McMurray, and the Cold Lake Sands on the eastern borders of Alberta. Total ultimate in place reserves in the four deposits may be as high as 1,000 billion barrels of crude bitumen; the recoverable potential is 330 billion barrels, of which approximately 75% could be upgraded to 250 billion barrels of synthetic crude — more than two-thirds as much as the entire reserves of the Arab oil-producing countries.

Using an overburden limit of 45 m, it is estimated that about 80 billion barrels (or approximately ten times Canada's conventional crude oil reserves), are recoverable by the mining and extraction method used at Fort McMurray. The extracted bitumen is upgraded on the site and piped to Edmonton and from there to customers via an interprovincial pipeline system.

In 1974 the estimate for a first-generation surface extraction plant at Athabasca ranged from $1.5 to $2 billion (100,000 barrels daily capacity); this compares unfavourably with present-day North Sea oil cost figures ($2 billion for 400,000 barrels per day output). Nevertheless, if all goes well, seven to ten plants could be in operation by 1984 producing between 700,000 to 1 billion barrels daily. Where the overburden at Athabasca is more than 45 m thick, the only efficient recovery mechanism is the in-situ method where steam is injected to heat the sands or a portion of the oil-impregnated unit is deliberately set on fire (fire-flood) to lower the viscosity of the bitumen to allow it to be brought to the surface. By far the greatest success to date has been with steam injection, although promising experiments are either underway or about to begin with underground combustion.

A major new in-situ pilot plant at Cold Lake is underway; by 1974 some 56 wells to penetrate a 150-m-thick deposit at depths of 450—550 m had been drilled. In-situ recovery methods require the constant drilling of wells in new patterns to maintain the sweep of underground heat generation. Up to 10,000 wells (ranging to depths of 610 m) might have to be drilled in the course of the 25 years normally taken as the economic life span of an oil sand plant. Nuclear stimulation is regarded as a fascinating, although remote, prospect by industry experts.

Another project is planned to perfect in-situ production methods in the Peace River area where the deposits are said to be equal to or perhaps better than the choicest portions of the main Athabascan oil sands. Eventually, if the spectacular tar matte belt is to be fully developed, most of the synthetic oil production will come from in-situ operations at much lower financial and environmental costs than present strip mining pits.

In Venezuela the huge Orinoco Tar Sand belt has an approximate length of 600 km with a width of 53 km; however, only 10% of the estimated 700 billion barrels of oil is regarded as recoverable with existing extraction technology. The heavy oil is rich in sulphur, vanadium, nickel and other elements, and there is no technology to remove them economically; according to industry estimates it would take at least five years and $4 to $5 billion to put the Orinoco belt into production. A number of oil companies are interested in obtaining concessions, but the Venezuelan government is determined that the exploitation of the area will benefit the Venezuelan economy and not the developed countries.

In other countries viscous oil occurs in numerous discontinuous and lenticular sandstone reservoirs in shallow and surface areas in Kansas (U.S.), in numerous tar sand belts in Madagascar, and even in England there is a · modest strip of tar-impregnated Pennsylvanian sandstones along the Severn banks.

Oil shales

The U.S. has reserves of shale in Colorado, Utah, and Wyoming capable of yielding 2,000 billion barrels of oil; this compares very favourably with the country's dwindling reserves of conventional crude. The 50,000-barrel/day plant at Parachute Creek in Colorado is in some of the richest shale deposits; a prototype, built at a cost of $100 million has confirmed extraction technology and now produces 800 barrels a day. Investigation of in-situ treatment of oil shales is underway in New Mexico; first steps will be to excavate a cylindrical chamber in the deposit by enlarging a drill hole and, using explosive charges in nearby drill holes, the surrounding shales will be fractured and collapsed into the cavity. Propane gas would then be pumped into the brecciated material and ignited; oxygen would be fed into the area to sustain combustion.

Nevertheless, the problems of oil shale development are immense. An output of 1 million barrels a day would involve the processing by surface mining of 5.7×10^8 tonnes of shale a year. Maximum possible output by 1980 is likely to be 300,000 barrels a day. Landscape and water scarcity problems mean that in-situ and nuclear techniques are the most hopeful.

Outside of the U.S., oil shales are known to occur in Canada, China, the U.K., Spain, France, Italy, Yugoslavia, Zaire, Chile, South Africa, Siberia, Turkey, Australia, New Zealand, and Brazil. The largest known deposit in southern Brazil contains possibly two-thirds as much as the western U.S. resources. A prototype plant is being built in Brazil, but similar development in most of the other countries listed above seems to be well in the future (the effect of price on world oil shale reserves is discussed in Chapter 2).

OTHER SOURCES OF ENERGY

There are a variety of other sources of energy which vary in stage of development from minor but increasing use — such as geothermal and solar sources — to exotic concepts such as warm ocean currents. These are briefly reviewed below.

Geothermal resources

The primary use of geothermal resources to date is for the generation of electricity; for this purpose (given existing technology) a permeable geothermal reservoir must have a temperature of at least 180°C (365°F) and lie at shallow enough depths (3 km or less). World electrical capacity from geothermal energy in 1971 was approximately 800 MW, equal to about 0.08% of total world electrical capacity. Italy leads in installed capacity, followed by the U.S. and New Zealand.

The production of geothermal power is obviously restricted to areas where geothermal energy is found in sufficient quantities; unlike coal, oil, and gas, geothermal steam cannot be transported long distances. In "dry-steam" fields the steam is fed directly from wellhead to turbine after removal of abrasive particles; "wet-steam" fields, where the steam and water are mechanically separated at the wellhead, are far more common.

The cost of geothermal power generation ranges from $2.50 to $7.00 per 1,000 kWh (Jaffe, 1971). In a developed country the cost of 1 kWh produced by a dry steam geothermal field is about three-fifths of the cost of 1 kWh produced by conventional fossil fuels for a power system of the same capacity; in the case of wet steam, the cost of 1 kWh produced by a geothermal field is competitive with the cost of 1 kWh produced by conventional methods. On this basis, the future of power production from geothermal energy appears promising.

Geothermal resources have other uses besides electricity generation, but to date they have been minor. Geothermal waters as low as 40°C are used locally for space heating and horticulture. Much of Reykjavik, the capital of Iceland, is heated by geothermal water, as are parts of towns in New Zealand, Hungary, the U.S.S.R. and in Idaho and Oregon in the U.S. Geothermal steam is also used in paper manufacturing in New Zealand and has potential use for refrigeration. Reservoirs like the Salton Sea in California contain immensely valuable by-products such as potassium, lithium, calcium, boron, and other elements. Use of geothermal energy to desalt geothermal water itself has been proposed. ,

Geothermal reservoirs are the "hot spots" of larger regions where the flow of heat from the depth of the earth is one and one-half to perhaps five times the world-wide average. Such regions of high heat flow commonly are zones of young volcanism and mountain building and are localized along the margins of major crustal plates. These margins are zones where either new material from the mantle is being added to the crust or where crustal material is being dragged downward and "consumed" in the mantle (subduction zones). The world's main geothermal zones are along the mid-Atlantic, the western Americas, the Pacific margins of the U.S.S.R. and Asia, in Oceania down through New Zealand, the Red Sea and east Africa, along the belt of mountains extending from Italy through Turkey to the Caucasus, along the coast of North Africa, and in Antarctica. Geothermal fields are absent from the stable continental shields. Although there are no known shallow geothermal reservoirs in the non-volcanic continental areas bordering the shields, hot water has been found at depths of 3—6 km in the U.S.S.R., Hungary, and in the Gulf Coast of the U.S.

Modern exploration techniques — including chemical thermometry, electrical resistivity, electromagnetic depth soundings, micro-earthquake seismology, gravimetrics, and remote (infrared) sensing — have increased the success ratio of discovery of geothermal resources significantly. In the U.S. geothermal resources could supply 132 million kW of electricity by 1986 and 395 million kW by the year 2000. In Mexico the first liquid phase geothermal system in the western hemisphere has been put into production.

Solar energy

The approximately 2 trillion kWh of electrical energy consumed in the U.S. in 1970 were equivalent to solar radiation falling on only some 1,360 km^2 of desert. If this solar energy could be tapped with only 5% efficiency, 27,200 km^2 of desert alone (10% of U.S. arid lands) could supply the amount of electrical energy consumed in 1970. The world potential of solar radiation has been estimated at about 60 times actual conventional energy supply. Development of solar energy has not yet become an economic proposition and will not become so until research funds and energy prices increase considerably.

The three non-biological classes of solar energy utilization are terrestrial, space, and marine. Terrestrial and space systems would use incident solar energy, while marine systems would use both incident and natural solar energy stored in sea thermal gradients. Essentially two schemes have been proposed for terrestrial systems: one involves the use of solar cells and the direct conversion of solar energy into electrical energy; the other involves the absorption of solar energy as heat which is either used directly or converted into some other energy form. With the present enormous fabrication costs for solar cells, the terrestrial solar heat systems offer the best prospects for supplying economically a significant portion of the world's energy supply.

The use of solar energy for space heating, air conditioning, and hot water heating is one of the extremely attractive possibilities for conservation of non-renewable energy resources. Annual incidence of solar energy on the average built-up areas of the U.S. is six to ten times the amount required to heat the buildings. The cost of using solar domestic hot water heating in the 60 million dwelling units in the U.S., which would result in a 1.5% reduction in demand on other fuels, would be approximately $4.5 billion (Berg, 1973).

Space solar cell systems would employ a number of satellites in synchronous orbits about the earth's equator. Batteries of solar cells would collect the solar energy and convert it to electricity; this energy would then be transmitted to collecting antenna arrays on earth.

Two types of marine thermal systems have been proposed: one would use sea thermal gradients and extract the solar energy that is stored in the surface layer of the ocean; the other would use the oceans to support a floating platform and a system of concentrating mirrors to focus the incident solar energy on a boiler. This, and the other systems discussed above, could provide a significant fraction of the world's energy in the future. Sea thermal gradient systems are already at the engineering design and construction phase.

Wind power

Wind energy is an ideal no-cost fuel, providing a non-polluting energy supply on a self-financing basis. Based on work begun in the 1950s, the U.K. leads in wind energy research, aerogeneration, and hill-wind dynamics; with its maritime climate and hilly terrain the U.K. is well suited for generating electrical power from wind. Even with a 25% load factor, aerogeneration coupled to a pumped storage scheme, would be economically feasible today.

Unpredictability of capricious air currents, and hence the need to provide some kind of back-up system, is the greatest weakness of wind power. However, an experimental system installed in Scotland in 1958 suggested that in some circumstances a back-up plant of anything more than the occasional few kilowatts of off-peak grid power would be unnecessary.

Australia, with the world's largest pumped storage complex might be considered for a major use of aerogeneration. Japan could also use aerogenerators on its numerous dams with short river flows. Other potential areas are New Zealand, the western U.S., Canada, southern Chile, Norway, and Faroes and Falkland Isles (see Bruckner, 1974).

Tidal power

At present there are only two power-producing tidal barrage schemes in operation. One is in Brittany (across the estuary between Dinard and St. Malo) and the other is in the U.S.S.R. (in the Kislaya Inlet, 80 km northwest of Murmansk). Both are essentially pilot schemes.

Along the coastal areas of the Bay of Fundy between Maine (U.S.) and Nova Scotia (Canada) and along the coast of Alaska there are many proposals for the construction of huge dams to capture high tide waters and release them at low tide. In the Bay of Fundy it has been estimated that 2 million kW could be developed at a cost of about $5.60 per 1,000 kWh; in comparison new thermal, nuclear, and hydroelectric generating plants costs range between $9.00 and $15.00 per 1,000 kWh.

Waste heat from power stations

An electricity generating station produces "low-grade heat" in the form of hot water as well as electricity. Low-grade heat is perfectly suitable for many purposes; one of these is the district heating of towns and cities. A generating station producing 1,000 MW of electricity produces 2,000 MW of low-grade heat which could warm and wash a million people.

The Central Electricity Generating Board in the U.K. already has a district heating scheme in London where low-grade heat is supplied by the Battersea generating station. Increased installations of district heating in other cities would cause physical dislocation, but nothing like what people have been prepared to suffer in the past to accommodate private motor cars. A decision to exploit low-grade heat from future nuclear electricity-generating stations would mean that they would have to be sited within ten miles or so of cities; the implications of this are severe.

Warm ocean currents

Another scheme for producing electricity proposes the use of floating heat engines that are fed on one side by warm water from the Gulf Stream's surface and on the other side by near-freezing water piped upward from the seabed nearly 1 km below the surface. This ocean-generated electricity could be used to produce hydrogen for fuel cells ashore.

CONCLUSIONS

While Boyd (1973, p. 7) is correct in his conclusion that the world's re-sources are "... large enough to stagger the imagination", man's ability to develop these resources is limited at any given time. The events of the first half of this decade have emphasized that the continued growth of energy demand throughout the world has alarming implications both in terms of in-creasing costs and pressures on resources — not to mention environmental problems. Only the most optimistic expect the necessary funds and brain-power to be expended on the world's energy problem in the next few decades unless there is a major change in government's and industry's priorities and in the life-styles in the developed countries. In spite of its serious disruptive effect, the four-fold increase in the price of oil this decade may have had the effect of helping to bring about these changes — by making energy a high-cost commodity, conservation measures could be easier to implement, ex-ploration for new sources of energy may be stimulated, and the development and exploitation of non-conventional sources of energy may become eco-nomically feasible.

Hubbert (1973) has estimated that the peak of world oil production may occur by the year 2000 and that even for the world's most abundant fossil fuel, coal, production could peak by 2150. McCulloh (1973) is more optimistic, but like most observers feels that the future for fossil fuels is necessarily limited. Nuclear energy could be the panacea, but it also has tremendous hazards. Solar energy and geothermal energy are less hazardous, but the technical problems involved in their development have only begun to be faced.

An upper limit to global energy use suggested by Weinberg and Hammond (1970) of 600 million Btu per year per capita (about twice the U.S. per capita level in 1970) in a world populated by 20 billion people is beyond comprehension today. Given present rates of world population growth, this population could be achieved by the middle of next century (although it is hard to imagine that the prospects of a world so crowded would not result in massive efforts to make sure it does not happen); even with the greatest optimism, it is difficult to imagine the less developed parts of the world achieving a level of per capita energy consumption even approaching the present U.S. level, let alone surpassing it. Even if this theoretical limit is accepted, there is a danger that there will be a substantial air temperature rise within the next century if the present energy growth rate continues (Budyka, 1972); thermal pollution or waste heat introduced into the environment can alter weather on large as well as small scales (Patterson, 1973).

Measures to control pollution inevitably cause price increases. A major source of air pollution has been high-sulphur coal, and measures to prohibit its use directly reduce the usable reserves of coal; those who advocate increasing the use of coal to replace dwindling oil supplies must face the fact that its use will either increase pollution or result in vastly increased costs. The problems of surface mining and the disposal of wastes — a serious problem with coal mining — will also be a problem in the exploitation of the tar sands and oil shales. Marine pollution from oil spillage is an increasing hazard; even the use of geothermal power is attended by disposal problems caused by a high mineral content of the hot water.

The greatest potential source of danger to man and the environment is the development of nuclear power — the most optimistic source of future energy supply. The radioactive waste material cannot be casually buried; it must be safely stored for hundreds to a thousand years (depending on the product); obsolete reactors must be sealed and protected for 50—100 years before they can safely be dismantled. The costs of these protective measures are of a continuing nature and must be borne by present, and far distant future, production of energy.

The approach to the energy problem today must be two-sided. The supply side of the problem is essentially technological in nature — new exploration and the development of new techniques and non-conventional sources of energy are critical. The demand side is more complex since it deals with human nature. Efforts to reduce the rate of growth of demand are clearly necessary, and much can be done to reduce waste and to develop new low-

218

energy requirement techniques without much disruption in the present economic structure of the developed countries, although some changes in life-style may be necessary. This does not imply that the rate of economic growth has to be slowed down — particularly in the developing countries — but rather that the world's economies will have to adjust to higher-cost energy and develop better ways to use it.

REFERENCES CITED

Averitt, P., 1973. Coal. In: D.A. Brobst and W.P. Pratt (Editors), *United States Mineral Resources. U.S. Geol. Surv., Prof. Paper* 820, pp. 133—142.
Berg, C.A., 1973. Energy conservation through effective utilization. *Science*, 181: 128—138.
Boyd, J., 1973. Minerals and how we use them. In: E.N. Cameron (Editor), *The Mineral Position of the United States, 1975—2000.* University of Wisconsin Press, Madison, Wisc., pp. 1—8.
Bruckner, A., 1974. Taking power off the wind. *New Sci.*, 61: 812—814.
Budyka, M.L., 1972. The future climate. *Trans. Am. Geophys. Union*, 53: 868—874.
Cheney, E.S., 1974. U.S. energy resources: limits and future outlook. *Am. Sci.*, 62: 14—22.
Darmstadter, J., 1971. *Energy in the World Economy.* Johns Hopkins, Baltimore, Md., 876 pp.
Darmstadter, J., 1972. Energy. In: R.G. Ridker (Editor), *Population, Resources and the Environment.* U.S. Government Printing Office, Washington, D.C., pp. 103—149.
Dickey, P.A., 1972. Geology of oil fields of West Siberian Lowlands. *Am. Assoc. Pet. Geol. Bull.*, 56: 454—471.
Finch, W.I., Butler, A.P., Armstrong, F.C. and Weissenborn, A.E., 1973. Uranium. In: D.A. Brobst and W.P. Pratt, (Editors), *United States Mineral Resources. U.S. Geol. Surv., Prof. Paper* 820, pp. 456—467.
Geological Survey of Canada, 1973. *An Energy Policy for Canada.* Phase I, 2: 31—54.
Halbouty, M.T. (Editor), 1970. *Geology of the Great Petroleum Fields. Am. Assoc. Pet. Geol., Mem.* 14, 575 pp.
Hubbert, M.K., 1969, Energy resources. In: National Academy of Sciences and National Research Council, *Resources and Man.* W.H. Freeman, San Francisco, Calif., pp. 157—242.
Hubbert, M.K., 1973. Survey of world energy resources. *Can. Inst. Min. Metall. Bull.*, 66: 37—53.
Jaffe, F.C., 1971. Geothermal energy. *Bull. Ver. Schweiz. Petrol-Geol. Ing.*, 38: 17—40.
McCulloh, T.H., 1973. Oil and gas. In: D.A. Brobst and W.P. Pratt (Editors), *United States Mineral Resources. U.S. Geol. Surv., Prof. Paper* 820, pp. 477—496.
McKelvey, V.E., 1972. Mineral resource estimates and public policy. *Am. Sci.*, 60: 32—40.
Meadows, D.H., Meadows, D.L., Randers, J. and Behrens, W.W., III, 1972. *The Limits to Growth.* Universe Books, New York, N.Y., 205 pp.
Meyerhoff, H.A. and Meyerhoff, A.A., 1973. Arctic politics and geology. In: *Proceedings 2nd International Symposium on Arctic Geology. Am. Assoc. Pet. Geol., Mem.* 19, pp. 646—670.
OECD, 1966. *Energy Problems, Policy and Objectives.* Paris, 187 pp.
OECD, 1974. *Energy Prospects to 1985.* Paris, 2 volumes.
Oil and Gas Journal, 1973. 71(53): 83—88.
Patterson, J.T., 1973. Energy and the weather. *Environment*, 15: 4—9.
Roberts, K.V., 1974. Thermonuclear fusion power. *Environ. Change*, 2: 329—338.

Shimkin, D.B., 1964. Resource development and utilization in the Soviet economy. In: M. Clawson (Editor), *Natural Resources and International Development,* Johns Hopkins, Baltimore, Md., pp. 155—205.

Strishkov, V.V., Markon, G. and Murphy, Z.E., 1973. Soviet coal production. Clarifying the facts and figures. *Min. Eng.,* 25: 44—49.

Sutulov, A., 1973. *Mineral Resources and the Economy of the U.S.S.R.* McGraw-Hill, New York, N.Y., 192 pp.

Theobald, P.K., Schweinfurth, S.P. and Duncan, D.C., 1972. Energy resources of the United States. *U.S. Geol. Surv., Circ.* 650, 27 pp.

Thomas, T.M., 1973. World energy resources: survey and review. *Geogr. Rev.,* 63: 246—258.

United Nations, 1973. *World Energy Supplies,* 1968—1971. New York, N.Y., 187 pp.

U.S. Bureau of Mines, 1970. *Mineral Facts and Problems.* U.S. Government Printing Office, Washington, D.C., 1291 pp.

U.S. Department of the Interior, 1972. *United States Energy, a summary review.* U.S. Government Printing Office, Washington, D.C., 72 pp.

Warman, H.R., 1972. The future of oil. *Geogr. J.,* 138: 287—297.

Weeks, L.G., 1972. Critical interrelated geologic, economic, and political problems facing the geologist, petroleum industry, and nation. *Bull. Am. Assoc. Pet. Geol.,* 56: 1919—1930.

Weinberg, A.M. and Hammond, R.P., 1970. Limits to use of energy. *Am. Sci.,* 58: 412—418.

Chapter 8

MARINE MINERAL RESOURCES

FRANK F.H. WANG and V.E. McKELVEY

INTRODUCTION

The oceans comprise about three-fourths of the area of the earth. Together with their seabed and subsoil — as the rocks and mineral resources on and beneath the ocean floor have been described in the legalistic terms of international conventions — they represent a largely unexplored but potentially huge source of a variety of minerals important for future supply. Although accurately described as the world's principal unexplored frontier, the marine environment has a long history of mineral productivity — salt from evaporation of seawater and recovery of fresh water from submarine aquifers going back to ancient times; subsea mining of coal and other minerals from extensions of onshore deposits as long as 300 years ago; and in recent decades recovery of oil and gas, sulphur, sand, gravel, lime, coral, tin, diamonds, gold, platinum, and titanium minerals from relatively shallow near-shore waters. The prospects for additional supplies of these minerals from the subsea are bright, and recent studies indicate that the deep-ocean floor is also an important future source of metals associated with basaltic igneous activity, such as nickel, copper, cobalt, and manganese.

Perhaps more difficult than the development of the knowledge and technology necessary to realize the marine mineral resources potential is international agreement on the area of national jurisdiction over these resources and the international arrangements for resources development that are to prevail beyond. Negotiations on these subjects are currently in progress in the Law of the Sea Conference sponsored by the United Nations, the outcome of which may influence not only the extent to which these resources become available, but also world peace and the benefits that may accrue to mankind from their development.

GEOLOGIC AND PHYSIOGRAPHIC PROVINCES AND THEIR BEARING ON POTENTIAL MARINE MINERAL RESOURCES*

The sea floor of the earth consists of two greatly contrasting physiographic divisions: the continental margins which represent the submerged edges of

---

* Much of this section has been drawn from McKelvey and Wang (1970) and McKelvey et al. (1969a).

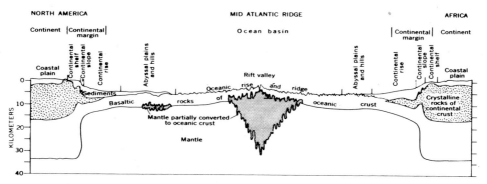

Fig. 8-1. The major physiographic provinces of the ocean basins, illustrated by a trans-Atlantic profile from New England to the Spanish Sahara (modified from Heezen, 1962).

the continental masses, and the deep-ocean basins (Fig. 8-1, Table 8-I). The physiographic contrast between the continents and ocean basins reflect fundamental differences in the underlying crust, in the geological processes that operate within the crust and mantle, and in the sedimentary processes that have deposited diverse types of sedimentary sequences as governed by the sedimentary environments and geologic history. The mineral resources in the various subsea physiographic provinces are also the products of the geologic processes which have formed and concentrated varieties of mineral resources in the sedimentary strata and in the underlying crystalline rocks (Fig. 8-2).

Continental shelf and slope

The submerged part of the continental mass includes the continental shelf and slope and, in many places, the continental rise where its thick apron of detrital sediments spreads over and conceals the transition between the continental and oceanic crust. The continental shelf is the relatively flat or gently inclined area of the sea floor between the low-water line and the change in the inclination of the sea floor that marks the shelf edge and the beginning of the continental slope. This marked change in inclination occurs at depths usually between 130 and 200 m, but exceptionally as shallow as 50 m or as deep as 500 m. The width of the shelf ranges from about 1 km up to 1,300 km. The inclination of the continental slope varies from less than 3° along coasts with large rivers and deltas to over 45° off supposed faulted coasts; an inclination of about 5°, however, is the most common.

The continental crust of the continental margins contains diverse varieties of igneous rocks but consists mainly of granitic-type rocks which are richer in silica and the alkalis and poorer in iron and magnesia than the oceanic crust. The continental crust is comparatively lighter in density and thicker than the oceanic crust (which averages about 5 km thick and is composed

TABLE 8-I

Areas of subsea physiographic provinces (after Menard and Smith, 1966)

Province	Area (10^6 km^2)	Percent of total area
Continental margins		
Continental shelf and slope [1]	55.4	15.3
Continental rise	19.2	5.3
Great ocean basins		
Abyssal plains and hills	151.5	41.8
Trenches and associated ridges	6.1	1.7
Oceanic ridges and rises	118.6	32.7
Volcanic ridges and cones and other features	11.2	3.2
	362.0	100.0
Small ocean basins (included above in continental rise and abyssal plains)	7.5	2.1

[1] The extent of the physiographic shelf has never been estimated, but is often taken to approximate the area landward of the 200-m isobath. Menard and Smith estimate this area to be 27.1 million km^2, about half that shown above for the shelf and slope combined.

largely of basalt and related rocks). The continental and oceanic crusts are in flotational equilibrium with the underlying mantle, and the lighter continents rise above the oceanic basins much like icebergs in the sea. The continental margins are generally overlain by sediments many kilometres in thickness compared with typically a few hundred metres of sediments overlying the oceanic crust of the ocean basins.

Known and expected mineral resources of the continental margins generally resemble those found in adjacent coastal areas and include: petroleum, sulphur, salt, potash, coal and other minerals in sedimentary basins of predominantly Tertiary and Mesozoic age; metallic deposits such as copper, lead, zinc, nickel, gold, silver, mercury, fluorspar, beryllium, tin, tungsten and many other minerals that have been concentrated by the igneous and hydrothermal processes operating within the continental crust; unconsolidated surficial placers of heavy minerals and precious metals, sand and gravel, lime shells and similar deposits that have been laid down in shallow marine environments or under subaerial conditions; and authigenic deposits of phosphorite, glauconite and associated minerals, and rarely ferromanganese oxides that have been formed by chemical or biochemical processes operating in the modern and ancient seas.

With respect to petroleum potential, the shelf and slope areas of the world as a whole appear more favourable than the exposed subaerial parts of the

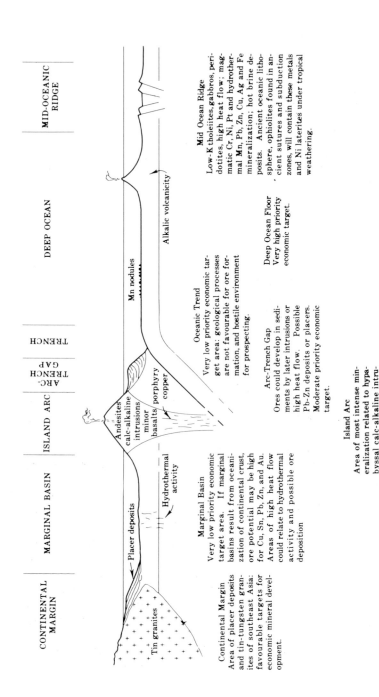

Fig. 8-2. Economic significance of the various plate tectonic elements in the marine domain (after CCOP/IOC, 1974, p. 15).

continents, because the shelves and slopes in general contain a greater thickness of marine Tertiary and/or Mesozoic sediments, from which comes the greater part of the world's petroleum production.

Continental rise and small ocean basins

Lapping against the base of the continental slope in many areas is the broad, smooth-surfaced, gently sloping continental rise which is underlain by a thick apron of sediments derived from the continent. Water depths at the landward edge of continental rises range from about 1,500 to 3,500 m, and depths at their seaward edge range from about 3,500 to 5,500 m where they merge into the adjacent abyssal plains as the sediments comprising them become gradually thinner seaward. The widths of continental rises vary from 100 km to as much as 1,000 km, and their lengths range up to several thousand kilometers, roughly parallel to the coasts of the nearest continent (Emery, 1969). The thickness of the wedge-shaped body of continental rise sediments is generally more than 3 km (Heezen and Tharp, 1965; Emery, 1969) and ranges to as much as 10 km (Drake et al., 1968; Weeks, 1971). Sediments of the continental rises may be deposited along the border of both the great ocean basins and small ocean basins and are mainly developed off continents wherever marginal deep-sea trenches are absent. They are particularly well developed off tectonically stable coasts having huge supplies of detrital sediments, opposite the deltas of the world's giant rivers, and around some of the semi-enclosed small ocean basins. The continental rises of the world occupy about 5% of the total area of the world's oceans, equivalent to about one-third the combined areas of the world's continental shelves and slopes, and contain extremely large volumes of clastic sediments largely of Cenozoic and Cretaceous age (Emery, 1969).

The relatively small, restricted basins that are partially enclosed by continents or island arcs are known as "small ocean basins", "mediterraneans", or "semi-enclosed seas". Menard (1967) grouped the few dozens of basins, already known, according to the tectonic characteristics of their boundaries. For example: the Aleutian and Okhotsk basins are separated by simple volcanic island arcs; the basins of the Indonesian archipelago, Caribbean, Sea of Japan and South China Sea are separated by highly complex island arcs; and the Mediterranean Sea, Black Sea, and the Gulf of Mexico are surrounded by continents.

They range in width from a few hundred kilometres to about 2,000 km. Because of their sediment fill, the water depths of the floor of the small ocean basins are less than those of large oceans and range from a few hundred to more than 5,000 m. Most of them include at least some central flat areas of abyssal plains below a depth of 2,000 m. The crust beneath typical small ocean basins is either the "oceanic" type or the "intermediate" type; the latter has seismic velocity and crustal thickness midway between continental and oceanic crust (Menard, 1967). Regardless of the origin of the small ocean basins and the characteristics of the underlying crust, most of these basins contain a great thickness of sedimentary fill, which commonly amounts to several kilometres and may occasionally reach as much as 20 km (Hedberg, 1970). The small ocean basins and continental rises as a whole appear favourable for petroleum, sulphur, potash, salt, and other sedimentary deposits that are closely associated with marine sedimentary basins of Tertiary and Mesozoic age.

Great ocean basins

Constituting the largest part of the world's seabeds and occupying an area about double the total land surface, the great ocean basins typically comprise the following physiographic features (Figs. 8-1 and 8-2): mid-oceanic ridges and rises, characterized by volcanic fields and a rift valley along the ridge crest; abyssal plains and abyssal hills lying on both sides of the oceanic rise; individual volcanoes and volcanic ridges scattered over the ocean basins, but often clustered to form groups of islands or seamounts; and deep-sea trenches, commonly present along volcanic island arcs or young mountain chains bordering the ocean basins.

In contrast to the continental crust, the crust beneath the deep oceanic basins is thin (averaging about 5 km) and relatively uniform in composition, consisting of oceanic basalt and typically associated with underlying sheeted dyke complexes which pass down into gabbros and into the peridotites of the upper mantle. The origin of some of the features of the great ocean basins as well as the continental margins is related to the processes of sea-floor spreading and plate tectonics. Thus, the earth's outer shell or lithosphere is regarded as consisting of a number of lithospheric plates of oceanic or continental crust. The lithospheric plates, apparently floating on viscous mantle material, tend to move relative to each other as semi-rigid units, allowing divergence, convergence and parallel slip, with most of the deformation arising from interactions along the boundaries between plates. Interpretation of magnetic lineations and dating of the oldest sediments overlying basaltic crust of the ocean floor have substantiated the spreading and void-filling processes that bring basic igneous rocks to the surface along the mid-oceanic ridges, and carry the new oceanic crust away from the divergent oceanic ridges at the rate of a few to as much as 15 cm a year. The older crust far from the mid-oceanic ridges may be thrust back and consumed within the mantle beneath the adjacent continental or oceanic plate, forming deep-sea trenches, island arcs and deep-seated tectonic and volcanic activities, or it may collide with, override, or extrude from beneath an adjacent plate, depending on the juxtaposition of the two (for a review see Dewey and Horsfield, 1970; Le Pichon et al., 1973). Much is still to be learned about the fundamental geological processes and geological framework of the world's oceans. Nevertheless, ocean basins are known to contain large concentrations of manganese, iron, nickel, copper, cobalt and other metals in the widespread ferromanganese nodules and encrustations on the surface of the ocean floor; the bedrock beneath the ocean basins may also contain large concentrations of the same suite of metals and base and precious metals (such as zinc, lead, mercury, nickel, chromium, platinum, gold, silver, copper, molybdenum, etc.) in deposits possibly formed from differentiation of the molten mantle material from which the basalt itself was derived or from melting of the crustal plates that were thrust into the mantle.

Potential subsea mineral resources

The continental shelves, continental slopes, continental rises and small ocean basins appear to contain the greater portion of potential subsea mineral resources, both in terms of variety of minerals and the value of those likely to be recovered within the next few decades. In contrast, the floors of the great ocean basins contain a much smaller variety of minerals, chiefly metalliferous deposits genetically related to mafic igneous rocks, but the total amount of some of these metals may be large.

Offshore oil and gas production has dominated the history of profitable mineral development off many countries of the world. The total value of world offshore crude oil and gas production in 1974 reached nearly $40 billion (thousand million), which exceeds that of all other marine non-living and living resources combined, including fisheries, ocean shipping and other commercial uses of the ocean space (Tables 8-II and 8-III).

A number of other mineral materials are produced in scattered offshore and beach localities in the world — nearly all in coastal waters less than 35 m deep. In 1972, the annual world production of non-hydrocarbon minerals, from the sea floor and the rocks beneath the sea floor, totaled about $740 million, which represented no more than 2% of the on-land production of these minerals. During the next few decades, production of some subsea hard mineral resources may expand, particularly:

(1) Extremely large and high-grade offshore deposits in easily accessible coastal waters (such as the tin deposits off the coasts of Indonesia, Malaysia and Thailand).

(2) Minerals in short supply (such as diamonds, gold, platinum, nickel and perhaps phosphorite locally).

(3) Bulk materials (such as sand and gravel, lime shells and lime mud, coal, phosphorite, and iron ore) in short supply in adjacent countries where transportation cost is the main factor in the market price.

(4) Deep-seabed manganese nodules desired as a new source of copper, nickel, cobalt, manganese and other metals.

(5) Possibly other offshore minerals desired by individual countries to reduce balance-of-payments deficits, to become less dependent on foreign supplies, and to conserve land resources and preserve land environment.

TABLE 8-II

World annual production of minerals from oceans and beaches (estimated raw material value in U.S.$ million)

	Year 1972		Year 1980
	production value	percentage by value from ocean	projected production value
Subsurface soluble minerals and fluids			
Petroleum (oil and gas)	10,300	18	90,000 [3]
Frasch sulfur	25	3	
Salt	0.1	nil	
Potash (production expected in 1980s)	none	—	
Geothermal energy	none	—	
Fresh-water springs	35 [1]	—	
Surficial deposits			
Sand and gravel	100	<1	
Lime shells	35	80	
Gold	none	—	2,000 [4]
Platinum	none	—	
Tin	53	7	
Titanium sands, zircon, and monazite	76	20	
Iron sands	10	<1	
Diamonds (closed down in 1972)	none	—	
Precious coral	7	100	
Barite	1	3	
Manganese nodules (production expected by early 1980s)			
Phosphorite	none		
Subsurface bedrock deposits			
Coal	335	2	
Iron ore	17	<1	
Extracted from seawater			
Salt	173	29	
Magnesium	75	61	
Magnesium compounds	41	6	
Bromine [2]	<20	30	2,000 [4]
Fresh water	51	—	
Heavy water	27	20	
Others (potassium salts, calcium salts, and sodium sulphate)	1	—	
Uranium	none	—	
Total			94,000

Non-petroleum commodities total production value $ 694 million

Total from seawater $ 388 million

Sources: Cruickshank (1969, 1974, 1976); Archer (1973); U.S. National Council on Marine Resources and Engineering Development (1970, table V-2, p. 67).
[1] More than 200 million gallons of fresh water are recovered per day from submarine

PETROLEUM

Present status of offshore development

Although oil was produced as early as 1923 from wells in Lake Maracaibo, Venezuela, true offshore exploration did not begin until 1947 when the Ship Shoal field was discovered in the open waters off Louisiana. From this beginning, exploration of the continental shelves has been extended to many parts of the world. Since 1969 the offshore search for petroleum has much accelerated and this trend was further stimulated in 1974 by the increases in crude oil price and by the desire of many countries to become less dependent on foreign petroleum sources. Commercial offshore activities have now been undertaken off the coasts of more than 80 countries, encompassing all continents of the world except Antarctica; in two-thirds of these countries petroleum activities are primarily or exclusively offshore.

Although exploitation of offshore petroleum is still in an early stage of development, the world's offshore production has increased ten-fold since 1960. In 1973 the world's offshore oil production averaged 10.4 million barrels per day and offshore gas production was about 16.8 billion (16.8×10^9) cu. ft per day, which accounted for about 19% of the world's total oil production and 12% of the world's natural gas production. Proved offshore reserves have increased manyfold during the past decade, and as of the end of 1973 account for about one-quarter of the world's total oil reserves of 627 billion barrels and the world's total gas reserves of 2,033 trillion cu. ft (Gardner, 1973; Offshore, 1974, pp. 59—65). Total subsea petroleum resources of the world, including those not yet discovered, possibly match those of the continents.

Various projections indicate that by 1985 more than 35% (Birks, 1974), or possibly between 40 and 50% (Bynum and Lovie, 1975) of the world's oil production, or more than 25 million barrels per day (Birks, 1974), or possibly in excess of 31 million barrels per day (International Management and Engineering Group of Britain, 1972) will come from beneath the sea. The increase in gas production is expected to be even larger, possibly reaching the daily rate of over 52 billion cu. ft by 1985. Development of offshore petroleum in new areas, where there is little or no production at present, could change significantly the outlook for supply and demand for many countries and regions.

---

springs in Argolis Bay, Greece, but only a small portion of the produced water is utilized.
[2] Seawater plant at Freeport, Texas, closed down in late 1969 (U.S. Bureau of Mines, 1973, Mineral Year Book 1971, Vol. 1, p. 233).
[3] Projections indicate that offshore production by 1980 will probably at least triple the 1972 daily output rate of 9.5 million barrels of oil and 17 billion (10^9) cu. ft of gas per day and crude oil price will probably be stabilized around $10 per barrel.
[4] Also assuming an average 30% price increase of raw minerals.

Petroleum in the continental shelf and slope

The unique tectonic setting of the continental shelves and slopes, together with the adjacent coastal plains (constituting the dynamically active border realm between the continent and the ocean) and the past geological events

TABLE 8-III

World annual uses of marine resources and ocean space (estimated value of raw material production or cost of operations)

	U.S. $ million		Ultimate total capacity or quantity
	1971	1980	
Mineral resources			
Petroleum (oil and gas)	10,300 [1]	90,000	finite and depletable
Solid minerals and fluids	700	2,000	
Extraction of seawater con-stituents and desalination	388	2,000	unlimited
Living resources			
Fisheries (fish catch)	12,000	20,000	
Aquaculture for food (in marine, brackish, and fresh waters)	600	2,000	
Aquaculture for pearls	80	120	replenishable
Drugs from marine organisms	5	50	
Energy resources			
Tidal energy resources	50	200	energy inexhaustible but economic potential now limited to 20 power sites
Energy from ocean waves, currents and water temperature gradients	none	none	energy inexhaustible but present technology inadequate and uneconomical
Transient uses of the ocean space			
Ocean shipping	18,000 to 20,000	60,000	coastal sealanes already badly congested
Military activities	25,000 to 30,000	*	
Supporting maritime, services (navigation, charting, sea and weather routing of ships, coastal traffic control, search and rescue)	400	1,000	
Scientific research	1,000	2,000	
Underwater archaeological excavation	5	*	
Marine pollution control and abatement (by industries and governments)	1,000	5,000	

TABLE 8-III (continued)

	U.S. $ million		Ultimate total capacity or quantity
	1971	1980	
Prolonged uses with modification of the ocean space			
Harbour and channel dredging	200	500	
Offshore mining	300	2,000	
Offshore petroleum structures and production facilities (including pipelines and storage tanks)	2,000	10,000	
Submarine cables (communications and power transmission)	*	*	
Coastal and engineering construction (harbours, artificial offshore ports, floating airports, floating power plants, submarine tunnels, sea canals)	*	*	suitable ocean space for specific uses is limited
Coastal and offshore recreation (artificial beaches, underwater parks, underwater hotels), habitation and industrial development (reclamation of coastal lands, floating residential structures and sea cities)	*	*	
Waste disposal	*	*	semi-enclosed waters have very limited capacity to absorb and disintegrate waste materials
Pollution accidents	negative benefit (damage to living resources and recreation)		

Sources: Lecture presented by Frank F.H. Wang to the Group Training Course in Offshore Prospecting in Japan, October 1972; Pirie (1973); Vetter (1974).
[1] 1972 daily output averaged about 9.5 million barrels of oil and 17 billion (10^9) cu. ft of gas, and crude oil price was around $3 per barrel.
* Not estimable.

that took place both on land and under the oceans, have made the continental margins the ideal locale for development of large sedimentary basins with extremely thick Tertiary and Mesozoic sediments favourable for petroleum (Klemme, 1972; Weeks, 1972; Wonfor, 1972). However, knowledge of the world's continental margins as a whole, particularly the slope areas and beyond, still is meagre. In view of the lack of reliable basic data, it is not yet possible to make a complete world-wide assessment or a meaningful revision of L.G. Weeks' 1969 estimate of over 1,000 billion equiv. barrels of undiscovered hydrocarbons beneath the shelves and his 1973 estimate of 1,800 billion equiv. barrels of undiscovered oil and gas beneath the shelves and slopes (United Nations, 1973). These figures are based on broad geologic fa-

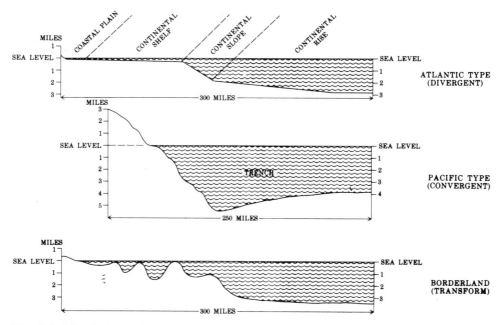

Fig. 8-3. Morphology of divergent-, convergent- and transform-type continental margins (after Thompson, 1974).

vourability factors and assumptions and are useful only in providing relative orders of magnitude. Taking into account the technological difficulties and high costs of petroleum development in increasingly deep waters, it is possible that not more than one-quarter of the total undiscovered hydrocarbon resources would be commercially recoverable.

Three fundamental types of continental margins are generally distinguished (Fig. 8-3) and each is characterized by its unique types of petroliferous sedimentary basins (Beck and Lehner, 1974; Thompson, 1974). The *Atlantic-type* margins ("divergent" or "trailing-edge" continental margins) are developed along most parts of the periphery of continents and islands that have been moving away from the rising and spreading mid-oceanic ridges in the expanding ocean basins (Inman and Nordstrom, 1971). These margins represent the transitional zones between coupled oceanic and continental crusts and are characterized by lack of intense seismicity, absence of widespread volcanism, an uninterrupted long history of sediment accumulation and consequent development of a typically broad margin, which consists of a coastal plain, broad continental shelf, gentle continental slope and prominent continental rise. As the ocean basin expands and the continental masses separate, the divergent margins subside and a broadly predictable sequence of sediments accumulates. The sequence includes, from oldest to youngest, non-marine clastic sediments, euxinic sediments and evaporites (deposited in early stages of basin development), marine carbonates, and marine clastic sediments (Thompson, 1974). Continental margins of this type are typically developed along the present coasts of both sides of the Atlantic and are also found along the coasts of East Africa, the Indian subcontinent, Australia, and most parts of Antarctica. Many large Tertiary deltas, Mesozoic salt basins, and carbonate platforms on the divergent margins have already been drilled and proven to be highly petroliferous provinces.

Recent studies (Hedberg, 1970; Pegrum et al., 1973) indicate that many crustal blocks have been separated from the outer edges of the Atlantic-type margins and drifted far away from the parental continents as a result of tensional rifting related to the opening of ocean basins. Such isolated "caps" or "detached shelves" are present in various parts of the Atlantic as well as the Indian Ocean, the Arctic Ocean and the South China Sea. Some of these, for example the Rockall and Hattan Banks now separated by deep troughs from the British Isles and Iceland, contain thick sedimentary basins and appear likely to be future petroleum provinces.

In the Atlantic and other oceans, crustal blocks broken off from the continental margins have been downdropped to oceanic water depths, mostly after Middle Cretaceous or during Neogene times (Sander, 1970). These blocks commonly form "marginal plateaus" which appear as huge steps down from the normal level of the continental shelves to that of the ocean deeps. Some of these marginal plateaus are underlain by several kilometres of sediments and may have good petroleum possibilities; an example is the Voring Plateau off Norway, where a core hole recently drilled by the Deep Sea Drilling Project encountered natural gas and oil saturation at 120 m to about 450 m below the sea floor under 1,439 m of water.

The *Pacific-type* margins (or convergent continental margins), like those bordering the Pacific Ocean, are extremely complex, irregular and orogenically active, often displaying intense seismicity, volcanism, magmatic intrusion and mountain building which are active today and in the geologic past (Fig. 8-4). These activities are largely related to the crustal dynamics of the convergent plate boundaries, involving collision, subduction or obduction of major crustal plates, with attendant deformation or destruction of the crust. The most conspicuous occurrences of these orogenically active margins are represented by the island arc—continental margins, such as the Western Pacific arc—trench systems from Alaska to Indonesia along the convergent boundaries between the continent and ocean (CCOP/IOC, 1974). The Tonga arc—trench system represents another type of convergent margin involving two oceanic crustal plates.

In the typical island arc—continental margin in Southeast Asia, a great oceanic basin (such as the Pacific and Indian Oceans) is separated from the continental mass by a major arc system, which is bounded by a deep elongate trench on the oceanic side and accompanied by foreland-type shelves,

234

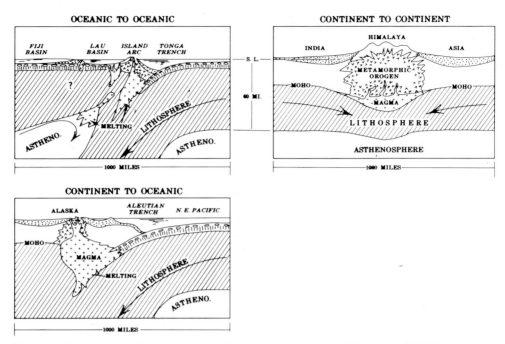

Fig. 8-4. Major types of convergent continental margins (after Thompson, 1974).

marginal seas and small ocean basins on the landward side. The trench regions are regarded as the main loci of crustal subduction, where one lithospheric plate descends beneath another plate and into the asthenosphere along the Benioff seismic zone and is eventually consumed. Intensive petroleum exploration in the Indonesian archipelago during the past few years indicates that the collision and subduction of crustal plates might have resulted in repeated tectonic activities (involving vertical block faulting, transcurrent faulting, crustal compression and extension) and pronounced phases of magmatic intrusion and volcanism, which have fragmented the foreland regions into complex arrays of relatively short-lived Tertiary sedimentary basins separated by tectonic ridges and arches (Ben-Avraham and Emery, 1973; Beck and Lehner, 1974; Pulunggono, 1975). Several of the foreland-type Tertiary basins occurring beneath the shelves and coastal lowlands have already proven to be highly petroliferous.

Many of these margins along the island arcs have been in a long-continued stage of development, having been repeatedly modified, destroyed, and renewed by crustal processes along the collision or subduction zone and by uplift and subaerial erosion. However, the significant geothermal conditions in these basins may greatly enhance the early generation, migration, and accumulation of hydrocarbons in shallow reservoirs, possibly within a short geologic period (possibly as young as Pleistocene sediments) and may relatively increase the overall magnitude of hydrocarbon recovery (Klemme, 1972; Thompson, 1974).

Another important variety of continental margins associated with large-scale *transform faulting* (dominantly strike-slip faulting in transformation from one sense of convergent or divergent displacement to another) is shown by the southern California borderland, the borderland off southeast Alaska (between Queen Charlotte Island and Juneau) (Thompson, 1974), the Venezuela borderland (Maloney, 1968) and the Sahul shelf off the northwest coast of Australia (Fairbridge, 1966a). These borderlands differ from the normal continental margins in having a very irregular topography (often consisting of a number of deep basins separated by submarine ridges, banks and offshore islands), in lacking the well-defined topographic break between the shelf and the slope, and in exposing extremely complicated structures. Much of the California continental margin, including its southern extension to Baja California, now strongly reflects the tectonics related to the San Andreas system of strike-slip faults, which represent the transformation from the divergence at the East Pacific Rise to the convergence in the western Gulf of Alaska (Thompson, 1974).

With crustal stretching and widening of the continental margin, the borderland basins were progressively enlarged and continuously subsided (Crowell, 1973). Consequently, great volumes of organic-rich Neogene sediments were deposited in the deep basins. The topographic configuration often created pronounced upwelling and high organic productivity in the surface waters, yet it favoured the preservation of organic matter under the stagnant bottom waters. These sedimentary strata periodically underwent deformation to form petroleum traps and were also affected by Miocene volcanism in the basin floors. The significantly high geothermal conditions in the pull-apart grabens may greatly enhance the early generation, migration and accumulation of hydrocarbons. The borderland tectonic setting is thus highly favourable for prolific petroleum accumulations.

Dewey and Burke (1974) recognized that some of the formerly divergent plate boundaries may ultimately converge and that the setting of ancient divergent margins may influence orogeny and basin history during continental collision. *Mild continental collision*, involving moderate deformation of thick sedimentary prisms of *formely divergent margins*, sometimes creates the most favourable conditions for concentration of hydrocarbons in giant petroleum fields. For example, thick sedimentary piles of the formerly divergent margins of the Tethys Seaway were affected by the convergence between the Arabian and African plates and consequent closing of the Tethys during Mesozoic and Cenozoic times. The oil-prolific Persian Gulf apparently was saved from collision destruction by transform movement of India past Arabia (Thompson, 1974). Giant petroleum accumulations also exist in other remnants of the Tethys ocean. Continuing efforts of exploration and

236

research in the world's ocean may reveal similar types of convergent tectonics of Tertiary of even younger age that may be favourable for petroleum accumulations.

Petroleum potential of seabed provinces beyond the continental slope

Although there has been no petroleum exploration in oceanic areas beyond the continental slope, marine geological and geophysical research throughout the world's oceans during the recent years, including deep-sea drilling from the *Glomar Challenger*, has provided a great deal of new geologic information helpful in evaluating and inferring the hydrocarbon potentials beneath the continental rise, deep-sea trenches and troughs (Fig. 8-5), small ocean basins and the deep seabed beyond. These seabed provinces as a whole have considerable potential for undiscovered hydrocarbon resources, particularly the small ocean basins (Hedberg, 1970; McKelvey and Wang, 1970). Recent studies by the National Petroleum Council (1974) suggest that most of the world's ultimately recoverable subsea petroleum is likely concentrated in

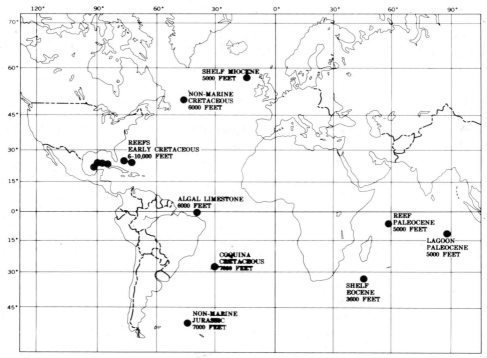

Fig. 8-5. Cores of shallow-water rocks recovered from the deep seabed by the Deep Sea Drilling program (after Thompson, 1974).

the continental margins and small ocean basins. It is further inferred that possibly over one-half of the world's total subsea petroleum potential lies in shelf areas, and another one-third probably lies in the continental slopes, continental rises and the small ocean basins. In contrast, the deep-ocean floor probably contains only a minute fraction (not more than 2%) of the total subsea petroleum potential. However, the petroleum potential of the seabed provinces beyond the upper part of the continental slope has been little investigated, mainly because of the great water depths and the present economic and technological limit for petroleum production.

A feature of great interest to petroleum geologists is the *continental rise*, the apron of sediments lapping against the base of the continental slope and sloping gently oceanward to the abyssal floor. Characteristics of the continental rises that seem particularly significant include: rapid deposition of great thickness of largely clastic sediments of Cenozoic and Cretaceous age; a considerable proportion of displaced organic-rich deltaic and other shallow-water sediments brought down by turbidity currents and mass movements and re-deposited as potential source rocks of hydrocarbons (Heezen et al., 1966; Schneider et al., 1967; Emery, 1969); buried ancient submarine channels and other facies variations of sandy turbidites as possible stratigraphic traps (Hedberg, 1967); and thick Mesozoic sedimentary basins that were originally deposited under shelf and slope environments and possibly foundered to deeper waters and buried by younger sediments of the rise. Recent geophysical research conducted by the petroleum industry in the eastern Atlantic (Beck and Lehner, 1974) reveals that a continental rise with more than 4 km of Mesozoic sediments occurs on the west side of the Canary Islands tectonic trend, which separates the rise from the large Triassic salt basin beneath the present shelf and slope. The study also shows that the continental rises off the Niger and Congo deltas contain 4—6 km of Cretaceous-Tertiary sediments above an oceanic basement (interpreted as Cretaceous flood basalts). Both the Cenozoic and Mesozoic sedimentary basins beneath the rises seem to be generally favourable for petroleum accumulations, particularly the foundered sedimentary basin originally deposited under shelf and slope conditions.

Many of the Pacific-type or convergent continental margins and virtually all the island arcs are bordered by deep-water arcuate features of huge elongate *trenches and troughs*. They are often characterized by some of the greatest negative gravity anomalies (suggesting a deficiency of mass) and by frequent earthquakes located along the Benioff zone.

Peripheral trenches, located near a source of sediments eroded from a continent or large island, provide the natural repository for the deposition of organic-rich terrigenous sediments, alternating with beds of slowly deposited pelagic oozes. Although the western part of the Aleutian Trench, for example, is a typical deep empty depression, its eastern part has become a partially filled sedimentary trough and the trench axis can be followed eastward to the potentially petroliferous province of the Gulf of Alaska, where the topographic entity disappears completely beneath thick terrigenous sediments (Shor, 1962; Menard, 1964). Similarly, the southern half of the Peru-Chile Trench has a substantial sedimentary filling (Scholl et al., 1968), the Hikurangi Trench is partially filled with sediments derived from New Zealand (Menard, 1964), the Antillean Trench is filled near Barbados, and other examples exist along the Scotia Arc (Ewing and Landisman, 1961). Many former marginal trenches and troughs have been completely filled with sediments and now are overlain by continental rises (Hedberg, 1970).

Sediment-filled trenches involved in a mild or "early stage" collision between a continent and an island arc may be very favourable for petroleum accumulations, possibly similar to the closing of ancient Tethys oceans by continental collision. A good example is the thick sedimentary basin under the Timor trough, where a series of northeast-trending anticlines have been formed by the initial approach to collision between Australia and Indonesia (Thompson, 1974).

A newly recognized potentially petroliferous sub-province of the island arc—trench systems is the so-called "inter-arc trough" or "inter deep" (or sometimes called outer-arc submarine basin) that occurs parallel to and on the landward side of the major deep-sea trenches, whenever the trenches are bounded on their landward side by double island arcs — an inner volcanic arc (forming an island mountain range with andesitic volcanoes on the island arc proper) and an outer non-volcanic arc or ridge (a submarine ridge or belt of smaller islands, consisting largely of deformed flysch or pelagic sediments) (Fairbridge, 1966b, CCOP/IOC, 1974). Situated between the outer non-volcanic ridge and the island arc proper, the inter-arc trough is never as deep as the water depth of the major trench. Tectonic damming of the trough between the two arcs not only increases the sedimentary fill, commonly 3—7 km of Neogene sediments along the basin axis, but also favours the preservation of organic-rich sediments under the waters of restricted circulation. Geophysical prospecting in the Indian Ocean off Java and Sumatra shows the existence of several potentially petroliferous Neogene sedimentary basins in the inter-arc trough that lies between the volcanic arc of Java and Sumatra and the submarine ridge north of the deep Java Trench (Beck and Lehner, 1974). Recent gas discoveries off the west coast of Sumatra (Meulaboh) and off the east coast of central Japan (40 km offshore east of Iwaki) suggest that the inter-arc troughs offer new opportunities for petroleum exploration (Ranneft, 1972).

Oceanic trenches associated with convergent boundaries between oceanic crustal plates may result in geological characteristics similar to the island arc—continental margins in the sense of a subduction zone, and similar favourability from the standpoint of organic production in the waters. Being far removed from continents and large islands, the oceanic trenches, such as the Tonga (Fig. 8-4) and other trenches in the southwest Pacific, generally received less land-derived sediments but may be favourable for development of large carbonate reefs of Cenozoic age, which would provide potential petroleum reservoirs close to known oil seeps on the Tonga Islands (Thompson, 1974).

The partially enclosed small ocean basins located proximate to continents or large islands have served as the largest sedimentary sinks during their long history of rapid deposition, and trapped huge volumes of detrital sediment, commonly ranging from several kilometres to as much as 20 km in thickness (Hedberg, 1970). Many of the world's largest rivers discharge not into the open oceans but into these restricted basins; 35% of the world's continental areas drain into the small ocean basins which occupy only about 2% of the total area of the world's oceans. Therefore, it is hardly surprising that about one-sixth of all identifiable subsea sediments may lie in these basins (Menard, 1967). The restricted configuration of these basins favours the accumulation and preservation of organic matter in sediments as potential source rocks of hydrocarbons; it is also conducive to the deposition of salt and related evaporites. The Gulf of Mexico, Red Sea, and the Balearic Basin in the western Mediterranean already are known to be underlain by extensive evaporite sequences which include salt diapir structures (Hedberg, 1970).

Recent studies indicate that in some of the small ocean basins, such as the Gulf of Mexico, western Mediterranean, and the Indonesian archipelago, portions of the abyssal areas were broken off and down-faulted from pre-existing continental margins, possibly by crustal extension and subsidence, and that the buried older sedimentary units (originally deposited under shallower waters) appear particularly favourable for petroleum accumulations (Sander, 1970; Beck and Lehner, 1974). The new evidence of liquid and gaseous hydrocarbon shows encountered by the Deep Sea Drilling Project (Fig. 8-6) in the western Mediterranean and the Gulf of Mexico (McIver, 1972, 1974) and the oil seeps and presence of tar masses in the abyssal floor of the Gulf of Mexico (Pepper, 1958; Rezak et al., 1969; Sweet, 1974) show that hydrocarbons have generated and migrated in sediments at abyssal water depths, thus enhancing the possibilities of commercial petroleum accumulations beneath some of the small ocean basins. Furthermore, drilling from the *Glomar Challenger* has also encountered gaseous hydrocarbons (largely methane with traces of ethane and propane generated from sediments with terrigenous organic matter by bacterial biochemical processes) in many core holes (about 20% out of some 300 holes drilled until early 1973) in small ocean basins, continental rises and trenches (Peterson and Turrentine, 1972; McIver, 1974). In these regions, the high pressure of the overlying water column (if greater than 1,500 m) and the relatively low temperature (2—5°C) of the sediments at the bathyal and abyssal water depths tend to combine the gaseous methene (as well as ethane and propane) with water molecules to form solid gas hydrates (crystalline ice-like clathrate compounds) generally within the upper 500 m of sediment (Trofimuk et al., 1973; Claypool and Kaplan, 1974; Stoll,

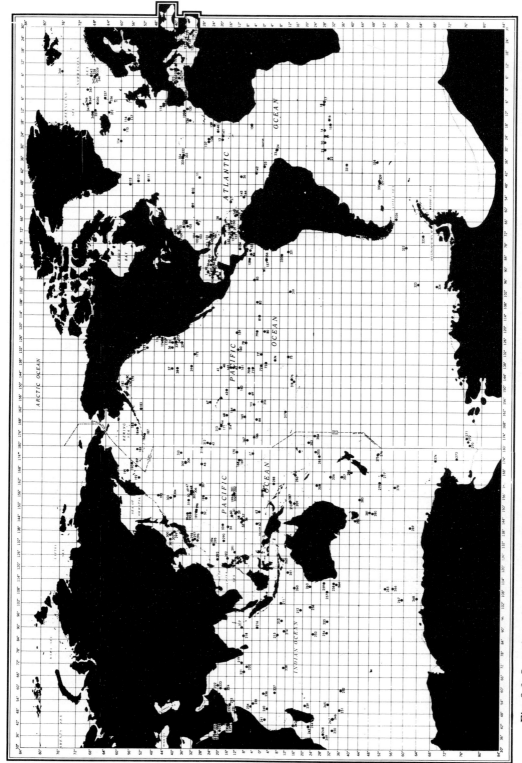

Fig. 8-6. Location of core holes drilled until the end of 1974 during the Deep Sea Drilling Project of the JOIDES program sponsored by the National Science Foundation.

1974). Many of the layers of solid gas hydrates have been detected during seismic surveys as "irrational" reflecting horizons (Peterson and Turrentine, 1972; Bryan, 1974). Below the burial depths of about 500 or 700 m, the solid gas hydrates are decomposed and re-converted to methane and other gaseous hydrocarbons by the increasing temperature in the subsurface sediments as shown by geothermal gradients. The widely occurring layers of solid gas hydrates may be an additional factor in the migration and accumulation of natural gases in the deep seabeds by reducing the loss of gaseous methane from the sediments to the overlying oceans waters, allowing upward migration of gases at a pace controlled by the sedimentation rate, and favouring an enrichment of gases immediately below the layer of solid gas hydrates which may act as an overlying seal (Claypool et al., 1973).

Beneath the abyssal floor of the *great ocean basins,* such as the Pacific, Atlantic, and Indian Oceans, the sedimentary section is usually relatively thin, less than 1 or 2 km thick, consisting primarily of pelagic sediments with very few beds of finer-grained sands in the flysch-type deposits derived from land, and with relatively little structure (National Petroleum Council, 1969). Most evidence suggests that the great ocean basins — including abyssal plains, volcanic seamounts and mid-oceanic ridges — are far less favourable than the continental margins and small ocean basins with respect to petroleum possibilities. Nevertheless, geological and geophysical research has already shown that many large crustal blocks had been rifted off from continental margins and drifted and subsided in deep oceanic areas. The sedimentary sequences which were originally deposited under shallow-water environments but are now under deep waters hundreds of kilometres from present shorelines may be favourable for generation of hydrocarbons. Occurrences of such shallow-water sedimentary rocks have been encountered in a number of holes drilled by the Deep Sea Drilling Project in the Atlantic and Indian Oceans (Fig. 8-5). These and other findings suggest that small portions of the deep-sea floor may be favourable for petroleum.

The rapid pace at which offshore oil well drilling has been moving into increasingly deeper waters and more hostile ocean environments reflects the rapid advance in offshore petroleum technology. Today commercial operations in up to a hundred metres of water are common and wildcat drilling has recently been conducted in water depths over 640 m; full-scale exploration is being initiated in many areas under 610—915 m of water. By 1977 at least one exploratory well will probably be drilled in 1,830 m of water (*Offshore Service,* 1974, p. ii). As of 1973 exploration concessions and permit areas have been granted off 16 countries beyond 182 m water depth to as much as 2,750 m (Adye et al., 1974, p. 139).

As a result of the continuing evolution of increasingly sophisticated and versatile mobile drilling rigs, commercial exploratory drilling is now technically and economically feasible in any foreseeably desired water depth. However, the water-depth limit of offshore production operations still lags some

500 m behind exploratory drilling (Fig. 8-7), because it is more complicated and difficult to develop the oil and gas production facilities which must be able to withstand stress, corrosion and tremendous extremes of weather and sea conditions during the life of the producing fields (often extending from 20 to 40 years). With recent technological advances, fixed production platforms for above-water production systems, now already feasible in 260 m of water in the earthquake-prone environment of the Santa Barbara Channel, will probably be installed in 305 m water depth in the Gulf of Mexico in the early 1980s (Fig. 8-8; *Oil & Gas Journal*, 1974) and will be eventually ex: tended to as deep as 450 m water depth in moderate climates (National Petroleum Council, 1974). Engineering research and development in progress indicate that by the early 1980s new techniques, including greatly refined ocean-floor completion and production systems, can be perfected to permit commercial exploitation operations in 915 m of water. This technical capability can be rapidly extended to water depths of as much as 2,440 m in the 1990s (Fig. 8-9), provided that sufficient economic targets in the ultra-deep waters are demonstrated in the next few years by at least a few extremely large discoveries in and beyond the 915 m water depth (Geer, 1973a, 1974; Adye et al., 1974, pp. 129—141). Given a highly favourable governmental policy and economic incentives, even a moderately large discovery (100—200 million barrels) in 1,000 m water depth could be developed economically in nearly all but the severest of the ocean environments (National Petroleum Council, 1975).

However, technical difficulties, costs and the required magnitude of investment will increase considerably with water depth and with severe condi-

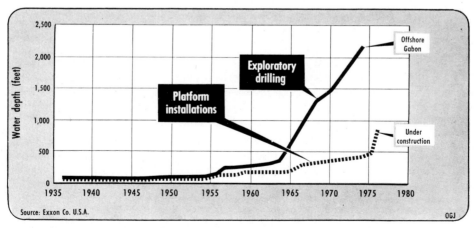

Fig. 8-7. World water-depth records for commercial wildcat drilling and installations of fixed platforms for above-water petroleum production systems (from *Oil & Gas Journal*, April 7, 1975, p. 39; data from Exxon Co., U.S.A.).

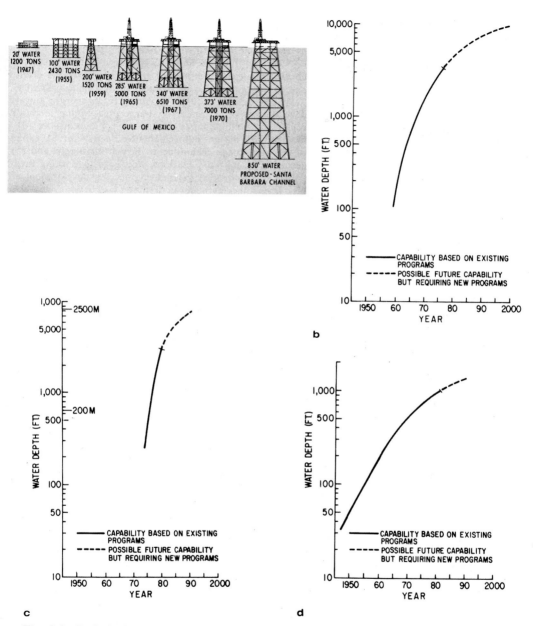

Fig. 8-8. Technical capability of fixed platforms for above-water petroleum production systems 1947—1990 and capability of underwater production systems (after Geer, 1973b). (a) Twenty-five years of platform development. (b) Industry mobile drilling and underwater completion capability projection. (c) Industry underwater production and manifold system capability projection. (d) Industry fixed platform capability projection.

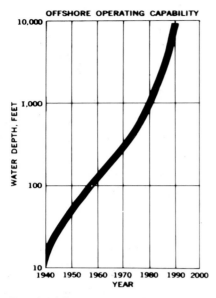

Fig. 8-9. Technical capability of complete commercial wildcat drilling and production systems, 1940—1990, including under-water production systems (after *Ocean Industry*, September 1974, p. 141).

tions of the ocean environment, particularly the ice-laden polar seas (Tables 8-IV, 8-V, and 8-VI). Given the extremely high cost of ultra-deep water petroleum development, the availability of petroleum from the continental

TABLE 8-IV

Comparative capital cost of energy resources development (per barrel/day production capacity in 1974 dollars) (from Birks, 1974)

Energy resource	Cost
Oil	
Alaska onshore (north slope)	3,750
North Sea	3,000
Mideast offshore	1,025
Mideast onshore	375
Gas	
Offshore	2,250
Onshore	1,250
Tar sands	8,750
Shale oil	6,250
Coal oil	7,200

TABLE 8-V

Offshore petroleum exploration drilling cost for various water depths and climatic conditions (from National Petroleum Council, 1974)

Offshore exploration expenditure index: 1.0 = $2.7 million per well in 1974 dollars

Water depth (m)	Climatic conditions [1]				
	mild (1)	moderate (2)	severe (3)	ice laden	
				75% (4)	100% (5)
200	0.8	1.0	1.8	2.3	4.6
500	1.0	1.3	2.1	2.8	5.4
800	2.3	2.6	3.3	4.0	6.1
4,000	3.8	4.0	4.3	5.6	7.5

[1] Typical of the various climatic conditions are:
 (1) Senegal, Gabon, Honduras, Mediterranean, Java Sea, Persian Gulf.
 (2) Gulf of Mexico, South Atlantic, South Pacific *, Northwest Australia *, Sea of Japan *, Yellow Sea.
 (3) North Sea, Bay of Biscay, South Australia, Gulf of Alaska *, North Atlantic, North Pacific, West Coast of Canada, Nova Scotia.
 (4) Bristol Bay *, West Greenland **.
 (5) Arctic Ocean, Chukchi Sea **.

 * Earthquakes; ** icebergs.

TABLE 8-VI

Expenditure for offshore petroleum development and production for various water depths and climatic conditions (from National Petroleum Council, 1974)

Development and production estimated expenditure requirements: 1.0 = $95 million per system in 1974 dollars

Water depth (m)	Climatic conditions [1]				
	mild (1)	moderate (2)	severe (3)	ice laden	
				75% (4)	100% (5)
200	0.9	1.0	2.8	unknown but estimated to be substantially greater than severe	
300	—	—	6.2		
500	2.7	3.0	—		
1,000	4.3	4.8	10.2		

[1] See footnote of Table 8-V.

shelves and slopes, and the increasing efforts in utilizing alternative energy sources, opportunities for petroleum production from the small ocean basins and the continental rises are likely to be restricted during the next few decades to a small number of the very largest, highly productive fields in the most favourable locations.

POTASH, SALINE MINERALS, AND SULPHUR ACCESSIBLE TO DRILL-HOLE EXTRACTION METHODS

Most of the world's deposits of common salt, potash-bearing minerals, anhydrite and gypsum were formed in the geologic past by evaporation of seawater and other natural brines in saline sedimentary basins of restricted circulation. Important deposits of magnesium-bearing salts were also deposited in such basins, and elemental sulphur deposits formed in some of them by complex biogenic processes involving the alteration of anhydrite in the evaporite sequence and in the salt-dome cap rocks. Such ancient saline basins occur extensively and are commonly found beneath the continental shelf, slope and rise as well as beneath the abyssal floor of small ocean basins, such as the Gulf of Mexico and the Mediterranean. As a result of offshore petroleum exploration and the research efforts of the Deep Sea Drilling Project, the known distribution of sub-seafloor deposits of saline minerals has recently been expanded significantly. A number of salt basins have been located in the Mediterranean and the Red Sea and off the coasts of Africa, the Middle East, Europe, eastern Canada, Arctic Canada, northwest Australia, and other regions.

The offshore saline sedimentary basins are known to contain enormous quantities of salt, gypsum, anhydrite, and possibly also potash, magnesium and other saline minerals. For example, potash deposits together with anhydrite, gypsum and salt in the Upper Permian evaporite sequence extend extensively off the British coasts of northeast Yorkshire and southeast Durham (Dunham and Sheppard, 1969). Petroleum exploration in the North Sea shows that great thicknesses of salt deposits occur in three separate salt basins and potash seams are likely to be present (Dunham, 1969). Large potash deposits and thick beds of magnesium salts are high-value resources, and strategically located subsea deposits, such as those off the Yorkshire Coast, are likely to be developed in the future, particularly where they are amenable to solution mining or to underground mining from a land entry.

A significant portion of the world's production of sulphur comes from elemental sulphur deposits associated with anhydrite, either in bedded evaporites or salt-dome cap rocks. Production of sub-seafloor deposits, however, has been limited to two salt domes offshore Louisiana, which during the peak production in 1968 yielded about 20% of U.S. sulphur production. The first offshore sulphur deposit was discovered accidentally in 1949 during the petroleum drilling about 13 km seaward of Grande Isle, Louisiana. The Grande Isle deposit, with an average sulphur content ranging from 15 to 30%, covers an area of several hundred acres and varies from 70 to 130 m in thickness, making it the third largest known sulphur deposit in the U.S. The sulphur is recovered in a molten state from bore holes by injecting heated seawater under pressure.

The presence of elemental sulphur in a salt dome encountered by drilling from the *Glomar Challenger* in the Sigsbee Deep in the Gulf of Mexico and in another bore hole drilled in the Mediterranean (Pautot et al., 1970) suggests that some of the salt domes beneath continental margins and small ocean basins may contain commercial quantities of elemental sulphur. Potential world resources of subsea sulphur in salt domes have not been estimated, but are likely to amount to scores of millions of tons or more, and some of these undiscovered deposits may be similar to the known recoverable deposits off Louisiana (McKelvey and Wang, 1970).

HEAVY-MINERAL PLACERS AND OTHER SURFICIAL DEPOSITS CURRENTLY MINED BY DREDGING METHODS

Tin, rutile, ilmenite, zircon, monazite, magnetite, precious coral, barite, lime shells, and sand and gravel are the chief surficial materials economically recovered from the marine environment (all from less than 35 m of water and within 10 km of shore). Marine placer deposits, both offshore and along the modern beaches, constitute a major category of surficial deposits (Table 8-VII). Among the marine placers, the heavy metal deposits (tin, gold and platinum with specific gravities ranging from 6.8 to 21) occur chiefly in submerged extensions of stream channels, generally within several kilometres distance from their primary sources; deposits of the lighter heavy minerals (rutile, ilmenite, zircon, monazite and magnetite with specific gravities ranging from 4.2 to 5.3) are virtually confined to present beaches and submerged beaches at distances as far as a few hundred kilometres from their continental sources; diamonds and other gems (specific gravities range from 2.9 to 4.1) occur in streams as well as in beaches (Emery and Noakes, 1968); and sands of glass and foundry quality are concentrated in some modern and submerged beaches. Most of the offshore placers were formed during the Pleistocene Ice Ages when the sea level was as much as 160 m lower than today (Donn et al., 1962), and because of the cyclic nature of the sea-level changes during the glacial and interglacial periods, a succession of beaches and extensions of stream channels probably developed anywhere from the present 160-m isobath to the present shoreline or landward to a slightly higher elevation. Offshore placers may also locally include relict or reworked deposits which may have been concentrated by offshore currents from pre-

TABLE 8-VII

Typical on-going and possible future marine mining of unconsolidated surficial deposits (modified from Cruickshank, 1973)

Commodity	Ore mineral	Specific gravity	Grade of ore [1]	Equivalent ppm	Price range (U.S. $) [2,3]
Non-metallic					
Silica	quartz sand	2.65	76.93	25,644.4	29–92/st
Lime	shells and shell sands	2.7	1,428.6	476,142.9	1.40/st [4]
Magnesite	magnesite	2.9–3.1	33.33	11,110.0	60–100/st
Sand & gravel	various	3.0	1,923.1	640,961.5	1.04–7.87/st
Phosphate	phosphorite nodules and sand	3.0	307.7	102,553.8	6.50–1,020/st
Topaz	topaz	3.4–3.6	1 ct/cu. yd	0.1469	1–5/ct
Spinel	spinel	3.5–4.5	0.2 ct/cu. yd	0.0294	5–100/ct
Corundum	corundum	3.9–4.1	28.55	9,522.86	70–130/st
Heavy mineral sands					
Beryllium	beryl	2.75–2.8	66.67	22,220.0	30–35/st
Titanium	rutile	4.18–4.25	11.43	3,809.1	175/st
	ilmenite	4.7	101.8	33,936.0	22–24/lt
Chromium	chromite	4.6	93.33	31,108.0	24–56/lt
Zirconium	zircon	4.68	40.0	13,332.0	56–70/lt
Manganese	hausmannite braunite	4.72–4.84	36.72	12,239.2	61–68/lt
Iron	magnetite	5.18	203.64	67,872.0	11/lt

Thorium	monazite	5.0–5.3	12.44	4,147.73	180–200/lt
Columbium	columbite (10:1)	5.4–6.2	1.25	416.63	1,600–1,700/st
Rare Earths	group of 15 Me oxides		7.15	2,383.1	0.14–3,000/lb [5]
Tin	cassiterite	6.8–7.1	0.7463	248.73	2,680/st [6]
Mercury	cinnabar	8.10	0.498	166.13	152.5/fl(76 lb)
Precious and rare metals					
Diamonds	diamond (indust.)	3.5	0.25 ct/cu. yd	0.0367	4–50/ct
Copper	native metal	8.9	2.04	680.27	0.49–0.53/lb
Silver	native metal	10.5	0.6369 oz/cu. yd	13.268	1.57/oz
Gold	native metal	15–19.3	0.0203 tr.oz/cu. yd	0.464	49.26–49.46/tr. oz
Platinum group	native metal	14–19	0.0091 tr. oz/cu. yd	0.1894	110/tr. oz
Deep-sea authigenes					
Mn/Fe/Co/Ni/Cu	manganese nodules	2.6	66.84	22,279.4	29.92–56.61/st [7]
Au/Ag/Cu/Pb/Zn	metalliferous oozes	1.6 [8]	481.93	160,626.5	4.15–12/st [9]

1 Lb per cu. yd unless otherwise specified (to convert to kg/m³ multiply by 0.342).
2 Low value in price range used.
3 Prices from *Engineering & Mining Journal* of May 1972, and U.S. Bureau of Mines (1970).
4 Calculated from per-ton value of calcium in shell at 40% of this value, from atomic weights.
5 Using $ 1.50 as the average value, rather than the low price in the price range.
6 Calculated from per-pound value (1972) of tin at 79% market value, from atomic weights.
7 Varies from average Atlantic to highest Pacific (data from Mero, 1967, and others).
8 From Walthier and Schatz (1969).
9 As above, wet value.
Abbreviations: st = short ton, lt = long ton, tr. oz = troy ounce, ct = carat.

existing alluvial deposits or glacial till, such as the gold-bearing gravel covering a glaciated area of the Bering Sea shelf off Alaska (Nelson and Hopkins, 1972).

Offshore placers of potential economic importance are likely to be restricted to local shelf areas, either adjacent to the primary source rocks on land or cropping out on the sea floor, and are not expected to concentrate in areas beyond the shelf. Because of the very special geological conditions required for their formation in the marine environment, the world's overall potential for offshore placers is not likely to be as abundant as on land and modern beaches. However, there are important exceptions of very large and rich deposits formed in shelf areas under exceptionally favourable geological conditions. For example, the broad shelf areas off Indonesia, Malaysia and Thailand are long known for rich *tin* placers. In fact, offshore dredging for tin, first started in 1907 by an Australian sea captain in the Tongkan Harbor off southern Thailand, even preceded tin dredging in the nearby land area by three years. Offshore production in Thailand has been increasing continuously, and in 1973 yielded over 6,000 tons of tin concentrate. The average grade of offshore deposits mined is at least twice as high as the onshore deposits (Archer, 1973).

Titaniferous magnetite sand deposits are known to occur in near-shore waters and along modern beaches in many parts of the world, but they are generally lower in grade and smaller in size than the types of iron ores being exploited on land. There are important exceptions, however, in local offshore areas such as off southern Kyushu, Japan, off the west coast of Luzon (Philippines), off the west coast of New Zealand, and along the beaches in southern Java.

Rutile, ilmenite, zircon, monazite and associated minerals have been mined for many years from beach and dune sands in Australia (Queensland and New South Wales), New Zealand, Ceylon, Africa, India and the southeastern U.S. (Florida and South Carolina). In Australia, economic deposits are known to exist along a 1,000-km strip of beach stretching from Fraser Island in the north to Newcastle in the south. In the late 1960s prospecting was conducted on the adjacent shelf areas, and drilling of over 1,000 shallow test holes revealed as many as five drowned ancient beaches, some of which appear to have considerable potential for heavy mineral deposits. A substantial offshore deposit containing rutile as well as zircon, ilmenite and magnetite was delineated in 30 m of water about 2.5 km off the coast at the Queensland — New South Wales border, but this deposit in the open sea has not yet been developed. World production of these minerals has increased considerably during recent years; continuing increase in demand may stimulate the exploitation of some of the high-grade deposits in shallow-water areas.

Raised beach sands containing *diamonds* in southwest Africa have been mined for decades, and in 1961 gem-quality diamonds were discovered in gravels in shallow shelf areas. Subsequent investigations indicated that several concession areas (from the coast to 4.8 km offshore, up to about 30 m of water) contain high-grade deposits, and one area is estimated to contain more than 10 million carats of diamonds in some thirty separate placer deposits. Although the average grade of offshore diamond deposits is at least three times higher than on land (Archer, 1973), offshore production cost has been much higher and operations were suspended in 1972. Nevertheless, marine diamonds and other offshore placer minerals, such as gold, platinum and chromite, still show some promise for the future.

Calcium carbonate is commercially recovered offshore in the form of shells and shell sands, aragonite mud, coral and calcareous algae (*Lithothamnia*) and is primarily used in the manufacture of Portland cement, fertilizers, pulp and paper, and in extracting magnesium compounds from seawater. The largest production of lime shells comes from San Francisco Bay, off the Gulf coast in the U.S., and off Arkanes in Iceland. Extensive deposits of aragonite mud and sand, possibly between 50 and 100 billion tons, have been found on the broad shelf areas around the Bahama Islands; one of the areas has been developed by a large-scale dredging operation, and now produces 1 million tons per year. Coral sands have been dredged in Hawaii and Fiji for local uses as fill and aggregate. Similar calcareous reefs and sands are abundant off the coasts of many tropical areas where local limestone sources may be sparse or totally lacking. The high cost of shipping these bulky materials from distant sources suggests that nearby calcareous reefs and sands will be increasingly utilized in the future.

Offshore sand and gravel are important construction materials now being mined in shallow-water areas off the U.K., Denmark, Sweden, Japan, the U.S. and several other countries (Cruickshank, 1969; Archer, 1973). In terms of the tonnages dredged and in terms of annual production value, sand and gravel are by far the most important of the sea-floor surficial deposits now being commercially recovered.

Though still in its infancy, the U.K. marine sand and gravel mining industry is the largest and most advanced offshore operation of its type in the world (Hess, 1971). About 85 suction dredges are employed in numerous unprotected offshore areas in waters up to 30 m as well as in tidal estuaries (Fig. 8-10). Offshore production of sand and gravel increased from 3.4 million tons in 1955 to 13.5 million tons in 1971; it was valued at about $32 million and represented about 12% of the total U.K. production in that year (Archer, 1973). Since 1967, an export trade of washed sea gravel of construction quality has been developed, principally to the Netherlands and France. Production of marine aggregates in the U.K. has been increasing about 6% per annum.

A geological investigation on the Atlantic shelf of the U.S. (Schlee, 1968; Schlee and Pratt, 1970) indicates that the offshore sand deposits cover extremely large areas and contain enormous quantities of sand that will be useful in the future. Fairly extensive surface gravel layers are found off New

Fig. 8-10. Gravel and sand mining operations in the U.K. waters (from Chapman and Roder, 1969).

Jersey and New York, but most gravel beds are covered by 12—45 m of sand so that exploitation may not be feasible unless the less valuable sand can also be marketed. Marine sand and gravel have also been periodically recovered elsewhere in the world (e.g. in Thailand and Hongkong). Dredging for offshore gravel and sand is expected to grow rapidly, particularly near the densely populated coastal metropolitan areas and some oceanic islands, where the continuously rising demand for construction materials can no longer be supplied by the dwindling on-land sources available in areas adjacent to construction.

SUBMARINE PHOSPHORITE ACCESSIBLE TO DREDGING

Phosphorite, a complex calcium phosphate rock, has long been used as a source of phosphate for fertilizers and chemicals. The most common phosphate mineral is carbonate-fluorapatite, which usually contains 3.5—4% fluorine and small but important quantities of uranium (0.005—0.05%), vanadium (from 0.01 to over 0.03%), and rare earth elements bound up in the crystal lattice. Submarine phosphorite was first recognized among samples dredged from the Agulhas Plateau off South Africa by the *Challenger* Expedition in 1873 (Murray and Renard, 1891). In 1937 phosphorite nodules were found in samples from tops of submarine banks off southern California (Dietz et al., 1942). Subsequently, oceanographic expeditions have reported phosphorite nodules in many scattered localities off the coasts of North and South America, Africa, Australia and New Zealand, mostly in the outer shelf and upper slope areas and on tops and sides of submarine banks and seamounts. Occurrences are known from closely near shore to more than 300 km offshore, in water depths ranging from 20 m to more than 3,000 m, but most commonly in waters less than 400 m. Occurrences of sparsely scattered phosphorite nodules and phosphatized rocks are likely to be anywhere in middle latitudes (between 40°N and 40°S) in the shelf and slope areas without excessive modern detrital sedimentation. In addition to these scientific investigations of submarine phosphorite, some preliminary prospecting has been undertaken during the past decade in areas off Baja California and southern California, off northwest Africa, and off Australia and New Zealand. To date there has been no commercial phosphorite production offshore.

Phosphorite found on the sea floor is of four major types: phosphorite nodules, phosphatic pellets and sands, phosphatic muds, and consolidated phosphatic beds in Tertiary and older strata exposed on the sea floor; all these are of marine origin. In addition, Pleistocene or older guano deposits, which were formed from droppings of birds (or droppings of bats in caves) under sub-aerial conditions, may possibly have been submerged as a result of eustatic sea-level changes or tectonic subsidence of the Guano Islands (Cook, 1974). Offshore deposits possibly of the submerged guano-type have recently been reported from the vicinity of Christmas Island in the Indian Ocean (Bezrukov, 1973).

Nodular phosphorite, the most common form found on the ocean floor, includes nodules as well as flat slabs, irregular masses, and coatings on rocks; they are usually sparsely distributed but occasionally occur as extensive blankets. They range in size from small pebbles to large slabs nearly 1 m in length.

Phosphorite nodules have been found in bottom samples recovered from several shelf and slope areas adjacent to the Atlantic coast of the U.S., particularly off Florida and on the Blake Plateau (Pratt, 1971; Emery and Uchupi, 1972), off southern California and Baja California (Inderbitzen et al., 1970; Dietz et al., 1942), off northern Chile and Peru and off the southeast coast of South America. Occurrences of phosphorite nodules have been long known off the coasts of Spanish Sahara, Morocco and Guinea, and on the Agulhas Plateau off South Africa, and recently some of these areas have been systematically investigated (Tooms and Summerhayes, 1968; Summerhayes, 1970; Parker and Simpson, 1972; Rogers et al., 1972). In the southwest Pacific, scientific investigations have also been made on the phosphorite nodules occurring on the Chatham Rise and Campbell Plateau east of New Zealand (Waters, 1968; Summerhayes, 1969; Buckenham et al., 1971), and on the upper continental slope off northern New South Wales (Von der Borch, 1970). In the Indian Ocean, phosphorite nodules have been recently recog-

nized among samples taken off the North Andaman Island. Phosphorite specimens containing about 10% P_2O_5 were found in 1967 by a Russian oceanographic expedition at a sampling station off Quilon, southwest India (British
Sulphur Corporation, 1971).

Most of these reported occurrences consist of thinly scattered nodules too
sparse to be of any economic significance, and only a small number of deposits are expected to consist of abundant nodules. An example is the relatively
abundantly distributed phosphorite nodules off southern California that
have been investigated for more than 30 years; an evaluation based on all
phosphorite samples collected in this region (Inderbitzen et al., 1970) indicates that nodules are probably present in about 10% of the total area of
93,000 km², but only a few areas are suggested as prospecting targets, where
at best only marginal grade deposits (possibly containing a total of about 45
million tonnes of phosphatic nodules and 11 million tonnes of phosphatic
pellets) may be present.

Concerning the quality of phosphorite off southern California, the P_2O_5
content of 53 selected phosphate-rich nodules and pure portions averages
about 27%, but the overall average of unselected impure nodules are of much
lower grade. An economic analysis (Sorensen and Mead, 1969) showed that
the quality and quantity of phosphorite off southern California are too low
to warrant commercial exploitation in the foreseeable future.

By comparison, most of the phosphate deposits being mined on land contain considerably higher P_2O_5 content (generally 31—36%), which seems to
be due to secondary enrichment by removal of part of the carbonate and
other impurities during weathering and by other geological processes taking place under subaerial conditions. It appears that the best nodular phosphorite deposits on the sea floor will not be much higher than 29% P_2O_5,
which apparently corresponds to the "proto-ore" of on-land phosphate
deposits before enrichment by weathering.

Phosphatic strata of Tertiary and older age are known to extend offshore
from on-land phosphate deposits and crop out at places in the shelf and
slope areas, mostly underlying or adjacent to the occurrences of phosphorite
nodules. For example, off Florida and Georgia, Miocene phosphate strata are
probably exposed as discontinuous thin beds on the sea floor (Emery and
Uchupi, 1972), and where present, may contain 10—25% P_2O_5 (McKelvey et
al., 1969b). Sea-floor outcrops of Tertiary and Upper Cretaceous phosphatic
strata also occur off Morocco and Spanish Sahara (Tooms and Summerhayes,
1968; Summerhayes, 1970). Off southern California, low-grade phosphatic
shales within Miocene strata often crop out on the submarine banks. Generally speaking, the phosphatic bedrock exposed on the sea floor appears to
have no economic potential at present, even if deposits of sufficient grade
and quantity could be found. The mining costs of consolidated rock, even on
land, are higher than those of unconsolidated loose material.

The *phosphorite in the shallow-water shelf areas* and the near-shore zone are generally sand-sized phosphate pellets and oolites, ranging from 0.1 to over 1 mm in size. Low-grade deposits of *phosphatic sands* have been found off Cape Fear, North Carolina (Pilkey and Luternauer, 1967), off the South African coast between Luderitz and Walvis (Rogers et al., 1972), off Baja California, as well as in several beach and backshore areas along the Baja California coast (British Sulphur Corporation, 1971).

Unconsolidated low-grade deposits of phosphatic sands occur widely in the shelf areas west of Baja California and the deposits where investigated in the 1960s by both oceanographic institutions and mining companies. In one large area under 50—130 m of water, phosphatic sands are extensive and contain 15—40% apatite grains (D'Anglejan, 1967), equivalent to about 4.5—12% P_2O_5. It has been estimated that about 2,000 million tons of phosphate material may be present in this area (Mero, 1967), but it has not been determined which patches of deposits, if any, may have economic value. Shallow water depth, unconsolidated nature and virtually no overburden covering the deposits are favourable factors. But the possibility of profitable exploitation in the future depends upon the costs of offshore dredging, and of beneficiation and processing.

Phosphatic mudbanks are known in shallow waters off the southwest coast (Malabar) of India. Between monsoons, the mudbanks often contain significant amounts of phosphate (Seshappa, 1953), with a few patches assaying 1—5% P_2O_5 on a dry basis, together with some amounts of nitrate and potash. The highest P_2O_5 content sampled is 18%. During the monsoon season, the bottom sediments are disturbed and the phosphate is stirred up into the water.

Exploitation of the phosphatic muds is favoured by proximity to shore, shallow-water occurrence, and the finely divided nature of the material which would facilitate pumping. Possibly the raw phosphate material may eventually be used in direct application to improve the local acid soils.

Most of the occurrences of submarine phosphorite known thus far do not appear to be commercially exploitable in the near future. However, they may prove of interest in some areas far removed from other sources.

COAL AND OTHER SUB-SEAFLOOR DEPOSITS ACCESSIBLE TO UNDERGROUND MINING AND OTHER METHODS

Conventional underground mining of sub-seafloor deposits has long been an accomplished fact. One of the first undersea coal mines was operated more than 350 years ago off the coast of Scotland, where a shaft was sunk through an artificial island built up in shallow water (Austin, 1967a). To date, more than 100 subsea underground mines with shaft entry from land, islands, or artificial islands, have recovered minerals such as coal, iron ore,

256

nickel-copper ores, tin, gold, copper, mercury, and limestone off the coasts of Australia, Canada, Chile, Finland, France, Greece, Japan, Poland, Spain, Taiwan, Turkey, the U.K., the U.S. and other areas (Austin, 1967a,b,c). Some of the larger mines have working depths below sea level ranging from 30 m to more than 2,400 m, water cover as much as 120 m and are as much as 8 km from shore (Cruickshank, 1969).

Coal is the most important sub-seafloor-bedded deposit known to occur in large quantities in some of the ancient non-marine sedimentary basins now buried beneath the continental shelves and beyond. Under certain geological circumstances, portions of the coal-bearing basins might have subsided to-gether with the underlying crustal blocks and now occur beneath the conti-nental slope and adjacent areas. One of the new findings was coal in non-ma-rine Jurassic sedimentary strata by the Deep Sea Drilling Project in a hole at Orphan Knoll about 600 km off Newfoundland. A coal seam of Mesozoic age was also encountered in offshore petroleum drilling on the Grand Banks more than 300 km off Newfoundland (Emery and Uchupi, 1972).

Extensions of large coal fields have been delineated in many shelf areas of the world, and for many years coal has been produced from subsea under-ground mines off the U.K., Japan, Turkey, Chile, and several other countries. In 1972, undersea coal production accounted for 38% of total coal produc-tion in Japan and 10% in the U.K.; the world's annual subsea production was about 30 million tons. In Japan, subsea coal mining dates back to 1860 when the Takashima coal mine was first extended off the coast of western Kyushu. To date, many subsea coal fields have been discovered off eastern Hokkaido and western Kyushu, with over 300 million tons of proven coal reserves re-coverable under current economic conditions. The six existing subsea coal mines have been continually expanded and in 1972 produced a total of 10.8 million tons (including 5.3 million tons of coking coal). The Miike mine, the largest subsea coal mine in Japan, has galleries extending more than 5 km off the coast and now has more than 5,000 underground workers. Offshore in-vestigations in the U.K. have delineated large coal reserves off the Durham coast and also indicated that enormous quantities of coal probably lie fur-ther offshore beneath the western flank of the North Sea Basin, where the deposits may be beyond the present economic limit of subsea underground mining with entry from land but might become recoverable by in-situ gasifi-cation in the future.

Subsea coal deposits are widely known in many shelf areas of the world including off Israel, Spain, Arctic U.S.S.R., Norway, Greece, Brazil, Argen-tina, Australia, Canada, Alaska, and possibly off New England, Washington and Oregon (Tokunaga, 1967; McKelvey et al., 1969b). An appreciable por-tion of the increasing demand for coal may well come from future subsea mines, since subsea underground mining is more environmentally acceptable and involves no serious consequence of any surface subsidence above the mines. Continuing engineering research to develop new automated techniques of underground excavation and tunnelling through hard rocks may eventually extend the economic limit of subsea mining of coal and other large deposits (beneath a cover of impervious rocks) to more than 35 km from the coasts and islands (Wang and Cruickshank, 1969). Moreover, the world's increasing demand for energy may accelerate development of in-situ gasification and other new methods that would permit commercial in-situ extraction of not only the offshore coal deposits presently beyond the limit of subsea underground mining, but also the low-grade coal seams and organic-rich shales offshore as well as on land.

In addition to coal, many other metallic and non-metallic deposits within bedrock (such as tin ore, iron ore, nickel-copper ores, gold, mercury, and limestone) have been recovered from many subsea underground mines in widely scattered areas. Off the Cornwall coast in the U.K., for example, one of the early subsea underground tin mines was actively operated in the late 18th century until the shaft entry was accidentally hit by a ship and flooded. Geological evidence indicates that the shelf off northwest Cornwall may be extensively mineralized, and that the belt of granitic plutons, some of which may be tin-bearing, is likely to extend westward to the Scilly Islands and possibly continue 160 km further offshore (Dunham, 1969; Dunham and Sheppard, 1969).

Subsea iron ore of commercial grade has been actively mined off Jussaro Island, in Finland, off Elba Island in Italy, and off Cockatoo Island in the northwest part of Australia. In Newfoundland, the subsea iron-ore deposits, which were previously mined from shaft entries on Bell Island, are estimated to have several billion tons of reserves (Pepper, 1958). Generally, any mineral that is worth more than $15 to $20 a ton and extends offshore from the coastal fringes as a large thick deposit (several hundred million dollars worth of reserves) beneath a cover of impervious rock may be economically recoverable by conventional underground mining. Under favourable circumstances, consolidated lode deposits exposed on the sea floor may be amenable to open-pit bedrock mining, involving underwater blasting and recovery of the fragmented ore by dredging (Fig. 8-11). The first operation of this type was initiated in 1969 to recover a high-grade barium sulphate deposit off Castle Island, southeast Alaska.

Theoretically speaking, in-situ extraction of sub-seafloor solid minerals through bore holes offers the distinct advantage that production is not dependent on physical access to the subsurface deposits, thus avoiding the great mechanical effort and expense of excavation, haulage, milling, and waste disposal required in conventional mines. Several seemingly feasible methods of extracting subsurface metallic deposits on land have been proposed and are being investigated, including chemical and bacterial leaching of ore, and fracturing the underground orebody by chemical explosives or nuclear explosion (Fig. 8-11). But these experiments are still in their beginning stage; many new engineering developments are needed before any attempt to

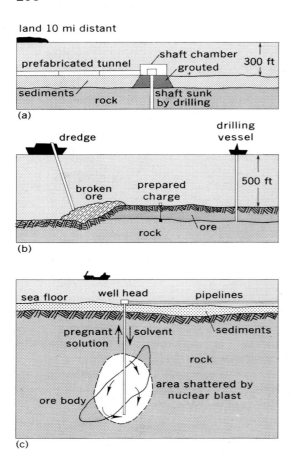

land 10 mi distant

prefabricated tunnel

shaft chamber
grouted
300 ft

sediments
rock

shaft sunk
by drilling

(a)

dredge

drilling
vessel

broken
ore

prepared
charge

500 ft

rock

ore

(b)

sea floor

well head

pipelines

pregnant
solution

solvent

sediments

rock

ore body

area shattered by
nuclear blast

(c)

Fig. 8-11. Possible future methods of mining consolidated mineral deposits under the sea (from Cruickshank, 1973).

develop the field technology of in-situ extraction, and it will be a slow and expensive business requiring enormous financial resources and engineering ingenuity.

In summary, the underlying sedimentary and crystalline rocks appear to contain many types of metallic and non-metallic minerals in large quantities similar to those found on land. Among these, mineral deposits of sedimentary origin probably dominate, because of the generally thick sedimentary section beneath the continental margins. Mineral deposits occurring in ancient sedimentary strata and buried at various depths below the sea floor may include, in addition to oil and gas, minerals such as sulphur, geothermal energy, brines, bedded salt and potash deposits, gypsum and anhydrite, coal, bauxite, limestone, phosphatic strata, bedded iron ore, and many other types known on land. Known and expected mineral resources occurring as lode or disseminated deposits associated with crystalline rock masses beneath the continental shelf and slope may include various metallic deposits as well as non-metallic minerals such as fluorite. Geological evidence indicates that metallogenic belts extend offshore in many shelf areas as well as in inland seas; extensively exposed bedrock areas, some of which might be mineralized, are found in several shelf areas, such as off the coasts of eastern Canada and northeast U.S. (Emery and Uchupi, 1972) and off the west coast of South Africa. But the generally thick sedimentary sections in most shelf and slope areas of the world tend to conceal any mineral deposits that may be present within the older bedrock; moreover, deep burial makes these deposits inaccessible to the offshore recovery techniques available today.

MANGANESE OXIDE NODULES ACCESSIBLE TO DREDGING

The occurrence of manganese oxide nodules on the deep-sea floor was first reported approximately 100 years ago by the *Challenger* Expedition of 1872—1876. During the past century, these unusual deposits have been sampled and photographed by oceanographic expeditions in broad abyssal areas. The early studies were largely of scientific nature, but in the early 1960s manganese nodules became recognized as the largest and most exceptional of the known mineral resources on the deep seabed.

Since the late 1960s, the manganese nodules of the deep-ocean floor have become the object of accelerated investigations throughout the world to study their distribution, nature, and potential value; to develop and test experimental mining systems for the recovery of the nodules; to develop pilot processing plants for commercial extraction of the metallic constituents of the nodules; and to initiate deep-ocean environmental base-line and impact studies that will be required as environmental guidelines for future mining operations. The possibility of commercial exploitation of the deep-sea manganese nodules, which had been considered highly uncertain until the late 1960s, is now approaching a reality as a result of the recent development of deep-ocean mining technology and the knowledge of the distribution, concentration and values contained in the nodules in several prospective mining sites; there seems to be little doubt that the nodules are destined to be an important future source of copper, nickel, cobalt and other metals (Table 8-VIII).

Manganese oxide deposits on the sea floor have a wide variety of shapes; they take just about any external form, including nodules, smaller grains, larger slabs and coatings over rocks, as well as fillings and impregnations in porous rocks. They most commonly occur as somewhat spherical concretions, with either smooth or knobby exteriors. The size of nodules is highly variable, but generally ranges in diameter from less than 1 to 25 cm, with an overall average of about 5 cm. Occasionally, an agglomerate slab of manga-

TABLE 8-VIII

World demand in 1980 for manganese, nickel, copper and cobalt and hypothetical supply from the seabed

	Manganese	Nickel	Copper	Cobalt
Mineral production in 1968 (short tons)	8,624,000	614,992	6,129,760	22,624
Assumed annual growth rate to 1980	5%	6%	6%	5%
Estimated world demand in 1980 (short tons)	15,456,000	1,232,000	12,147,692	40,656
Percentage of world demand in 1980 which could be supplied from:				
one mining operation	2.0	1.3	0.13	7.9
two mining operations	4.0	2.6	0.26	15.9
three mining operations	6.1	3.9	0.38	23.8
five mining operations	10.1	6.5	0.64	39.7
ten mining operations	20.2	13.1	1.28	79.3
twenty mining operations	40.4	26.2	2.56	158.7
fifty mining operations	101.1	65.5	6.41	396.7

Source: United Nations (1971).

nese nodules may be longer than a metre; one of the largest recovered (by a telegraph cable-laying ship from a depth of 5,200 m in the Philippine Trench) weighed some 750 kg. The colour of nodules varies from light brown to earthy black, generally reflecting the iron content of the nodules. The nodules generally have a specific gravity of 2.1—3.1 and are somewhat porous and friable; their friability seems to be inversely proportional to the amount of carbonate cementing material present. Internally, most nodules show concentric growth layers, frequently around a nucleus such as a shark's tooth, whale's earbone, micrometeorite (cosmic particle), volcanic glass, red clay, grain of a detrial mineral, or fragment of pumice, basalt, or other rock.

Sand-sized manganese oxide grains or "micro-nodules", around 0.5 mm in diameter, occur as a common constituent of ocean-bottom red clays and bio-genic deep-sea oozes. The manganese encrustations covering bedrock surfaces on seamounts generally are about 2 cm thick but in places reach 10 cm or more. Ferro-manganese nodules, especially rich in iron but poor in other metals, have been found in several shallow-water areas of the Arctic, on pla-teaus along continental margins, and in epi-continental seas, such as the Bal-tic (Manheim, 1965).

The mineralogy of manganese nodules is relatively complex, consisting generally of a mixture of manganese dioxide minerals, goethite (amorphous iron hydroxide), clay minerals, and minute amounts of apatite, barite, celes-

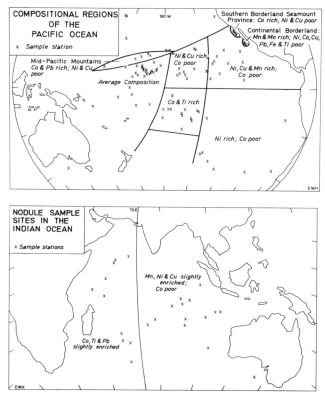

Fig. 8-12. Salient chemical characteristics of manganese nodules from different oceanic regions (see Table 8-IX) (modified from Cronan, 1972, p. 23).

tite, various silicates, and detrital minerals. The dominant manganese minerals in the nodules are identified as birnessite, for occurrences in bathyal water depths generally less than 3,000 m, and todorokite (a hydrous manganese oxide) in occurrences at greater water depths. Several other ferromanganese oxide minerals have been recognized; a few are not similar to any previously known mineral of this element. Various metallic elements and cations (such as nickel, copper, cobalt, zinc, molybdenum, cadmium, strontium, sodium and calcium) can substitute in rather high concentrations for manganese and iron in the disordered layers of the crystal structures of these manganese minerals. The chemical composition and physical characteristics of nodules vary considerably, not only from one location to another, but within a single nodule. Since the late 1960s, knowledge of the distribution, concentration, and properties of deep-seabed nodules has increased significantly as a result of accelerated investigations, involving many industrial and research organizations throughout the world. Nodules have been sampled, photograph-

TABLE 8-IX

Average composition of surface nodules from different areas within the Pacific and Indean Oceans[1] (modified from Cronan, 1972)

	1	2	3	4	5	6	7	8	9	10	11
Mn	33.98	15.85	22.33	19.81	15.71	16.61	16.87	13.96	12.29	13.56	15.83
Fe	1.62	12.22	9.44	10.20	9.05	13.92	13.30	13.10	12.00	15.75	11.31
Ni	0.097	0.348	1.080	0.961	0.956	0.433	0.564	0.393	0.422	0.322	0.512
Co	0.0075	0.514	0.192	0.164	0.213	0.595	0.395	1.127	0.144	0.358	0.153
Cu	0.065	0.077	0.627	0.311	0.711	0.185	0.393	0.061	0.294	0.102	0.330
Pb	0.006	0.085	0.028	0.030	0.049	0.073	0.034	0.174	0.015	0.061	0.034
Ba	0.171	0.306	0.381	0.145	0.155	0.230	0.152	0.274	0.196	0.146	0.155
Mo	0.072	0.040	0.047	0.037	0.041	0.035	0.037	0.042	0.018	0.029	0.031
V	0.031	0.065	0.041	0.031	0.036	0.050	0.044	0.054	0.037	0.051	0.040
Cr	0.0019	0.0051	0.0007	0.0005	0.0012	0.0007	0.0007	0.0011	0.0044	0.0020	0.0009
Tl	0.060	0.489	0.425	0.467	0.561	1.007	0.810	0.773	0.634	0.820	0.582
L.O.I	21.96	24.78	24.75	27.21	22.12	28.73	25.50	30.87	22.52	25.89	27.18
Depth (m)	3003	1146	4537	4324	5049	3539	5001	1757	4990	3793	5046

(1) Continental Borderland off Baja Calfornia, (2) Southern Borderland Seamount Province, (3) Northeast Pacific, (4) Southeast Pacific, (5) Central Pacific, (6) South Pacific, (7) West Pacific, (8) Mid-Pacific Mountains, (9) North Pacific, (10) West Indian Ocean, (11) East Indian Ocean.

[1] Data in weight percent, air dried weight, from Cronan (1967); Cronan and Tooms (1969).

ed, or viewed by sea-floor towed television in many parts of the world's oceans between 50°N and 60°S. Data indicate that the nodules vary greatly in composition and concentration per unit area on the sea floor. Major provinces of nodules have been recognized (Cronan, 1972) in the Pacific, Atlantic and Indian Oceans (Fig. 8-12, Table 8-IX), and there appears to be a strong correlation between the content of minor elements and sediment type. Regionally average concentrations of manganese, nickel, and copper tend to be high in samples dredged from the eastern parts of the oceans, decreasing westward as the amounts of iron, cobalt, titanium, and lead increase (Cronan, 1972); this relationship of average metal contents is distinct within the Pacific, less well defined for the Indian Ocean, and obscure in the Atlantic. Similar relationships in metal contents also appear to apply to water depths at which the nodules occur and to rate of sedimentation; nodules from abyssal regions that have depths exceeding about 4,500 m and slow rates of sedimentation consistently have much higher nickel and copper contents than those from continental terraces, submarine ridges, and seamounts.

Investigations in the Pacific indicate that surficial manganese nodules are abundant on the sea floor in several large areas but are rare or absent in other areas. Zenkevich and Skorniakova (1961) estimated that the Pacific Ocean as a whole may contain about 90,000 million tons of manganese nodules. Mero (1967) estimated that the Pacific may contain about 1.7 trillion (10^{12}) tons of manganese nodules, which are now being continuously accumulated at an annual rate of about 10 million tons. Whatever the total amount of nodules proves to be, the amount of suitable quality, abundance, and environmentally favourable setting to warrant future exploitation is likely to be much smaller than the total.

Among the major provinces of surficial manganese nodules found to date, those in the Pacific at depths of 4,500—6,000 m appear to be most extensive and most favourable (Fig. 8-13). The important nodule provinces fall in two broad belts that extend east-west across the Pacific Ocean, from Central America to the Mariana Trench north of the equator and from the Peru-Chile Trench to the Tonga Trench south of the equator; a zone of relatively rapid pelagic sedimentation along the equator separates these two belts. A nodule-rich belt, similar to that of the southern Pacific, extends across the central Indian Ocean from northwestern Australia to the Seychelles-Mauritius Ridge (Mascarene Plateau). In the North and South Atlantic, manganese nodule provinces are segmented by the sea-floor relief and linear zones of high sedimentation along transverse fracture zones.

In a review of nodule distribution and composition, Horn and Horn (1973) show that nickel- and copper-rich deposits of nodules are largely confined to the broad belt north of the equator in the equatorial North Pacific. A region east of the Line Islands between the Clipperton and Clarion Fracture Zones (Fig. 8-14) is generally considered most promising for the future mining of manganese nodules and has therefore become the focus of recent

264

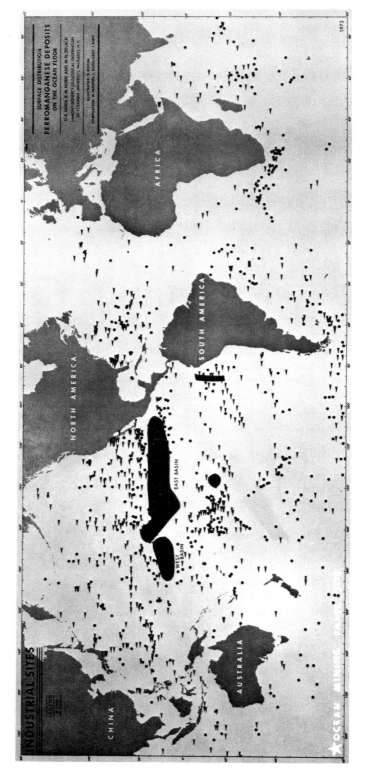

Fig. 8-13. Promising ocean areas as potential sites of manganese nodule mining; most potential is offered by mining sites located from 8°31′N and 150°W to 10°N and 131°30′W (see Fig. 8-14) (from Horn et al., 1973b, p. 53). Black dots are cores, arrows indicate dredges.

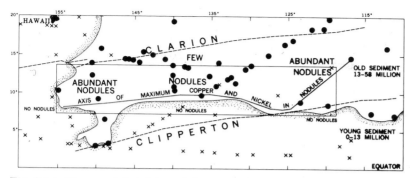

Fig. 8-14. Copper-nickel-rich nodules occurring with east-west band of radiolarian ooze between the Clipperton and Clarion fracture zones (from Horn et al., 1973c, p. 4). Black dots denote camera stations where nodules occur; ×'s, where photographs indicate seafloor is barren.

industry exploration and scientific studies. The southern Clipperton Fracture Zone forms a well-defined boundary separating the province of abundant nodules to the north from the equatorial zone of sparse nodules and high biological productivity and rapid deposition of biogenic oozes (dead skeletal material) to the south. Manganese nodules between the two fracture zones and in areas north of the Clarion rest on nearly flat to gently sloping and rolling surfaces of siliceous oozes and red clays. In most places, sediment cores have few or no nodules beneath the 2- to 6-cm layer of the surficial nodule deposits. Average water depths in which nodules appear to be most abundant range from about 5,000 to 5,500 m. At such great depths, the dead skeletal material from the surface waters settles through a "zone of carbonate compensation", where carbonate material is dissolved and only small amounts of siliceous remains reach the bottom. This large region provides nodules in sufficient abundance and high nickel-copper content, relatively smooth bottom surface, and generally favourable weather and other operating conditions.

Extensive nodule provinces have also been found in other regions, and new areas of abundant nodules are indicated west of Peru and northeast of the Tuamotu Islands in the equatorial South Pacific, south of Australia, northeast of the Hawaiian Islands, southeastern part of the Atlantic, and several areas in the Indian Ocean (Corwin and Berryhill, 1973; Horn et al., 1973a). However, nodules from many of the oceanic areas of the world are little investigated, and these areas may have individual deposits comparing favourably with those in the equatorial eastern Pacific. It is generally considered that a nodule density of at least 7.3 kg per square metre of ocean floor and a combined copper-nickel content of at least 2% are the prerequisite factors.

With the current intensive efforts of private corporations and governmental organizations of many countries in pursuing exploration and technological development, and with the successful testing of pilot-scale mining and processing systems and favourable results of economic feasibility studies during the past few years, it is expected that the first commercial exploitation of the nodules will become a reality as early as 1980 (Kaufman, 1974b). However, some experts believe that there are numerous practical problems to be solved during operations in the early years, and that deep-ocean mining and processing technology must be further developed for more efficient exploitation operations — less down-time, reduced mining costs, and the capability to withstand extremes of sea and weather conditions. The level of economic production cannot be estimated at present, but it could be in the range of a few thousand tons to more than 20,000 tons of nodules dredged per day. It is generally thought that manganese nodule mining and processing would be of moderate scale during the early years, and that the world's nickel market would play the predominant role in limiting the production level and growth of the new industry (see Chapter 4). The deep-sea-produced nickel, copper, and manganese will represent only minor shares of the world's total output, possibly meeting the projected increase in demand of 3.5—6% per year (Rigg, 1974). However, the cobalt output could represent a significant share of the world market, at least in the early stage. From the mid 1980s to the end of the century, it is not possible to project what proportion of world output of the four metals will come from the deep-sea bed.

METALLIFEROUS SEDIMENTS AND BRINES

Abnormally hot (more than 40°C), highly saline (240 parts per thousand) brines with abnormally high metallic contents have been discovered along the central rift valley of the Red Sea in a series of deep basins at depths of 1,900—2,200 m, the largest being the Atlantis II Deep with an area of about 12 km × 5 km (Miller et al., 1966; Degens and Ross, 1969; Backer and Schoell, 1972). In the Atlantis II Deep, the Discovery Deep, and a few other deep basins, metal-enriched sediments occur widely on the bottom, apparently associated with the overlying brine pools. Recent studies have reported new occurrences of metal-enriched sediments in widely separated localities in the Red Sea, as much as 320 km north and 160 km south from the Atlantis II Deep (Tooms, 1970; Backer and Schoell, 1972).

Geological and geochemical evidence, including elemental composition of the brines and oxygen and hydrogen isotope ratios, suggest that they were formed as a result of leaching of metals from basaltic bedrock and the evaporite beds beneath the Red Sea by heated circulating seawater brine, essentially along the rift and fracture zones which are also characterized by high heat flow. Where fractures in the bedrock permit seawater to enter, the high heat flow along the fracture zones is likely to set up a convection system, drawing cold seawater into the crust and heating it. The higher temperature of the water increases its capacity to extract metals from the country rock. On re-emergence to the bottom surface, precipitation of the metals from the hot hydrothermal anoxic brine is probably caused by mixing and reaction with the overlying cooler, oxidizing, weakly alkaline and less saline water (possibly also caused by the introduction of H_2S generated from dissolved sulphates by bacterial action). Ponding of the emerging relatively heavy metal-enriched brines in bottom depressions, such as the Red Sea deeps, forms the stagnant brine pools and allows localized precipitation of concentrated metalliferous constituents on the bottom.

The layer of hot brines, which is about 160 m thick in the Atlantis II Deep and 200 m in the Discovery Deep, has a total salt content about 10 times higher than normal seawater, and contains heavy metals, such as iron, manganese, zinc, lead, copper, silver, and gold in extraordinary concentrations of from 1,000 to 50,000 times the amounts found in normal seawater. The metalliferous muds may have a greater economic potential than the overlying brine pools. In the Atlantis II Deep, for example, the upper 10 m of the metal-rich mud are estimated to be over 50 million tons, and the zinc, copper, lead, silver, and gold they contain are estimated to have a potential in-situ value of about $2,500 million; this does not take into account the iron, manganese, mercury, and other metals in the mud (Bischoff and Manheim, 1969).

It is clear that the Red Sea brine pools, metalliferous muds, and associated geothermal energy in the subsurface represent a very large and valuable potential resource. The possibility of exploiting these resources will be critically considered in the future when suitable technology is developed for handling these corrosive media at such great depths and distances from the shore and for processing and extracting the colloidal metalliferous material contained in the muds in unique physical and chemical forms.

In 1972 drilling from the *Glomar Challenger* in the Red Sea deeps encountered at depth metal-rich shales and sandstones (above layers of anhydrite and salt) in two drill holes. These metalliferous strata may be one of the sources of the metals in the surficial muds, and they may also constitute a new potential sub-seafloor resource of copper, zinc, and vanadium (Ross et al., 1972, p. 26).

Metal-enriched sediments, similar to those in the Red Sea deeps, are known to exist in a few areas of anomalously high heat flow at the crest of the East Pacific Rise (Boström and Peterson, 1966, 1969) and they may also occur well away from the active rifts, such as adjacent to the submarine Banu Wuhu volcano off Indonesia (Zelenov, 1964). The Deep Sea Drilling Project has encountered these types of metal-rich sediments in the Pacific just above the igneous rock basement over a wide area along the rise crests. Although the geological significance of these findings relating to the spreading of the rift valleys and associated hydrothermal emanations have not yet been fully evaluated, it has been inferred that metal-rich sediments may be found at the base of sedimentary columns in some areas adjacent to the axial zone of spreading and the outer edge of the spreading rift systems (Peterson, 1970).

FRESH GROUNDWATER AND GEOTHERMAL ENERGY

Fresh-water aquifers are known to extend offshore beneath the continental shelf in many parts of the world, and submarine springs have been observed for many years off the Atlantic coast (Florida, South Carolina), and Pacific coast (California) of the U.S., the Bahamas, Barbados, Cuba, the Yucatan Peninsula of Mexico, Chile, Samoa, Guam, Australia, Japan, and many areas in the Mediterranean (Kohout, 1966; Manheim, 1967; Stefanon, 1972). There is also an unconfirmed occurrence of large submarine springs off the Rumanian coast in the Black Sea. Utilization of submarine spring waters dates back many centuries; they supplied fresh water to the ancient Phoenician city of Aradus on a rocky island off Syria, and water from springs in the Persian Gulf was recovered by Arab boats (Fye et al., 1968). Details of the Florida springs, which are well documented, are found in JOIDES (1965), Manheim (1967), Brooks (1961), and Emery and Uchupi (1972); Cruickshank (1974) has reported a spring in the Argolis Bay in Greece which emits 10 m^3 per second, or almost 228 million gallons of fresh water each day, representing an annual value of $36 million.

Fresh water, as well as brackish waters, occurring in extremely large quantities in aquifers beneath many shelf areas represent an undeveloped resource, which is particularly important in semi-arid and arid coastal areas and offshore islands where on-land fresh-water sources may be sparse or totally lacking. While subsea fresh-water aquifers can be exploited for many uses for the coastal community, brackish waters (less than 1,500—2,000 ppm) may be suitable for agricultural and industrial purposes or, after blending with fresh water from a land source, for domestic purposes.

The continuing release of geothermal energy from the earth's crust is relatively uniform over large areas of the stable continental masses but often reaches abnormally high levels in some of the tectonically active regions, which probably cover 10% of the land surface and much greater areas under the oceans. Nearly all the known geothermal fields and potential geothermal provinces occur in the tectonically active belts characterized by excessive geothermal flux. They represent a replenishable source of energy as long as the deeply seated sources of volcanic or magmatic heat persist and as long as the geothermal energy stored in the thermal fluid reservoirs can be transferred by geologic processes to balance its natural discharge and artificial recovery. Man is beginning to realize that the present utilization of depletable fossil fuels will prove to be merely a short transitory phase over a long span of human history, and that energy from this source must be gradually supplanted by large-scale utilization of geothermal energy as well as by other new types of clean energy, such as solar energy, controlled thermonuclear fusion, hydrogen fuel, wind power, tidal power, ocean currents, and ocean thermal gradients (see Chapter 7). Moreover, certain types of geothermal brines may provide new sources of industrial and agricultural chemicals. In fact, commercial extraction of minerals from geothermal fluids has been actively conducted in Larderello, Italy, since 1777; at present this geothermal area continues to produce borax, sulphur, boric acid, and other industrial minerals.

Search for geothermal energy on land during recent years has already indicated potential geothermal sites in about 80 countries and there is a vast potential for commercial geothermal development in many of these areas. A highly important type of geothermal energy recently discovered (Jones, 1969, 1970) is the regional geo-pressured system, where heat flow of the earth is trapped by insulating and impermeable clay beds in some of the offshore and coastal geosynclinal basins characterized by rapid deposition of great thicknesses of young sediments and by crustal subsidence. Enormous amounts of residual heat are often stored in the porous water-logged sand beds at subsurface depths of 1,500—5,000 m, resulting in super-heating of the fluids and creating abnormally high interstitial fluid pressure (geo-pressure) wherever natural compaction and escape of the fluids from a thick sedimentary sequence are impeded. The geothermal resources in such deep geosynclinal basins differ from other types of geothermal systems in that they are not dependent upon continuing recharge and deep circulation of meteoric water; their thermal waters are derived from the sediments themselves, and fluid depletion occurs with energy release. A prime example of this type of offshore geothermal resources is the Tertiary basin in the northern Gulf of Mexico as a broad geo-pressured belt, which extends some 1,300 km parallel to the shelf edge and the coast from the Rio Grande to the mouth of the Mississippi. Offshore petroleum drilling beneath the Gulf of Mexico has encountered many geo-pressured geothermal reservoirs containing super-heated salty water (at temperatures above 120°C) which would self-distill when pressure is released upon reaching surface. Many of the geo-pressured reservoirs also contain huge quantities of dissolved hydrocarbons. The enormous size of these geo-pressured reservoirs offers a large potential future energy source, as well as a possible new source of fresh water desalinized by self-distillation and recovery of the dissolved hydrocarbons. Similar geo-pressured geothermal reservoirs probably exist in many shelf areas where great thicknesses of young sediments were deposited over zones of high heat flow. Subsea sources of geothermal energy far from land have little economic value at present, but in coastal waters they may have a potential use in the generation of electric power, solution mining of sulphur and potash, and possibly most important of all, in repressuring offshore oilfields for secondary recovery of petroleum.

POTENTIAL OF MINERAL RESOURCES IN BEDROCK BENEATH THE DEEP SEA-BED

It is generally recognized that beneath a veneer of young sediments in the great oceanic basins at abyssal depths, much of the bedrock is composed of oceanic-type basalt which is remarkably uniform over large areas and seems to have little potential for concentrated metallic deposits (James, 1968; Dietz, 1969). However, new evidence suggests that the deep seabed probably contains enormous quantities of metallic minerals, such as nickel, copper, molybdenum, cobalt, zinc, lead, mercury, chromium, platinum, gold, silver, etc., in deposits possibly formed from differentiation of the molten mantle material, from which the basalt itself was derived, or from melting of the crustal plates that were thrust into the mantle.

Discovery of kimberlite (similar to the diamond-bearing rocks occurring on the African continent) and chromite by Soviet oceanographic cruises in the Indian Ocean, and the new indications of metalliferous mineralization associated with oceanic-type basalt that have been encountered by the Deep Sea Drilling Project (Fig. 8-2), indicate that oceanic-type magmas may differentiate irregularly and permit the concentration of ores commonly associated with basic and ultra-basic igneous rocks. Nevertheless, the deep seabed is not likely to contain, in any significant quantity, those primary ore deposits that are associated with the granitic rocks of the continental masses.

The metalliferous hot brines and metalliferous muds widely occurring and continually being formed on the sea bottom in the central rift valley of the Red Sea along the zone of divergence of spreading lithospheric plates (see p. 266) suggest that higher-grade sulphide ore might be present in the underlying strata (Walthier and Schatz, 1969). Although the major divergent plate boundaries, represented by the global system of mid-oceanic rifts, have not been investigated to any extent, sediments enriched in iron, manganese, copper, chromium, nickel, and lead have been found in several large areas along the crest of the East Pacific Rise, particularly at and just above the igneous rock basement (Boström and Peterson, 1966, 1969; Peterson, 1970). Similar metal-enriched cores have also been recovered by the Deep Sea Drilling Project in the Northwest Atlantic (iron, zinc, and manganese mineralization in the Tertiary sediments as well as a thin layer of elemental copper in Jurassic carbonate rocks overlying basalt at site 105 in 5,251 m of water) (Ewing et al., 1970), at places along the Mid-Atlantic Ridge (Scott et al., 1973), in the northeast Pacific (McManus et al., 1969), and at the crest of Ninety East Ridge in equatorial Indian Ocean (Von der Borch et al., 1972). A mineralized rock sample containing a vein of copper-iron sulphides has also been dredged from a depth of 3,500 m in the rift valley of the Arabian-India Ridge (Rozanova and Baturin, 1971).

Evidence of mineralization associated with oceanic rifts was recently investigated on land in southeastern Iceland along the Mid-Atlantic Ridge. This hydrothermal mineralization of copper, lead, and zinc (chiefly chalcopyrite, galena, sphalerite, and minor amounts of pyrite and molybdenum mineral) associated with Tertiary granophyre intrusions may possibly be derived from the differentiation of basaltic magma at the upper part of the mantle just beneath the oceanic crust (Jankovic, 1972). Recent studies on the long-known rich orebodies (copper, zinc, gold, silver, and other metals) in the Troodos Massif on the island of Cyprus have provided a new indication that extremely rich mineralization may possibly be associated with oceanic rifts of the geologic past. The Troodos Massif is now interpreted to be a slice of oceanic lithosphere (however, see Chapter 3, section *Oceanic crust*) that was originally generated from an oceanic rift in Cretaceous time and subsequently thrust upward to its present position (Moores and Vine, 1971; Gass and Smewing, 1973; Hudson and Robertson, 1973). Geological and geochemical evidence suggests that the large orebodies of metallic minerals occurring in a sequence of basic igneous rocks probably were formed beneath the oceans by hydrothermal processes related to sea-floor spreading (Corliss et al., 1972). The presence of base metal mineralization in Iceland and economic orebodies in Cyprus suggest that large, rich ore deposits might have been formed or are continuously being formed under certain conditions along the mid-oceanic ridges and rises. But it is unlikely that any of the present methods could be capable of detecting such orebodies beneath the deep seabed, let alone developing their resources (Rona, 1972).

Recent studies on the origin of ore deposits in relation to the processes of ocean-floor spreading and plate tectonics indicate that the consuming plate boundaries, both the continent—ocean boundary and the island arc—ocean boundary, are far more favourable for genesis of metallic ores than the accreting mid-ocean rifts (Guild, 1971, 1972a,b). In the typical case of convergence of two lithospheric plates, the denser oceanic crust is normally underthrusted beneath the continent or island arc along a relatively steep subduction zone and becomes consumed within the mantle. Seismic data show that the descending plate often extends as deep as 700 km before losing its identity completely; it is generally believed that partial melting of the descending plate at shallower depth yields basaltic magmas, then calc-alkaline magmas and finally alkaline magmas (Dewey and Bird, 1970). These magmas rise to produce various intrusive and extrusive igneous rocks in certain tectonic belts during successive episodes of crustal deformation, and the origin and emplacement of calc-alkaline magmas appear to be associated with highly important metallogenetic processes in concentrating metallic ores, such as copper, molybdenum, lead, zinc, gold, silver, tungsten, tin, and other metals. Good examples are the porphyry copper deposits (a highly important type of deposit accounting for more than half the world's present copper production and often containing molybdenum and gold), which are known to occur in the metallogenic provinces along the Andes in South America and along the West and Southwest Pacific island arcs, particularly Solomon Islands,

272

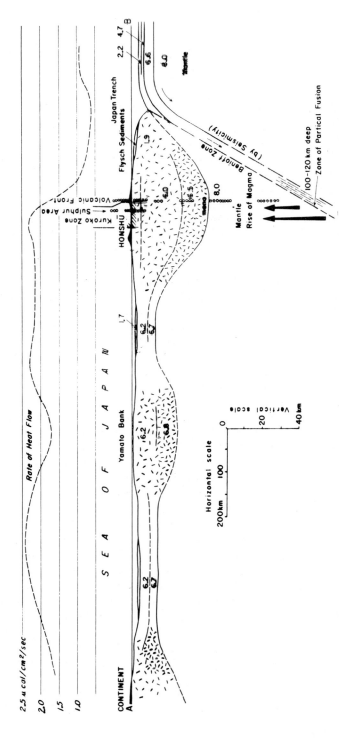

Fig. 8-15. Schematic section across the Japanese arc system, showing interpreted lithospheric structure, heat flow rate, and metal logenesis (from Nichiwaki, 1973).

Fiji, and from the Philippines to New Guinea. Volcanogenic stratiform deposits of massive sulphide ores (copper, zinc, lead, gold and silver), of which the Kuroko deposits of Japan are an example, also occur in the magmatic arcs (Fig. 8-15); other types of deposits in the volcanic rocks of the island arcs include gold in andesites and monzonites, and mercury in young volcanoes. The more acidic plutonic bodies of the magmatic arcs, such as the metallogenic belt from Indonesia through the Malay Peninsula and Thailand to Burma, apparently emplaced in pre-existing continental crust that might have been evolved from ancient arc systems, contain most of the world's tin and tungsten deposits with associated molybdenum, bismuth, and fluorite (CCOP/IOC, 1974). Inference made from such new understanding of ore genesis suggests that many undiscovered orebodies may possibly occur along uninvestigated segments of the island arcs in the western and northern Pacific, as for example, the Aleutians, Kuriles, Ryukyus, and behind the Mariana Trench and the Tonga-Kermadec Trench (Guild, 1972b). Most segments of these island arc—ocean boundaries are covered by water and young volcanic rocks, and even the exposed islands have not yet been explored to any degree.

According to a hypothesis developed by Sillitoe (1972), the new basaltic crust (containing thinly dispersed metallic minerals) that has been continuously formed along the mid-oceanic ridges would be carried away and eventually thrust beneath an island arc or a continent along a subduction zone, where partial fusion and complex distillation processes would concentrate the metallic elements into rich ore deposits. Such a complete cycle from the active mid-oceanic ridge like the East Pacific Rise to its active consumption zone beneath the Peru-Chile Trench, including the porphyry copper belt along the Andes, is being intensively studied.

MINERALS DISSOLVED IN SEAWATER

The world's oceans contain 1,370 million km^3 of seawater, and each cubic kilometre holds about 40 million tons of dissolved solids (Shigley, 1968; Gould, 1974). If all the dissolved substances in seawater could be extracted commercially, they would provide an inexhaustible resource for most of man's needs for minerals as well as fresh water. Seawater is perhaps the most complex solution that technology is ever likely to encounter. More than 77 elements have been identified at this time in seawater, and others can be inferred by their presence in marine authigenes or biota. It is likely that all naturally occurring elements are present in the ocean. This ancient solution, evolved since the beginning of the earth, contains simple mineral ions, complex aggregations of molecules, dissolved gases and a wide variety of organic molecules. Suspended in the solution are many kinds of colloidal and larger particulate minerals, insoluble organic compounds, and marine organisms ranging from the simplest to the most complex. Forchhammer (1865) made the classic statement that the quantity of various elements present in seawater is not proportional to the quantity of elements discharged by rivers, but inversely proportional to the facility with which the elements in seawater are made insoluble by the chemical processes and biological growth in sea. It could be further propounded that all the elements in the earth's crust are involved in a continuous temporal cycle, of which their residence time in the ocean is but a fraction. In the search for mineral resources, that time when the element is in its most concentrated or most easily isolated form is the most sought after (Cruickshank, 1974). For several elements, the residence time in solution in seawater is optimum, and for many others the solid forms as well as mineral fluids in rock strata are preferably sought. As an orebody, the ocean is unusual in that the minerals dissolved in seawater are being replenished by river runoffs at a rate equal to or greater than the present scale of their commercial extraction.

Water in the open sea appears to be relatively uniform in composition and its content of major elements reported by various workers does not differ significantly (see also Chapter 3). As for the minor elements, differences are generally less than one order of magnitude, but those of trace elements show wide variations, particularly in coastal waters (for example up to 2.4 orders of variations for copper and five orders for gold; Cruickshank, 1974). This is probably indicative of actual variations, for seawater is by no means homogeneous in its content of minor constituents. It implies that certain stagnant bottom waters in oceanic trenches and deep basins may contain greater concentrations of heavy metals and deuterium which may have economic significance in the future.

Despite the extreme complexity of seawater as a natural solution, its chemistry is essentially determined by a dozen constituents present in concentrations larger than 1 part per million (McIlhenny, 1969), and it is largely from these that mineral products are being recovered today. As of 1972 there were more than 300 near-shore operations in 60 countries which recovered seawater constituents, with a total annual value of about $415 million. The materials now commercially extracted directly from seawater (Table 8-II) include salt, magnesium metal, magnesium compounds, bromine, heavy water (deuterium oxide), and fresh water; additional small quantities of potassium and calcium compounds are produced from salt bitterns remaining after solar evaporation and salt crystallization. In fact, output from these operations accounts for about one-third of the total world salt production, two-thirds of the magnesium metal, and one-third of the bromine (Gould, 1974).

Attempts to locate economic concentrations of gold dissolved and suspended in seawater and to commercially recover gold have been unsuccessful. Research relevant to the possible extraction of uranium and other trace elements present in seawater has also been undertaken in several laboratories, but in no case has an economic commercial process been developed. Current intense interest in the production of fresh water from the sea has led to much additional research relevant to the extraction of minerals, and it is not unlikely that new commercial processes will eventually be developed to extract a few other constituents from seawater, from the concentrated effluent brines of desalination plants, or from the gently enriched recycle streams of solar salt extraction.

Several varieties of marine plants and animals can concentrate specific elements in their bodies by a factor of as much as 100,000. Iodine was commercially recovered from seaweeds in the 1840s and during World War I some 12,000 tons of potash were recovered from the giant kelp in California. Recently large-scale harvesting of seaweeds has been initiated or planned for the cold rocky coastal waters (such as the areas of prolific seaweed growth off Iceland and Norway) for mass production of supplemental fertilizers for agricultural uses. However, the possibility of utilizing the biochemical activities of some marine organisms in concentrating certain trace elements for possible commercial recovery still appears remote.

By far the most important development for the long-term seems to be the prospect of large-scale extraction of energy substances and fresh water. Deuterium (in heavy water) is one of the essential substances usable in the fusion process. The oceans contain 25 trillion tons of deuterium — the equivalent of 75×10^{26} Btu's. In comparison with fresh water and the cold polar seawaters, tropical seawater is somewhat enriched in heavy water, because of the fractionation occurring during evaporation. It is interesting to note that the heavy water extraction plant at Glace Bay in Nova Scotia was constructed to utilize the "Gulf Stream" waters, originating in the equatorial Atlantic. The seawater also contains abundant lithium; one of its isotopes is potentially usable in the fusion reaction.

The ocean has increasingly become a source for the world's growing freshwater needs, particularly in arid coastal regions, offshore islands and densely populated coastal cities, where fresh water is a necessary and valuable resource. Today there are several hundred desalination plants located throughout the world, with the capacity of individual plants ranging from several thousand gallons up to 7.5 million gallons per day. The world's installed desalination plant capacity has been increasing at an average rate of 30% per year (McIlhenny, 1969). By 1978 production of fresh water from the ocean is expected to reach 4 million m³ per day, worth over $250 million annually (Gould, 1974).

CONCLUSIONS

Geologically the earth's surface may be categorized into two great provinces — the continents composed of thick and comparatively light crust; and the ocean basins underlain by thin, comparatively heavy basaltic crust. The margins of most of the continents are submerged beneath the sea, and the rocks and minerals beneath the ocean are thus of continental origin and character around the edges of the ocean basins. The continental margins — including the shelf, slope, and, in some places, an apron of sedimentary debris derived from the continent called the continental rise — contain a diverse assemblage of minerals broadly similar to those produced on land. Among these, oil and gas are by far the most important in present commercial production. Placer deposits of heavy minerals, including gold, diamonds, titanium-bearing minerals, tin, and others are or have been produced from surficial deposits near shore, as have sand, gravel, shell, and lime-bearing muds. There are also phosphorite deposits and manganese oxide nodules and crusts present on the bottom surface in many areas which have not yet been mined commercially. Salt and sulphur deposits have been produced from drill holes, and coal, iron ore, and other minerals have been mined underground by conventional mining methods from shaft sunk on the land adjacent or from islands — natural or artifical.

The deep-ocean basins may contain some fragments of continents separated from the main continental mass in the course of continued drift, and they may contain some of these same kind of minerals. In the main, however, the minerals of the deep-ocean basins are those associated with basic igneous rocks. The manganese oxide nodules — rich in copper, nickel, and cobalt — which are extensive in two broad belts that extend across the Pacific near the equator are by far the most important in their potential for production in the near future.

The oceans and the rocks beneath, which cover about three-quarters of the earth's surface, represent a largely unexplored but potentially enormous supply of a great variety of minerals. The development of a technology to explore and exploit these resources — and agreement on the complicated issue of jurisdiction over them — is of critical importance. The conclusions of the Secretary General at the 51st Session of the United Nations Economic and Social Council (United Nations, 1971, pp. 1—2) summarize admirably the current problems of developing marine mineral resources. "It is clear from the accumulation of data on marine mineral resources and from the development of techniques for their exploitation that the marine minerals industry will continue to grow, possibly at a greater rate than has been achieved during the last few years. It is also clear that the rate of advance will continue to depend on demand, competition from other sources, the cost and effectiveness of marine operations and the potential productivity of deposits . . . the future development of marine minerals will depend to a large extent on scientific research related to the sea floor and on the future resolution of jurisdictional problems associated with the use of the seabed and the ocean floor."

REFERENCES CITED AND SELECTED REFERENCES

Adye, A.M., Gassett, P.L., Geer, R.L., Martin, R.G., Rupp, L.A. and Silcox, W.H., 1974. Experts discuss subsea production systems, 2. *Ocean Indus.*, 9(9): 129—141.

Ames, L.L., Jr., 1959. The genesis of carbonate apatites. *Econ. Geol.*, 54: 829—840.

Archer, A.A., 1973. Progress and prospects of marine mining. *5th Offshore Technol. Conf.*, Houston, 1973, Preprints, 1: 313—322. (OTC Paper 1757).

Austin, C.F., 1967a. In the rock — a logical approach for undersea mining of resources. *Eng. Min. J.*, 168(8): 82—88.

Austin, C.F., 1967b. Undersea drilling and production sites for petroleum. *J. Pet. Technol.*, 19: 1309—1316.

Austin, C.F., 1967c. Rock site — a way into the sea. *Sea Frontiers*, 13: 342—352.

Backer, H. and Schoell, M., 1972. New deeps with brines and metalliferous sediments in the Red Sea. *Nature (London), Phys. Sci.*, 240: 153—158.

Barnes, B.B., 1970. Marine phosphorite deposit delineation techniques tested on the Coronado Bank, southern California. *2nd Offshore Technol. Conf.*, Houston, 1970. Preprints, 2: 315—350 (OTC Paper 1259).

Baturin, G.N., Merkulova, K.I. and Chalov, P.I., 1972. Radiometric evidence for recent formation of phosphatic nodules in marine shelf sediments. *Mar. Geol.*, 13: M37—M41.

Beck, R.H. and Lehner, P., 1974. Oceans, new frontier in exploration. *Am. Assoc. Pet. Geol. Bull.*, 58: 376—395.

Ben-Avraham, Z. and Emery, K.O., 1973. Structural framework of Sunda Shelf. *Am. Assoc. Pet. Geol. Bull.*, 57: 2323—2366.

Berggren, W.A. and Hollister, C.D., 1971. Biostratigraphy and history of circulation of North Atlantic (abstract). *Am. Assoc. Pet. Geol. Bull.*, 55: 331.

Bezrukov, P.L., 1973. Principal scientific results of the 54th cruise of the R/V *Vityaz* in the Indian and the Pacific Oceans (February—May 1973). *Oceanology*, 13: 761—766. Also in *Okeanologiia*, 13: 921—926 (in Russian).

Birks, J., 1974. Proven reserves of offshore oil are produced at different rates. *Offshore*, 34(7): 66—72.

Bischoff, J.L. and Manheim, F.T., 1969. Economic potential of the Red Sea heavy metal deposits. In: E.T. Degens and D.A. Ross (Editors), *Hot Brines and Recent Heavy-Metal Deposits in the Red Sea.* Springer, New York, N.Y., pp. 535—541.

Blatt, H., Middleton, G. and Murray, R., 1972. *Origin of Sedimentary Rocks.* Prentice-Hall, Englewood Cliffs, N.J., 634 pp.

Boström, K. and Peterson, M.N.A., 1966. Precipitates from hydrothermal exhalations on the East Pacific Rise. *Econ. Geol.*, 61: 1258—1265.

Boström, K. and Peterson, M.N.A., 1969. The origin of aluminum-poor ferromanganoan sediments in areas of high heat flow on the East Pacific Rise. *Mar. Geol.*, 7: 427—447.

Braithwaite, C.J.R., 1968. Diagenesis of phosphatic carbonate rocks on Remire, Amirantes, Indian Ocean. *J. Sediment. Petrol.*, 38: 1194—1212.

British Sulphur Corporation, 1971. *World Survey of Phosphate Deposits.* London, 3rd ed., 180 pp.

Brooks, H.K., 1961. The submarine spring off Crescent Beach, Florida. *Q.J. Fl. Acad. Sci.*, 24: 122—134.

Bryan, G.M., 1974. In-situ indications of gas hydrate. In: I.R. Kaplan (Editor), *Natural Gases in Marine Sediments (Marine Science 3).* Plenum, New York, N.Y., pp. 299—308.

Buckenham, M.H., Rogers, J. and Rouse, J.E., 1971. Assessment of Chatham Rise phosphorite. *Australas. Inst. Min. Metall., Proc.,*

Bushinskii, G.I., 1964. On shallow-water origin of phosphorite sediments. In: L.M.J.U. van Straaten (Editor), *Deltaic and Shallow Marine Deposits (Developments in Sedimentology 1).* Elsevier, Amsterdam, pp. 62—70.

Bushinskii, G.I., 1966. The origin of marine phosphorites. *Lithol. Miner. Res.,* 1966(3): 292—311. Also in *Litol. Polezn. Iskop.,* 1966(3): 23—48 (in Russian).

Bynum, D. and Lovie, P.M., 1975. A thousand rigs could be needed to meet world goals. *Offshore,* 35(1): 53—60.

Cayeux, L., 1939. Phosphates sénoniens du Bassin de Paris. In: *Les Phosphates de Chaux Sédimentaires de France, 1* Imprimerie Nationale, Paris, pp. 202—259.

CCOP/IOC, 1974. *Metallogenesis, Hydrocarbons and Tectonic Patterns in Eastern Asia.* Sponsored by Committee for Co-ordination of Joint Prospecting for Mineral Resources in Asian Offshore Areas (CCOP) and International Oceanographic Commission, UNESCO (IOC), U.N. Development Programme (CCOP), Bangkok, 158 pp.

Chapman, G.P. and Roder, A.R., 1969. Sea-dredges sands and gravels. *Quarry Managers J.,* pp. 251—263.

Claypool, G.E. and Kaplan, I.R., 1974. The origin and distribution of methane in marine sediments. In: I.R. Kaplan (Editor), *Natural Gases in Marine Sediments (Marine Science 3).* Plenum, New York, N.Y., pp. 99—139.

Claypool, G.E., Kaplan, I.R. and Presley, B.J., 1973. Generation of light hydrocarbon gases in deep-sea sediments (abstract). *Am. Assoc. Pet. Geol., Bull.,* 57: 773.

Cook, P.J., 1972. Sedimentological studies on the Stairway Sandstone of Central Australia. *Aust. Bur. Min. Res., Geol. Geophys., Bull.,* 95, 73 pp.

Cook, P.J., 1973. The estuarine environment and phosphate deposition (unpublished manuscript).

Cook, P.J., 1974. *Prospects for Finding Offshore Phosphate Deposits in the Southwest Pacific.* U.N. Economic and Social Commission for Asia and the Pacific. Committee for Co-ordination of Joint Prospecting for Mineral Resources in the South Pacific Offshore Area, Apia, Western Samoa. 1974, 3rd Sess., Rep. NR/CCOP/SOPAC (3/CR.TAG/10).

Corliss, J.B., Graf, J.L., Jr., Skinner, B.J. and Hutchinson, R.W., 1972. Rare earth data from iron- and manganese-rich sediments associated with sulfide ore bodies of the Troodos Massif, Cyprus (abstract). *Geol. Soc. Am., 1972 Annu. Meet., Abstracts with Programs,* pp. 476—477.

Corwin, G. and Berryhill, H.L., Jr., 1973. Interim revision and updating of world subsea mineral resources. U.S. Congress. Senate. Committee on Interior and Insular Affairs. Subcommittee on Minerals, Materials and Fuels, *Mineral Resources of the Deep Seabed.* Hearings, 93rd Congr. 1st Sess. on S. 1134. Government Printing Office, Washington, D.C., pp. 716—747.

Cronan, D.S., 1967. *Geochemistry of some manganese nodules and associated pelagic deposits.* Ph.D. Thesis, Imperial College, University of London.

Cronan, D.S., 1972. Regional geochemistry of ferromanganese nodules in the world ocean. In: D.R. Horn (Editor), *Conference on Ferromanganese Deposits on the Ocean Floor.* National Science Foundation, Office for the International Decade of Ocean Exploration, Washington, D.C., pp. 19—30.

Cronan, D.S. and Tooms, J.S., 1969. The geochemistry of manganese nodules and associated pelagic deposits from the Pacific and Indian Oceans. *Deep-Sea Res.,* 16: 355—359.

Crowell, J.C., 1973. Origin of Late Cenozoic basins in southern California (abstract). *Am. Assoc. Pet. Geol. Bull.,* 57: 774.

Cruickshank, M.J., 1962. *The exploration and exploitation of offshore mineral deposits.* M.S. Thesis, Colorado School of Mines, 185 pp.

Cruickshank, M.J., 1969. Mining and mineral recovery. In: C.W. Covey et al. (Editors), *Undersea Technology Handbook-Directory*, 1969. Compass, Arlington, Va., pp. A45–A54.

Cruickchank, M.J., 1973. Unconsolidated deposits. In: *Marine Mining. SME Mining Engineering Handbook, Sect. 20.* SME AIME, New York, N.Y., pp. 20–114.

Cruickshank, M.J., 1974. Mineral resources potential of continental margins. In: C.A. Burk and C.L. Drake (Editors), *The Geology of Continental Margins.* Springer, New York, N.Y., pp. 965–1000.

Cruickshank, M.J., 1976. *Technological and environmental considerations in the exploration and exploitation of marine minerals.* Ph.D. Dissertation, University of Wisconsin (in preparation).

D'Anglejan, B.F., 1965. *Marine phosphorite deposits of Baja California, Mexico; present environment and recent history.* Ph.D. Dissertation, Scripps Institute of Oceanography, 214 pp.

D'Anglejan, B.F., 1967. Origin of marine phosphorites off Baja California, Mexico. *Mar. Geol.*, 5: 15–44.

Degens, E.T. and Ross, D.A. (Editors), 1969. *Hot Brines and Recent Heavy-Metal Deposits in the Red Sea.* Springer, New York, N.Y., 600 pp.

Dewey, J.F. and Bird, J.M., 1970. Mountain belts and the new global tectonics. *J. Geophys. Res.*, 75: 2625–2647.

Dewey, J.F. and Horsfield, B., 1970. Plate tectonics, orogeny, and continental growth. *Nature*, 225: 521–525.

Dewey, J.F. and Burke, K., 1974. Hot spots and continental break-up: implications for collisional orogeny. *Geology*, 1: 57–60.

Dietz, R.S., 1969. Ocean floor in the decade ahead. *Mar. Technol. Soc. J.*, 3(5): 68–69.

Dietz, R.S., Emery, K.O. and Shepard, F.P., 1942. Phosphorite deposits on the sea floor off southern California. *Geol. Soc. Am. Bull.*, 53: 815–848.

Donn, W.L., Farrand, W.R. and Ewing, M., 1962. Pleistocene ice volumes and sea-level lowering. *J. Geol.*, 70: 206–214.

Drake, C.L., Ewing, J.I. and Stockard, H., 1968. The continental margin of the eastern United States. *Can. J. Earth Sci.*, 5: 993–1010.

Dunham, K.C., 1969. Practical geology and the natural environment of Man, II. Seas and oceans. *Q.J. Geol. Soc. London*, 124: 101–129.

Dunham, K.C. and Sheppard, J.S., 1969. Superficial and solid mineral deposits of the continental shelf around Britain. *Proc. 9th Commonw. Min. Metall. Congr.*, Great Britain, 1969, 2: 3–25.

Emery, K.O., 1960. *The Sea off Southern California; A Modern Habitat of Petroleum.* Wiley, New York, N.Y., 366 pp.

Emery, K.O., 1969. Continental rises and oil potential. *Oil Gas J.*, 67(19): 231–243.

Emery, K.O. and Noakes, L.C., 1968. Economic placer deposits of the continental shelf. *U.N., Comm. Coord. Joint Prospect. Miner. Res. Asian Offshore Areas, Tech. Bull.*, 1: 95–111.

Emery, K.O. and Uchupi, E., 1972. Western North Atlantic Ocean: topography, rocks, structure, water, life and sediments. *Am. Assoc. Pet. Geol.*, Mem. 17, 532 pp.

Ewing, J., Hollister, C., Hathaway, J., Paulus, F., Lancelot, Y., Habib, D., Poag, C.W., Luterbacher, H.P., Worstell, P. and Wilcoxon, J.A., 1970. Deep Sea Drilling Project: Leg 11. *Geotimes*, 15(7): 14–16.

Ewing, M. and Landisman, M., 1961. Shape and structure of ocean basins. In: M. Sears (Editor), *Oceanography. Am. Assoc. Adv. Sci., Publ.*, 67: 3–38.

Fairbridge, R.W., 1966a. Sahul Shelf. In: R.W. Fairbridge (Editor), *Encyclopedia of Oceanography.* Reinhold, New York, N.Y., pp. 755–758.

Fairbridge, R.W., 1966b. Trenches and related deep sea troughs. In: R.W. Fairbridge (Editor), *Encyclopedia of Oceanography.* Reinhold, New York, N.Y., pp. 929–939.

Forchhammer, G., 1865. On the composition of sea-water in the different parts of the ocean. *Philos. Trans. R. Soc. London*, 155: 203—262.

Fuchs, R.L. and Westphal, W.H., 1973. Energy shortage stimulates geothermal exploration. *World Oil*, 177(7): 37—41.

Fye, P.M., Maxwell, A.E., Emery, K.O. and Ketchum, B.H., 1968. Ocean science and marine resources. In: E.A. Gullion (Editor), *Uses of the Seas*. Prentice-Hall, Englewood Cliffs, N.J., pp. 17—68.

Gardner, F.J., 1973. 1973: the year of major changes in worldwide oil. *Oil Gas J.*, 71(53): 83—88.

Gass, I. G. and Smewing, J.D., 1973. Intrusion, extrusion and metamorphism at constructive margins: evidence from Troodos Massif, Cyprus. *Nature*, 242: 26—29.

Geer, R.L., 1973a. Offshore drilling and production technology — where do we stand and where are we heading. *Am. Pet. Inst., 3rd Annu. Meet.*, API Paper 362-C, pp. C1—C44.

Geer, R.L., 1973b. Deepwater drilling: R and D and applied technology. *Ocean Indus.*, pp. 129—130.

Geer, R.L., 1974. Subsea production system technology: where it is and where it's headed. *Ocean Indus.*, 9(10): 45—54.

Goldberg, E.D. and Arrhenius, G.O.S., 1958. Chemistry of Pacific pelagic sediments. *Geochim. Cosmochim. Acta*, 13: 153—212.

Goldberg, E.D. and Parker, R.H., 1960. Phosphatized wood from the Pacific sea floor. *Geol. Soc. Am. Bull.*, 71: 631—632.

Gould, H.R., 1974. Minerals from the sea. In: R.C. Vetter (Editor), *Oceanography; The Last Frontier*. Basic Books, New York, N.Y., pp. 137—152.

Guild, P.W., 1971. Metallogeny: a key to exploration. *Min. Eng.*, (N.Y.), 23(1): 69—72.

Guild, P.W., 1972a. Massive sulphides vs porphyry deposits in their global tectonic settings. *Joint Meet. MMIJ-AIME*, Tokyo, 1972, Min. Metall. Inst. Japan, Tokyo, GI-3, 12 pp.

Guild, P.W., 1972b. Metallogeny and the new global tectonics. *24th Int. Geol. Congr.*, Montreal, 1972, Sect. 4 — Mineral Deposits, pp. 17—24.

Haber, F., 1927. Das Gold im Meerwasser. *Z. Angew. Chem.*, 40: 303—316.

Halbouty, M.T., 1967. *Salt Domes: Gulf Region, United States and Mexico*. Gulf, Houston, Texas, 425 pp.

Hamilton, E.L., 1956. Sunken islands of the mid-Pacific mountains. *Geol. Soc. Am.*, Mem. 64, 97 pp.

Hedberg, H.D., 1967. Why explore the deep offshore? In: V.S. Cameron (Editor), *Exploration and Economics of the Petroleum Industry*, 5. Gulf, Houston, Texas, pp. 61—84.

Hedberg, H.D., 1970. Continental margins from viewpoint of the petroleum geologist. *Am. Assoc. Pet. Geol. Bull.*, 54: 3—43.

Heezen, B.C., 1962. The deep sea floor. In: S.K. Runcorn (Editor), *Continental Drift (International Geophysical Series 3)*. Academic Press, New York, N.Y. and London, pp. 235—288.

Heezen, B.C. and Tharp, M., 1965. Tectonic fabric of the Atlantic and Indian Oceans and continental drift. *Philos. Trans. R. Soc. London, Ser. A*, 258: 90—106

Heezen, B.C., Hollister, C.D. and Ruddiman, W.F., 1966. Shaping of the continental rise by deep geostrophic contour currents. *Science*, 152: 502—508.

Hess, H.D., 1971. *Marine Sand and Gravel Mining Industry of the United Kingdom*. National Oceanic and Atmospheric Administration, NOAA Tech. Rep. ERL 213-MMTC 1, 190 pp. (COM 71-50585).

Hilde, T.W.C., Wageman, J.M. and Hammond, W.T., 1969. The structure of Tosa Terrace and Nankai Trough off southeastern Japan. *Deep-Sea Res.*, 16: 67—75.

Horn, D.R. and Horn, B.M., 1973. Worldwide metal content of ferromanganese deposits on the ocean floor (abstract). *Trans. Am. Geophys. Union*, 54: 338.

Horn, D.R., Delach, M.N. and Horn, B.M., 1973a. *Metal Content of Ferromanganese De-*

posits on the Oceans. U.S. National Science Foundation, Office for the International Decade of Ocean Exploration, Washington, D.C., NSF GX 33616, Tech. Rep. 3, 51 pp.

Horn, D.R., Horn, B.M. and Delach, M.N., 1973b. *Ocean Manganese Nodule Metal Values and Mining Sites.* U.S. National Science Foundation, Office for the International Decade of Ocean Exploration, Washington, D.C., Tech. Rep. 4.

Horn, D.R., Horn, B.M. and Delach, M.N., 1973c. *Factors Which Control the Distribution of Ferromanganese Nodules and Proposed Research Vessel Track North Pacific — Phase II Ferromanganese Program.* U.S. National Science Foundation, Office for the International Decade of Ocean Exploration, Washington, D.C., Tech. Rep. 8.

Hudson, J.D. and Robertson, A.H.F., 1973. Sedimentation on Cretaceous ocean ridge, Troodos Massif, Cyprus (abstract). *Am. Assoc. Pet. Geol. Bull.,* 57: 785.

Hutchinson, G.E., 1950. Survey of existing knowledge of biogeochemistry, 3. The biogeochemistry of vertebrate excretion. *Am. Mus. Nat. Hist. Bull.,* 96, 554 pp.

Inderbitzen, A.L., Carsola, A.J. and Everhart, D.L., 1970. The submarine phosphorite deposits off southern California. *2nd Offshore Technol. Conf.,* Houston, 1970. Preprints, 2: 287—304 (OTC Paper 1257).

Inman, D.L. and Nordstrom, C.E., 1971. On the tectonic and morphologic classification of coasts. *J. Geol.,* 79: 1—21.

International Management and Engineering Group of Britain, 1972. *Study of Potential Benefits to British Industry from Offshore and Gas Developments.* H.M. Stationery Office, London, 136 pp.

James, H.L., 1968. Mineral resource potential of the deep oceans. In: E. Keiffer (Editor), *Mineral Resources of the World Ocean. R.I. Univ., Narragansett Mar. Lab., Occ. Publ.,* 4: 39—44.

Jankovic, S., 1972. The origin of base-metal mineralization on the Mid-Atlantic Ridge (based upon the pattern of Iceland). *24th Int. Geol. Congr.,* Montreal, 1972, Sect. 4 — Mineral Deposits, pp. 326—334.

JOIDES (Joint Oceanographic Institutions: Deep Earth Sampling Program), 1965. Ocean drilling on the continental margin. *Science,* 150: 709—716.

Jones, P.H., 1969. Hydrology of Neogene deposits in the northern Gulf of Mexico Basin. *La. Univ. Water Resour. Res. Inst. Bull.* GT-2, 105 pp.

Jones, P.H., 1970. Geothermal resources of the northern Gulf of Mexico Basin. In: *U.N. Symposium on Development and Utilization of Geothermal Resources, Pisa, 1970. Geothermics,* 2(1): 14—26 (special issue).

Kash, D.E. et al., 1973. *Energy under the Oceans; A Technology Assessment of Outer Continental Shelf Oil and Gas Operations.* University of Oklahoma, Norman, Okla., 378 pp.

Kaufman, R., 1974a. The selection and sizing of tracts comprising a manganese nodule ore body. *6th Offshore Technol. Conf.,* Houston, 1974. Preprints, 2: 283—298 (OTC Paper 2059).

Kaufman, R., 1974b. The development of ocean mining continues to grow. *Offshore,* 34(13): 111, 114.

Kazakov, A.V., 1937. The phosphorite facies and the genesis of phosphorites. In: B.M. Himmelfarb, A.V. Kazakov and I.M. Kurman (Editors), *Geological Investigations of Agricultural Ores. U.S.S.R. Trans. Sci. Inst., Fert. Insecto-Fungic.,* 142: 95—113.

Kerr, J.W., 1970. Importance of continental drift to petroleum exploration (abstract). *Am. Assoc. Pet. Geol. Bull.,* 54: 855.

Klemme, H.D., 1972. Heat influences size of oil giants. *Oil Gas J.,* 70(29): 136—144; 70(30): 76—78.

Koenig, J.B., 1967. The Salton-Mexacali geothermal province. *Calif. Div. Mines Geol., Min. Inform. Serv.,* 20: 75—81.

Kohout, F.A., 1966. Submarine springs. In: R.W. Fairbridge (Editor), *Encyclopedia of Oceanography.* Reinhold, New York, N.Y., pp. 878—883.

282

Kolodny, Y., 1969. Are marine phosphates forming today? *Nature*, 224: 1017—1019.
Le Pichon, X., Francheteau, J. and Bonnin, J., 1973. *Plate Tectonics (Developments in Geotectonics 6)*. Elsevier, Amsterdam, 302 pp.
Lipps, J.H., 1971. History of circulation in Pacific Ocean (abstract). *Am. Assoc. Pet. Geol. Bull.*, 55: 348—349.
Luyendyk, B.P., Forsyth, D. and Phillips, J.D., 1972. Experimental approach to the paleocirculation of the oceanic surface waters. *Geol. Soc. Am. Bull.*, 83: 2649—2664.
McIlhenny, W.F., 1969. Ocean and beaches, a ready source of raw chemical and mineral materials. *Offshore*, 29(5): 56—62, 198.
McIver, R.D., 1972. Evidence of migrating liquid hydrocarbons in Deep Sea Drilling Project Cores (abstract). *Am. Assoc. Pet. Geol. Bull.*, 56: 639.
McIver, R.D., 1974. Hydrocarbon gas (methane) in canned Deep Sea Drilling Project core samples. In: I.R. Kaplan (Editor), *Natural Gases in Marine Sediments (Marine Science 3)*. Plenum, New York, N.Y., pp. 63—69.
McKelvey, V.E., 1963. Successful new techniques in prospecting for phosphate deposits. Science Technology and Development; United States papers prepared for U.N. Conference on Application of Science and Technology for Benefit of Less Developed Areas, 2: 163—172. Also in: *Aust. Min.*, 58(3): 8—13, 1966.
McKelvey, V.E., 1974. International report on ocean developments: United States. *Ocean Indus.*, 9(4): 208—213.
McKelvey, V.E. and Chase, L., 1966. Selecting areas favorable for subsea prospecting. *Trans. 2nd Mar. Technol. Soc. Conf. Exhibit*, Washington, D.C., 1966, pp. 44—60.
McKelvey, V.E. and Wang, F.F.H., 1970. World subsea mineral resources; preliminary maps. *U.S. Geol. Surv., Misc. Geol. Invest. Maps*, I-632, 4 sheets and accompanying text. Reprinted and slightly revised.
McKelvey, V.E., Swanson, R.W. and Sheldon, R.P., 1953. The Permian phosphorite deposits of western United States. *19th Int. Geol. Congr.*, Algiers, 1952, Sect. 11, pp. 45—64.
McKelvey, V.E., Stoertz, G.E. and Vedder, J.G., 1969a. Subsea physiographic provinces and their mineral potential. *U.S. Geol. Surv., Circ.* 619: 1—10.
McKelvey, V.E., Wang, F.F.H., Schweinfurth, S.P. and Overstreet, W.C., 1969b. Potential mineral resources of the United States outer continental shelf. In: Nossaman, Waters, Scott, Krueger and Riordan (firm), *Study of Outer Continental Shelf Lands of U.S.*, IV, appendix 5-A. U.S. Department of the Commerce, NBS, Washington, D.C., 117 pp. (PB 188 717).
McManus, D.A., Burns, R.E., Weser, O., Vallier, T., Von der Borch, C., Olsson, R.K., Goll, R.M. and Milow, E.D., 1969. Deep Sea Drilling Project: Leg 5. *Geotimes* (Sept.), 14(7): 19—20.
Maloney, N.J., 1968. Geomorphology of continental margin of Venezuela, 1. Cariaco Basin, Cumana, Venezuela. *Univ. Oriente, Inst. Oceanogr. Bol.*, 5: 38—53.
Manderson, M.C., 1972. Commercial development of offshore marine phosphates. *4th Offshore Technol. Conf.*, Houston, 1972. Preprints, 2: 393—398 (OTC Paper 1658).
Manheim, F.T., 1965. Manganese-iron accumulations in the shallow marine environment. *R.I. Univ., Narragansett Mar. Lab., Occ. Publ.*, 3: 217—276.
Manheim, F.T., 1967. Evidence for submarine discharge of water on the Atlantic continental slope of the southern United States and suggestions for further search. *Trans. N.Y. Acad. Sci., Ser. II*, 29: 839—853.
Menard, H.W., 1964. *Marine Geology of the Pacific*. McGraw-Hill, New York, N.Y., 271 pp.
Menard, H.W., 1967. Transitional types of crust under small ocean basins. *J. Geophys. Res.*, 72: 3061—3073.
Menard, H.W. and Smith, S.M., 1966. Hypsometry of ocean basin provinces. *J. Geophys. Res.*, 71: 4305—4325.

Mero, J.L., 1967. Ocean mining — a potential major new industry. *Proc. 1st World Dredging Conf.*, New York, 1967, pp. 625—641.

Miller, A.R., Densmore, C.D., Degens, E.T., Hathaway, J.C., Manheim, F.T., McFarlin, P.F., Pocklington, R. and Jokela, A., 1966. Hot brines and recent iron deposits in deeps of the Red Sea. *Geochim. Cosmochim. Acta*, 30: 341—359.

Moore, T.C., Jr., 1972. DSDP: successes, failure, proposals. *Geotimes*, 17(7): 27—31.

Moores, E.M. and Vine, F.J., 1971. The Troodos Massif, Cyprus, and other ophiolites as oceanic crust; evaluation and implications. *Philos. Trans. R. Soc. London, Ser. A*, 268: 443—466.

Murray, J. and Renard, A.F., 1891. Report on deep-sea deposits based on the specimens collected during the voyage of H.M.S. *Challenger* in the years 1872 to 1876. In: Gt. Br. Challenger Office, *Report on the Scientific Results, III. Deep-Sea Deposits*. 525 pp.

National Petroleum Council, 1969. Committee on Petroleum Resources Under the Ocean Floor, *Ocean Petroleum Resources under the Ocean Floor*. Washington, D.C., 107 pp.

National Petroleum Council, 1974. Committee on Ocean Petroleum Resources, *Ocean Petroleum Resources; An Interim Report*. Washington, D.C., 39 pp.

National Petroleum Council, 1975. Committee on Ocean Petroleum Resources, *Ocean Petroleum Resources; Report of the National Petroleum Council*. Washington, D.C., 98 pp.

Nelson, C.H. and Hopkins, D.M., 1972. Sedimentary processes and distribution of particulate gold in the northern Bering Sea. *U.S. Geol. Surv., Prof. Paper* 689, 27 pp.

Nichiwaki, C., 1973. Metallogenic provinces in Japan. In: N.H. Fisher (Editor), *Metallogenic Provinces and Mineral Deposits in the Southwestern Pacific. Aust. Bur. Miner. Resour., Geol. Geophys. Bull.*, 141: 81—94.

Niino, H., 1959. Manganese nodules from shallow water off Japan. *1st Int. Oceanogr. Congr.*, New York, 1959. Preprints, pp. 646—648.

Offshore, 1974. Offshore production rises despite the slowup in the wells drilled. 34(7): 59—65.

Offshore Service, 1974. August, p. ii.

Oil & Gas Journal, 1974. World's deepest-water platforms due in Gulf. 72(38): 102.

Parker, R.J. and Simpson, E.S.W., 1972. South African Agulhas Bank phosphorites. *Phosphorus Potassium*, 58: 18, 27.

Pasho, D.W., 1972. *Character and Origin of Marine Phosphorites*. Final report for Office Mar. Geol., U.S. Geol. Surv., Los Angeles, Calif. Univ. South. Calif., Dept. Geol., Rept. USC Geol., 72-5, 144 pp.

Pautot, G., Auzende, J.M. and Le Pichon, X., 1970. Continuous deep sea salt layer along North Atlantic margins related to early phase of rifting. *Nature*, 227: 351—354.

Pegrum, R.M., Rees, G., Mounteney, S.M. and Naylor, D., 1973. New geophysical and geological data on Northwest European shelf and their bearing on sea-floor spreading and oil and gas exploration (abstract). *Am. Assoc. Pet. Geol. Bull.*, 57: 800.

Pepper, J.F., 1958. Potential mineral resources of the continental shelves of the Western Hemisphere. *U.S. Geol. Surv. Bull.*, 1067: 43—65.

Peterson, M.N.A., 1970. Deep Sea Drilling Project; prelude to a decade of ocean exploration — the scientist evaluation. *Mar. Technol. Soc. J.*, 4(5): 9—13.

Peterson, M.N.A. and Turrentine, R.E., 1972. D/V *Glomar Challenger* and the Deep Sea Drilling Project. *Aust. Pet. Explor. Assoc. J.*, 12(1): 85—93.

Pevear, D.R., 1966. The estuarine formation of United States Atlantic Coastal Plan phosphorite. *Econ. Geol.*, 61: 251—256.

Pilkey, O.H. and Luternauer, J.L., 1967. North Carolina's Frying Pan phosphate sands. *Geo-Mar. Technol.*, 3(1): 24—25.

Pirie, R.G. (Editor), 1973. *Oceanography. Contemporary Readings in Ocean Sciences*. Oxford University Press, New York-London-Toronto, 530 pp.

Pratt, R.M., 1971. Lithology of rocks dredged from the Blake Plateau. *Southeast. Geol.*, 13: 19—38.

Pulunggono, A., 1975. Hydrocarbon potential of sedimentary basins, Indonesia. *Am. Assoc. Pet. Geol., Mem.* 25, in press.

Ranneft, T.S.M., 1972. The effects of continental drift on the petroleum geology of western Indonesia. *Aust. Pet. Explor. Assoc. J.*, 12(2): 55—63.

Rezak, R., Bouma, A.H. and Jeffrey, L.M., 1969. Hydrocarbons cored from knolls in southwestern Gulf of Mexico. *Trans. Gulf Coast. Assoc. Geol. Soc.*, 19: 115—118.

Rigg, J.B., 1974. Minerals from the sea. *Ocean Indus.*, 9(4): 213—219.

Rogers, J., Summerhayes, C.P., Dingle, R.V., Birch, G.F., Bremner, J.M. and Simpson, E.S.W., 1972. Distribution of minerals on the seabed around South Africa and problems in their exploration and eventual exploitation. In: Engineering Commission on Oceanographic Research, *Symposium on Ocean's Challenge to South African Engineers.* CSIR, Stellenbosch, S71, 8 pp.

Rona, P.A., 1972. *Exploration Methods for the Continental Shelf: Geology, Geophysics, Geochemistry.* National oceanic and Atmospheric Administration, NOAA Tech. Rep. ERL 238 and Atlantic Oceanographic and Meteorological Laboratory AOML 8: 47 pp.

Ross, D.A., Whitmarsh, R.B., Ali, S., Boudreaux, J.E., Fleisher, R.L., Matter, A., Nigrini, C., Stoffers, P., Coleman, R., Girdler, R., Manheim, F. and Supko, P., 1972. Deep Sea Drilling Project in the Red Sea. *Geotimes*, 17(7): 24—26.

Rozanova, T.V. and Baturin, G.N., 1971. Hydrothermal ore shows on the floor of the Indian Ocean. *Oceanology*, 11: 874—879. Also in *Okeanologiia*, 11: 1057—1064 (in Russian).

Sander, N.J., 1970. What's ahead for the international offshore? *World Oil.* 171(1): 83—88.

Schlee, J., 1968. Sand and gravel on the continental shelf off the northeastern United States. *U.S. Geol. Surv., Circ.* 602, 9 pp.

Schlee, J. and Pratt, R.M., 1970. Atlantic Continental Shelf and Slope of the United States-Gravels of the northeastern part. *U.S. Geol. Surv., Prof. Paper* 529-H, pp. H1—H39.

Schneider, E.D., Fox, P.J., Hollister, C.D., Needham, H.D. and Heezen, B.C., 1967. Further evidence of contour currents in the western North Atlantic. *Earth Planet. Sci. Lett.*, 2: 351—359.

Scholl, D.W., Von Huene, R. and Ridlon, J.B., 1968. Spreading of the ocean floor: undeformed sediments in the Peru-Chile Trench. *Science*, 159: 869—871.

Scott, M.R., Scott, R.B., Nalwalk, A.J., Rona, P.A. and Butler, L.W., 1973. Hydrothermal manganese in the median valley of the Mid-Atlantic Ridge (abstract). *Trans. Am. Geophys. Union*, 54: 244.

Seshappa, G., 1953. Phosphate content of mudbanks along the Malabar coast. *Nature*, 171: 526—527.

Sheldon, R.P., 1964a. Paleolatitudinal and paleogeographic distribution of phosphorite. *U.S. Geol. Surv., Prof. Paper* 501-C, pp. C106—C113.

Sheldon, R.P., 1964b. Exploration for phosphorite in Turkey — a case history. *Econ. Geol.*, 59: 1159—1175.

Shepard, F.P., 1973. *Submarine Geology.* Harper and Row, New York, N.Y., 3rd ed., 517 pp.

Shigley, C.M., 1968. Sea water as a raw material. In: E. Keiffer (Editor), *Mineral Resources of the World Ocean.* R.I. Univ., Narragansett Mar. Lab., Occ. Publ., 4: 45—50.

Shor, G.G., Jr., 1962. Seismic refraction studies off the coast of Alaska: 1956—1957. *Seismol. Soc. Am. Bull.*, 52: 37—54.

Sillitoe, R.H., 1972. A plate tectonic model for the origin of porphyry copper deposits. *Econ. Geol.*, 67: 184—197.

Simpson, D.R., 1964. The nature of alkali carbonate apatites. *Am. Mineral.*, 49: 363—376.

Sorensen, P.E. and Mead, W.J., 1969. A new economic appraisal of marine phosphorite deposits off the California coast. *Trans. 5th Mar. Technol. Soc. Conf. Exhibit*, Miami, 1969, pp. 491—500.

Stefanon, A., 1972. Capture and exploitation of submarine springs. *2nd Int. Oceanol. Equipment and Services Exhibition and Conf.*, Brighton, March, 1972. Conference Papers, pp. 427—430.

Stoll, R.D., 1974. Effects of gas hydrates in sediments. In: I.R. Kaplan (Editor). *Natural Gases in Marine Sediments (Marine Science 3)*. Plenum, New York, N.Y., pp. 235—248.

Strauch, F., 1968. Determination of Cenozoic sea-temperatures using *Hiatella arctica* (Linné). *Palaeogeogr., Palaeoclimatol., Palaeoecol.*, 5: 213—233.

Summerhayes, C.P., 1969. Marine geology of the New Zealand Subantarctic sea floor. *N.Z. Oceanogr. Inst., Mem.* 50, 92 pp.

Summerhayes, C.P., 1970. *Phosphate deposits on the northwest African continental shelf and slope*. Ph.D. Thesis, University of London.

Sweet, W.E., 1973. Natural hydrocarbon seepage on the continental shelf of the Gulf of Mexico. *Int. Symp. on Relationships of Estuarine and Continental Shelf Sedimentation*, Bordeaux, France, July, 1973, in press.

Sweet, W.E., 1974. Tar balls in the sea: a new source concept. *6th Offshore Technol. Conf.*, Houston, 1974. Preprints, 1: 651—658 (OTC Paper 2002).

Thompson, T.L., 1974. Plate tectonics in petroleum exploration of continental margins. Paper presented at *4th Explor. Sem.*, sponsored by the Egyptian General Petroleum Corp., Cairo, Nov. 18—22, 1974.

Tokunaga, S., 1967. Outline of offshore coal fields in Japan (Part I). Japan. *U.S. Geol. Surv. Bull.*, 18(9): 65—72.

Tooms, J.S., 1970. Review of knowledge of metalliferous brines and related deposits. *Trans. Inst. Min. Metall., Sect. B.*, 79: B116—B126.

Tooms, J.S. and Summerhayes, C.P., 1968. Phosphate rocks from the Northwest African continental shelf. *Nature*, 218: 1241—1242.

Tooms, J.S., Summerhayes, C.P. and Cronan, D.S., 1969. Geochemistry of marine phosphate and manganese deposits. *Oceanogr. Mar. Biol. Annu. Rev.*, 7: 49—100.

Trofimuk, A.A., Cherskiy, N.V. and Tsarev, V.P., 1973. Accumulation of natural gases in zones of hydrate-formation in the hydrosphere. *Dokl. Akad. Sci. U.S.S.R., Earth Sci. Sect.*, 212: 87—90. Also in *Dokl. Akad. Nauk S.S.S.R.*, 212: 931—934 (in Russian).

United Nations, 1971. General Assembly, *Possible Impact of Possible Seabed Mineral Production in Areas Beyond National Jurisdiction on World Markets with Special Reference to the Problems of Developing Countries: A Preliminary Assessment*. Report of the Secretary-General. Rep. A/AC. 138/36, May 28.

United Nations, 1973. General Assembly. Committee on the Peaceful Uses of the Sea-bed and the Ocean Floor Beyond the Limits of National Jurisdiction, *Economic Significance, in Terms of Sea-Bed Mineral Resources of the Various Limits Proposed for National Jurisdictions*. Report of the Secretary-General. Rep. A/AC. 138/87 and Corr. 1. 39 pp. and two annexes.

U.S. Bureau of Mines, 1970. *Mineral Facts and Problems*. U.S. Government Printing Office, Washington, D.C., 1291 pp.

U.S. Bureau of Mines, 1973. *Minerals Year Book 1971*. U.S. Government Printing Office, Washington, D.C., Vol. 1, p. 233.

U.S. National Council on Marine Resources and Engineering Developments, 1970. *Marine Science Affairs: Selecting Priority Programs*. Annual Report of the President to the Congress on Marine Resources and Engineering Development.

Van Andel, Tj.H., 1965. Marine phosphorite prospects in Australia and marine geology in the BMR. *Aust. Bur. Min. Resour., Rec.* 1965/188, 12 pp.

Veeh, H.H., Burnett, W.C. and Soutar, A., 1973. Contemporary phosphorites on the continental margin of Peru. *Science*, 181: 844—845.

Vetter, R.C., 1974. *Oceanography: The Last Frontier. Voice of America Forum Series.* U.S. Information Agency, Washington, D.C., 430 pp.

Von der Borch, C.C., 1970. Phosphatic concretions and nodules from the upper continental slope, northern New South Wales. *Geol. Soc. Aust. J.*, 16: 755—759.

Von der Borch, C.C., Sclater, J.G., Gartner, S., Jr., Hekinian, R., Johnson, D.A., McGowran, B., Pimm, A.C., Thompson, R.W. and Veevers, J.J., 1972. Deep Sea Drilling Project: Leg 22. *Geotimes*, (June), 17(6): 15—17.

Walthier, T.N. and Schatz, C.E., 1969. Economic significance of minerals deposited in the Red Sea deeps. In: E.T. Degens and D.A. Ross (Editors), *Hot Brines and Recent Heavy-Metal Deposits in the Red Sea.* Springer, New York, N.Y., pp. 542—549.

Wang, F.F.H. and Cruickshank, M.J., 1969. Technologic gaps in exploration and exploitation of sub-sea mineral resources. *1st Offshore Technol. Conf.*, Houston, 1969. Preprints, 1: 285—298 (OTC Paper 1031).

Watters, W.A., 1968. Phosphorite and apatite occurrences and possible reserves in New Zealand and outlaying islands. *N.Z. Geol. Surv., Rep.* 33, 15 pp.

Weeks, L.G., 1971. Marine geology and petroleum resources. *Proc. 8th World Pet. Congr.*, Moscow, 1971, 2: 99—106.

Weeks, L.G., 1972. Critical interrelated geological, economic and political problems facing the geologist, the petroleum industry and the nation. In: V.S. Cameron (Editor), *Exploration and Economics of the Petroleum Industry, 10.* Gulf, Houston, Texas, pp. 221—246.

White, D.E., 1965. Geothermal energy. *U.S. Geol. Surv., Circ.* 519, 17 pp.

White, D.E., 1968. Environments of generation of some base-metal ore deposits. *Econ. Geol.*, 63: 301—335.

Wonfor, J.S., 1972. New global tectonics: its exploration application to the eastern sector of the Australian plate. *Aust. Pet. Explor. Assoc. J.*, 12(2): 34—45.

Zelenov, K.K., 1964. Iron and manganese in exhalations of the submarine Banu Wuhu Volcano (Indonesia). *Dokl. Acad. Sci. U.S.S.R.*, 155: 94—96. Also in *Dokl. Akad. Nauk S.S.S.R.*, 155: 1317—1320 (in Russian).

Zenkevich, N.L. and Skorniakova, N.S., 1961. Iron and manganese on the ocean bottom. *Priroda*, 1961(2): 47—50 (in Russian).

Chapter 9

CONSERVATION OF MINERALS AND OF THE ENVIRONMENT

DAVID B. BROOKS

> "All of the resources we now take from the earth come at an enormous price. This price is in effect a distress signal, a warning that thoughtless exploitation can in the end lead only to tragedy . . . What we face now is not deprivation, but the challenge of sharing. We need not do without, but we must be good stewards of what we have. To ensure nature's continued bounty, we are not asked to suffer, but we are asked to be reasonable. We are asked to adjust our demands to nature's limitations, to realize that unrestrained consumption by individuals and economic sweepstakes among nations are not acceptable ideals."
>
> Pierre Elliot Trudeau
> (Prime Minister of Canada, 1974)

INTRODUCTION

Man's impact on the environment began with his use of tools. Well before recorded history, the growth of population together with inappropriate agricultural practices had made a significant impact in certain areas, some of which have not recovered to this day. Salination of soil in the Near East from irrigation, denuding of hillsides in Greece from over-grazing and current water shortages in Afghanistan (also related to centuries of mining and smelting) are cases in point.

The lesson of history is clear: whether one takes the old view of the environment as merely something to be dealt with as expeditiously as possible in obtaining "productive" resources, or the somewhat more modern view in which consideration of environmental "constraints" is regarded as an essential part of the productive process, or the ideal view that all resources have value so that production is as much a constraint on the use of the environment as the other way around, it is clear that conservation of mineral resources and conservation of the environment are inextricably linked. Hence, both will be treated here in a single chapter, even though mineral production is the *raison d'être* for the book.

THE NATURE OF THE CONSERVATION PROBLEM

Tradeoffs: the key to decisions

In many ways, the problem of mineral conservation is the conservation problem *par excellence*. Modern civilization is inconceivable without mineral extraction and consumption. Yet these involve not only depletion and eventual exhaustion in an economic if not a physical sense, but also, and simultaneously, environmental change since the minerals are themselves part of the environment.

In some cases, depletion and environmental change are necessary and acceptable; in others, one or the other or both must be regarded as waste and degradation; and, in most cases, either improved recovery of minerals requires a greater impact on the environment, or protection of the environment dictates lower levels or rates of mineral recovery. The analytic questions are when and where and for whom each case applies, and these questions can only be answered in terms of what economists call "tradeoffs".

In order to evaluate tradeoffs one must compare what has been gained and what has been lost because of the production or use of minerals. Such comparisons cannot be made in physical units alone but must also incorporate economic units, and in many cases other social units as well. Poor conservation practices are only partly measured by the tonnes or kilojoules wasted. At least as important are the potential alternative uses of those tonnes and kilojoules and the environmental impacts of their use or non-use. However, such issues involve values, which are seldom absolute, and this is why mineral-environmental issues must be treated in terms of social science rather than of engineering. As stated by Kneese (1966, p. 710):

"Values are preeminent in the planning process, and there must be some means of thinking systematically about them and quantifying if possible. Here the price and allocation theory of economics and its normative or prescriptive underpinning, welfare economics, come into the picture as planning tools. It is the preoccupation of economics with the problem of social welfare, and the relation of resource use and investment decisions to it, that make it a social science and capable of giving special and important insights and tools to resources engineering."

The goal here is not to present either the analytic framework for dealing with tradeoffs involving natural resources or the information needed to evaluate them in specific cases. (For full development of this framework, see Herfindahl and Kneese, 1974.) It is simply to arrange the often confusing array of mineral-environmental issues into a systematic set of questions and to identify some of the considerations that are needed to respond to those questions.

As is common practice, minerals will be referred to as non-renewable resources, and air, water and soil as renewable resources, even though this terminology can be misleading. Similarly, "environment" can be taken to include all natural resources associated with a commercially valuable mineral, including the space it occupies and gangue minerals. The term "mine" will refer to all primary mineral extraction operations including deep mines, pits, and even oil and gas wells. For the most part, the principles are transferable from one type of operation to another. The chapter focuses on the exploration and mining stages of production; reference to mineral processing and mineral manufacturing is made only at a few points.

The major issues

What, then, are the issues involving mineral and environmental conservation in a society that is committed to providing a high level of material well-being for its members? The most fundamental question is:

(1) How much of any non-renewable resource should be produced and consumed in any time period?

Then, assuming that some non-renewable resources will be produced and consumed, two further questions follow:

(2) How can best use be made of whatever non-renewable resources are produced in that time period?

(3) How can protection best be accorded to the renewable resources associated with the non-renewables in nature?

Each of the secondary questions can be further subdivided. The need for efficient use of the non-renewables involves: (a) higher recovery ratios for non-renewables at the mine and at each subsequent stage of processing, (b) less non-renewable inputs per unit of final product (lower heat losses, etc.), and (c) improved recycling of final products after use.

The need for protection of renewable resources involves: (a) decisions about specific sites for production in view of alternative (and often conflicting) land uses, (b) controls over the processes used for production, and (c) reclamation of the site after production is complete.

These questions are the basis for the organization of this chapter. The first major section treats the three questions at the top of the list, while the next two treat, respectively, the specific questions about conservation of minerals and of the environment. This organization reflects the existence of a major difference between the first three issues and the last six: the former are socio-political issues that must be dealt with by broad decisions at the regional, national or even international level, while the latter involve engineering economics that has to be applied at individual mineral operations.

MAN AND RESOURCES: THE SOCIO-POLITICAL ISSUES

The aggregate rate of growth: issue 1

At heart, the first question asks about how materially rich we wish to be and how we wish to live, and these are matters of the deepest societal values. For nearly 200 years few people in the West, apart from a few ascetics and philosophers (notably John Stuart Mill), had even questioned whether we ought to grow. The early conservation movements sought to eliminate the waste of natural resources to be used in production (that is, of such inputs as timber, minerals, water). The conservation movement of the 1960s shifted attention to the environmental resources that were also being consumed in the process, in effect as other inputs to production. It was only in the 1970s that output itself, the *result* of the production process, came to be questioned. Nevertheless, the tremendous attention received by books on the limits to growth (Meadows et al., 1972), on intermediate-scale technology (Schumacher, 1973) and on new life styles (Roszak, 1968), indicates that this issue has become credible to many people within just a few years. However, there is little recognition of how it differs from a simple fight against pollution.

The question of the rate of consumption is then the most fundamental issue of all involving minerals, for its resolution requires resolution of the conflicts between standard of living and quality of life. The answer must vary among nations. Until levels of per capita income well above subsistence are reached, higher standards of living will, and should, take precedence. Indeed, at this stage of development, it is reasonable to think that increases in the standard of living and in the quality of life are reasonably parallel. However, as development proceeds, the assumptions on which further growth is based become weaker. It cannot be denied that development continues to depend upon minerals, or that over time the benefits of mineral extraction have hugely outweighed the adverse effects. Both statements are true but miss the point, which is whether *additional* units of mineral production and consumption will be as beneficial — on a net basis — as past ones have been. It is hard to deny that much of the ever higher output of minerals goes to produce taller buildings, bigger automobiles and higher incomes for people whose buildings are already tall, whose automobiles are already big and whose income is already high.

Higher prices for energy minerals and other basic materials will force shifts in consumption patterns, and there is some evidence of a slowing down in the rate of growth of resource consumption in industrial nations (Malenbaum et al., 1973). However, active efforts to stop population growth and to moderate further growth in incomes could greatly increase the impact of these developments.

The relationship between the two forces was recognized by the 1974

World Population Conference which adopted, as part of the World Population Plan of Action, a specific principle urging that attention be directed to justice in the distribution of natural resources and that wasteful usage be minimized. An official report preparatory to the Conference had put the case more strongly and less diplomatically: "Long-term and lasting solutions to the environment-resources-population-development complex in rich countries will require a thorough redefinition of societal goals. In particular wasteful uses of resources and uncontrolled growth of material consumption should be drastically curtailed" (United Nations, 1974, p. 10).

Of course, both population growth and economic growth are responsible for mineral and environmental consumption. However, while most observers recognize the need to bring population growth to a halt, the intellectual debate about whether economic growth either must or should come to a halt has never been sharper (compare Beckerman, 1974, with Heilbroner, 1974). While economic growth provides the means to cope with poverty, pollution and other social problems (United Nations, 1974, p. 7), even if that growth is directed to such problems (which is by no means certain), the very technology that supports the growth is, in other ways, a part of the problem (Commoner, 1971, pp. 140—177).

It is uncertain whether the sudden attention in Europe and North America to various physical or social limits to growth will become widely enough shared to reduce the enormous drain on material resources for which richer nations are largely responsible. Problems of income distribution, political and military power, and even racism continue to bedevil most attempts to move away from economic sweepstakes measured in gross national product. Few countries have as yet shown much interest in experimenting with ways other than traditional industrialism to increase real incomes and promote a high qualify of life. Still, for the first time since economic progress became a political reality, alternatives are being seriously studied, and even achieving some political influence (Trudeau, 1974).

The non-renewable resources: issue 2

If one uses words literally, most natural resources are in fact non-renewable. The 150 species of birds and mammals that have become extinct during man's brief tenure as master of the earth are examples of "renewable resources" that have been exhausted. Given appropriate circumstances, man can equally make fish, forests and soil non-renewable resources. Open space, wilderness and special-purpose sites, as for hydro dams, can also be considered depletable in that they gradually get used up or devoted to other uses. Moreover, given the possibilities for creating dynamic disequilibria in the atmosphere by breaking down the ozone layer (Hammond and Maugh, 1974), little apart from the sunlight reaching the upper atmosphere — not even that reaching the earth — would appear to be absolutely renewable. Thus, it is hardly surprising that theoretical economists have generally concluded that, so far as conservation is concerned (or, more accurately, investment in conservation), there was no distinction to be made between renewable and non-renewable resources (Scott, 1955).

In common use, however, the term non-renewable resources can conveniently be restricted to those resources for which exhaustion in some sense is inherent in production and use. Only minerals — both fuel and non-fuel — readily fit this more limited definition because at any moment of geological time the quantities of elements in the crust, as well as the numbers of their deposits at each size and grade, are absolutely fixed. Exploration and geochemical research may alter our knowledge, but they do not make the quantities and numbers any less finite.

Looking now at the definition in more detail, it is possible to question whether in a modern world minerals are in fact so absolutely non-renewable. Exploration strategies and technologic advances have been such that exhaustion, while real enough at individual sites, has not been a very important problem on a world-wide basis. The most obvious economic effect of the advent of exhaustion should be a steady increase in cost, and for very few non-renewable resources can such a long-term increase be demonstrated (Herfindahl, 1961; Barnett and Morse, 1963). (The exceptions among minerals are major non-ferrous metals, for which costs per unit appear to have risen slightly since the 1930s.) What has happened is that technologic advances together with new discoveries have in most cases more than made up for depletion so that we can now mine lower and lower grades of material at roughly the same real cost (that is, the same inputs of goods and services per units of output) once required for much higher grades (Barnett and Morse, 1963; Newcomb, 1967; Dawson, 1971).

Mineral prices have risen, dramatically in some cases, since 1970. However, prices rise as a result of many factors unrelated to real costs of production in the above sense. Producer cartels, import restrictions, limitations on foreign investments and nationalization are among the non-real cost influences that play a major role in mineral pricing. Important as price increases may be to producing firms, and however justified they may be in the eyes of individual nations, they cannot be taken as evidence of declining primary resource availability.

The explanation for the apparent dilemma between a resource that is absolutely fixed in a geological sense and yet that appears to be renewable in an economic sense lies in the nature of mineral distribution in the earth. With some exceptions (notably oil and gas, which are discretely different from the rock masses that contain them), the tendency is for more and more material to be available as mineable grade is lowered (Brooks, 1973). As man has learned to find lower-grade deposits and to extract minerals from them, supply availability has increased enormously. Geological extensions of supply through discovery and technological extension of supply through the ability to work lower-quality and non-conventional sources are powerful forces in a world in which knowledge is power. Technology may well prove to be the more important force in the future because, in contrast to the discovery process which adds deposits one by one, a new technology can open up deposits across the world. Moreover, to the extent that technology permits more common deposits to be worked, it offers the potential of offsetting the real costs of exploration, which do appear to be rising, though not by as much as is sometimes asserted (Cranstone and Martin, 1973).

The life of non-mineral non-renewables such as open space and wilderness cannot be so easily extended by technology. While in some cases there are ways to ameliorate the conflicts, as by angle drilling for petroleum to avoid additional land occupation, in many cases there are not, and true physical exhaustion of one resource is inherent in the decision to exploit another. Only recently have economic techniques been developed to provide guidelines for evaluating the tradeoffs implicit in such decisions (Krutilla et al., 1972).

The question of whether the process described above can continue into the future is hotly debated (see the reviews of various opinions in Brooks, 1973; U.S. National Academy of Sciences, 1973; Heilbroner, 1974; and Govett and Govett, 1976). In terms of technological ability to obtain mineral resources *with no regard for any of the side effects*, the answer is probably: "Yes, provided enough energy is available" (and this too is largely a matter of mineral supply). However, very quickly such "side effects" as pollution, land-use changes, radioactive wastes, sabotage dangers, social impacts and international capital flows rise to levels of such importance that technological ability appears almost secondary. As stated elsewhere: ". . . while it is naive to jump from the premise that physical resources are physically finite to the conclusion that this limits their economic availability, it is equally naive to jump from the premise that mineral resources are economically infinite to the conclusion that they could be produced in enormous volume without major social or political problems." (Brooks and Andrews, 1974, p. 13).

Thus, mineral conservation is just as important (and maybe even more important) today when technology is capable of expanding supply as it was in the past when mineral supply and mineral demand were both much lower. Now, however, concern has to be focused on the problems associated with expanding mineral supply rather than on a naive fear of "running out".

The associated renewable resources: issue 3

If mineral exhaustion per se has not, so far, turned out to be a pressing conservation issue, what has become so is the environmental damage resulting from mineral extraction. Paradoxically, in his drive to find and consume non-renewable resources, man has made his most obvious impact on the renewable ones.

The catalogue of ill-effects on air, water and land stemming from mineral extractions is long and well documented. It includes sewage, sedimentation, acid drainage and heavy metals in watercourses; acid fumes, lead fumes and other noxious or odorous gases in air; outcrop and mine fires; dust from trucks, quarries and tailings ponds; landslides from the outslopes of strip mines and scars from other forms of overburden disposal; subsidence; and, perhaps saddest of all, decrepit and decaying mining towns. Table 9-I provides a catalogue of environmental impacts at primary stages of production in Canada; similar catalogues have been made for other stages of production and for other countries where different commodities are mined.

Many of the mining-environmental impacts extend for tens or even hundreds of miles from the mine site itself, which explains why environmentalists become so exasperated at the oft-repeated statistic about the insignificant fraction of the land surface that is occupied by mining, as if this were a measure of its environmental impact. Also, some of the effects are subtle or slow to appear. Relatively small amounts of copper and zinc can have major impacts on fish populations even if the fish are able to live successfully at these concentrations (Sprague et al., 1965).

Many mining-environmental effects are intangible, that is unmeasurable, either because there is no good measuring rod (for rural scenery, for example) or because the mine is so isolated that it is difficult to define the injured party. Studies have begun to bring to light ways to make measurements in such cases (Krutilla et al., 1972; Coomber and Biswas, 1974). Ideally, the methods are based on some understanding of the "damage function", the relationship among environmental damages (however defined), rates of mineral production, and, where appropriate, estimates of climatic conditions. Direct measurement of environmental damage is often possible by calculating changes in farm revenues, road repairs, water treatment costs and the like. If direct measurements are unavailable, substitutes such as the added distance people will drive to find unpolluted water or the costs of creating an equivalent environment elsewhere can sometimes be used. Nevertheless, environmental damages are difficult to quantify, particularly in the isolated situations that are typical of mining, so the determination of mining-environmental standards has become largely a political judgement about what environmental conditions are desired by the community at large.

In economic terms the key characteristic of most environmental problems resulting from production activities is that their effects are "external" to the producing firm. Unlike accounting costs, such as fuel and labour, there are costs that are paid not by the firm causing them but by some other economic unit — another firm, an individual, the public in general. The classic example of external costs involves wastes dumped into a river which raise treatment costs for those who live downstream of that point. The cost per individual may be small because total cost is spread among many. Moreover, some of the external costs are intangible. Both qualifications, however, are irrelevant to the general issue, which is: when external costs are present, the market

TABLE 9-I

Environmental impacts of primary mineral production in Canada (from Mineral Development Sector Information, Canada Department of Energy, Mines and Resources)

Area of impact	Stage of production			
	exploration	mining	milling, benefi-ciation and storage	transportation
Humans				
Problem		*Underground*: wet, dust, noise, gases *Open pit*: dust, noise, exhaust fumes	Dust, noise *Non-ferrous*: noxious fumes, handling problem with some toxic reagents *Asbestos*: dust, fibre	Noise, dust, exhaust fumes from heavy vehicles, dust from conveyor belts
Effect		Contributes to labour turnover. Possible respiratory disease, particularly for asbestos, fluorspar, gold. Stress and other physical ailments	Contributes to labour turnover. Possible respiratory and carcinogenic diseases, particularly for asbestos and other dry-milled minerals	*To consumer*: heavy vehicles are irritants and hazards in populated areas (e.g., aggregate and construction materials)
Land				
Problem	Trenching, drilling, access routes, line cutting, abandoned equipment, permafrost damage	Open pits and quarries, subsidence. Needless clearing of overburden and vegetation. Mine water contamination. Roads and access routes. Waste piles. Impact of poorly designed townsites	Tailings dams and ponds, contamination from seepage and spills. Unsightly stockpiles (e.g., sulphur). Waste dumps (e.g., salt tailings) from potash production. Alumina production (red mud)	Wide roads for heavy vehicles (and associated borrow pits). Dust. Unnecessary clearing. Spills from derailments and road accidents *Arctic*: problem of road, port, airstrip construction

TABLE 9-I (continued)

Area of impact	Stage of production			
	exploration	mining	milling, benefi-ciation and storage	transportation
Effect	Erosion, scarring (minor). Vagetation damage. Permafrost damage in Arctic. Disruption of natural drainage	Limited sequential land use possibilities. Aesthetics. Permafrost damage in Arctic	Waste lands created by tailings areas. Land contamination by leaching and run-off from tailings and waste dumps. Permafrost damage in Arctic	Opens wilderness areas to possible degradation. Heavy traffic destructive to highways. Permafrost damage in Arctic
Water Problem	Suspended solids from erosion. Brine from drilling into aquifer (salt exploration)	Suspended solids from mine water, heavy metals, pH from metal mines. Alteration of water table, degradation of water quality	Suspended solids, heavy metals, pH, toxicity from direct discharge and overflow tailings systems. Spills and seepage from tailings potash and salt mines, salt contamination of water bodies. Large water consumption	*Water transport*: suspended-solids colouration (iron ore) at shipping terminals. Spills from derailments and road accidents. Possible problem from slurry pipelines
Effect	Contamination of streams and groundwater	Harmful to aquatic life	Harmful to aquatic life. Produces ecological imbalance	Possible harm to aquatic life
Air Problem		Wind-blown dust. Gases from fuel-burning equipment. Dust from blasting and drilling. Asbestos dust and fibre	Dust, airborne particles (asbestos fibres), gases, odours, evaporation from tailings ponds. Iron ore pelletization (SO_2 from process drying fuel)	Airborne particles from loading material being transported, and road surface

TABLE 9-I (continued)

Area of impact	Stage of production			
	exploration	mining	milling benefi-cation and storage	transportation
			Concentrate drying (SO_2, heavy metals). Thermal power generation (hydrocarbons, SO_2, NO_x)	
Effect		Minor	Possible respiratory effects. Rain washing of particles affects vegetation and soil. Increased costs from corrosion, dirt, etc. If close to urban areas, health effects from asbestos fibres	Minor

system, on which we rely for most of our economic decisions, is no longer operating as it should. In effect the system has lost its claim to efficiency and equity because costs are not being minimized: those who have the power to reduce environmental costs have no incentive to do so, while those who have the incentive to reduce them have no power to do so. Generally, government has to intervene in order to place environmental costs back on the appropriate account books, or, in common terms, to make the polluter pay. The economics of this situation are described in most modern economics texts. (For an excellent and simple exposition see Herfindahl and Kneese, 1965, or U.S. National Academy of Sciences, 1973.)

External costs are the main, but not the only, source of mining-environmental problems. Mining may also conflict with native peoples' rights, create unstable economic and social conditions during the construction phase (and typically beyond) and cause extensive land-use changes, as when farm land is dedicated to surface mining or when mining moves into a wilderness area. Such effects differ not only in their origins from external costs but also in the fact that, whereas technology can reduce the volume or the toxicity of effluents, it offers relatively little hope for mitigating these other effects. Again, government intervention of one sort or another is required to establish equitable and efficient ways of reconciling the differing interests and differing values about uses of the environment that are partly incidental to, partly inherent in, mining activities.

Occupational health and safety

Occupational health and safety problems represent yet another external cost of mining operations. Mining remains one of the most dangerous occupations: not only are accident rates high (even in surface mines), but there is a set of industrial diseases — generally forms of pneumoconiosis or cancer — that are common among miners and other workers in certain sectors of the minerals industry.

From the economic point of view, the in-plant environment can be treated by an analysis similar to that for the ex-plant environment (Brooks and Berry, 1972; U.S. National Academy of Science, 1973). Accidents and disease impose a variety of costs on the individual and on society, which are not well reflected in the costs of production. Some of these external costs, such as hospitalization, are obvious; others, such as income losses and pain, are equally valid but more difficult to calculate. Together they are high enough to permit the assertion that occupational health and safety should never be neglected when cataloguing mineral-environmental issues at any level of aggregation.

EFFICIENT USE OF NON-RENEWABLE RESOURCES

Recovery ratios

At every stage of production from mining through fabricating, losses of material are inevitable, so that the amount of material appearing in the final product is less than the amount that existed in the ground. The recovery ratio in mining is described by the difference between "in-situ" or "geological" reserves and recoverable reserves. The former refer to minerals in place above a certain cut-off grade, whereas the latter take into account the effects of dilution and of materials that must be left behind in the mining process. The former are physically definable; the latter depend on economic and geologic conditions and on mining method. The difference between the two is relatively small for most metals and natural gas but of great importance for bedded minerals and petroleum.

Mining method is generally the most important determinant of recovery ratio at a given deposit (J. Lajzerowicz, personal communication, 1974). Surface mining permits more flexibility in production including the ability to obtain 100% extraction within the pit limits. In underground mining the ground support system limits recovery, especially for those deposits which are produced by selective mining methods such as room and pillar. There has been considerable improvement in recovery in many countries during recent decades simply because of the increase in surface mining. Gains have also been made underground with the shift from selective to mass mining techniques such as block caving and to new techniques, such as long-wall mining (mining techniques are discussed in Chapter 13). Table 9-II shows the mining methods used in Canadian mines in 1971 along with the corresponding general recovery level. Better planning of mining districts can also improve recovery. A new mining district should be fully explored before any part of it is developed so that regional solutions to recovery problems can be developed (Lajzerowicz and Mackenzie, 1971).

Recovery ratios in concentration are highly variable, but are generally high

TABLE 9-II

Percentage of ore broken and recovery ratios with various mining methods in Canada, 1971 (from Canada National Mineral Inventory Data Bank)

Method	Groups of commodities			Recovery percentage
	metal mines	non-metal mines *	coal mines	
Underground mining				
Shrinkage	4.30	n.a.	—	75—85 at depths greater than 750 m
Cut and fill; undercut and fill	11.10	0.20	5.27 **	near 100
Open stoping	8.84	0.02	—	60—80 at depths less than 600 m
Room and pillar	1.57	31.84	19.73	50—70 ***
Sublevel caving	5.84	—	—	90
Other caving	1.12	0.06	—	near 100
Other methods	0.52	—	—	—
Surface mining				
All methods	66.71	65.08	75.0	85—100

* Totals may not balance owing to individual rounding and to absence of data (noted n.a.) which cannot be published because of confidentiality provisions.
** Long walls.
*** In evaporite mines 60—90%; in Saskatchewan potash mines 30%.

300

TABLE 9-III

Recovery percentages at post-mining stages of production in Canada, 1971 (from Canada National Mineral Inventory Data Bank)

Commodity	Concentration stage (%)	Smelting/refining stages (%)	
		hydrometallurgical	pyrometallurgical
Copper from:			
copper concentrate	60—98	≥95	97
zinc concentrate		≥95	process not available
lead concentrate		process not available	not recovered
Zinc from:			
zinc concentrate	40—95	≥93	90—95
lead concentrate		process not available	>90
copper concentrate		>95	not recovered
Lead from:			
lead concentrate	70—98	process not available	92—95
zinc concentrate		process not available	>90
copper concentrate		process not available	not recovered
Nickel from:			
nickel concentrate	70—86	>95	>97
copper concentrate		>95	>95

in modern plants. Much the same can be said for recovery ratios at the metallurgical stages. Examples of each are listed in Table 9-III.

Non-renewable inputs to production

Conservation of non-renewable resources can be practiced during the production process itself. Two major opportunities involve the use of energy and the design of final products.

Recent increases in the cost of energy have brought concern for energy conservation to the fore. In mineral processing and manufacturing different technologies have differing rates of energy use. Within thermodynamic limits, the energy efficiency of a process depends both on the engineering of design and on the skill of operation. Considerable information is available on energy conservation in industry from both a theoretical and a practical point of view (Berg, 1974; Berry et al., 1974), and estimates of possible energy savings through better application of known construction technologies plus use of more efficient transportation modes (generally rail rather than road-based systems) run from 10 to 40% of current consumption. The potential with new energy-conserving technology is still higher.

Energy is not just a convenient symbol for all forms of resource materials conservation; to a considerable extent, data on energy consumption incorporate the use of other resources because it takes energy to produce materials. Detailed calculations can show the full energy burden of increased consumption of goods and services (Herendeen, 1974). Needless to say, tradeoffs are always present. It is *not* efficient to reduce the use of energy by cutting the additives in steel-making below a certain point because this will only increase the volume later lost to scrap; it *is* efficient to produce vehicles from energy-intensive light metals because of later energy savings in their operation. In most cases additions of energy or materials provide better results for a time but eventually yield smaller gains for each additional input. The appropriate tradeoff depends on specific technologic and economic conditions. For example, at each energy price the use of energy to increase speed will first improve economic efficiency (reduce costs per kilometre), next pass through an optimum (least cost) point, then become rapidly inefficient as speeds are increased further; however, the optimum speed for each engine will be reduced as the price of energy goes up.

More efficient use of mineral resources can also be accomplished through changes in product design. Smaller cars, for example, use considerably less metal than larger models and yet provide a similar service (Table 9-IV). Had one million more autos been of compact size, world reserves of non-renewable resources would now be higher by 12,000 tonnes of aluminium, 7,000 tonnes of copper, 530,000 tonnes of iron, etc.

Increases in product life are another means of conservation. A specific

TABLE 9-IV

Quantities of materials in U.S. standard and economy size automobiles (from Yasnowsky and Colby, 1974; recalculated to metric basis)

Material	Percentage of total weight	Kilogrammes per standard size car (1732 kg)	Kilogrammes per economy size car (1019 kg)
Aluminium	1.9	33	19
Copper	1.1	20	12
Iron and steel	74.8	1,296	762
Lead	0.7	12	7
Nickel	0.1	2	1
Zinc	2.0	35	20
Glass	2.7	46	27
Rubber and plastics	12.7	219	129
Miscellaneous	4.0	70	41
Total	100.0	1,732	1,019

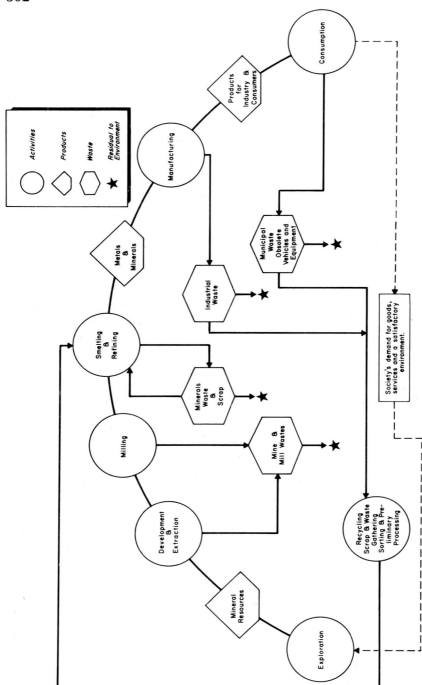

Fig. 9-1. The flow of materials through the economic system.

example of this approach is the returnable bottle, which depending upon size and type, is reused 12—25 times before being lost. The materials saving is thus enormous, and, even after netting for collection, washing and sterilizing, energy efficiency more than quadruples. Some current marketing practices, such as planned obsolence and advertising to promote unnecessary speed and power, severely obstruct this aspect of resource conservation. Because such anti-conservation practices are solidly rooted in modern corporate marketing, changes have been harder to effect than might seem likely. Nevertheless, the potential gains in material and energy savings are very large, as is now being recognized by consumer movements in many western nations.

Recycling

The flow of materials through the economic system from exploration to final demand can be shown graphically (Fig. 9-1). Note that there are "leaks" in the material flow system, and that materials which are unrecovered, lost or discarded become some form of waste. In Fig. 9-1, recycling appears as a redirection of such material so that it becomes another input to production. In this sense, "recycling" is a general term which includes *both* recycling of raw materials *and*, what is even better from a conservation point of view, re-use of the materials themselves. It neglects the best approach of all, namely, reduction of the volume of waste by eliminating unnecessary products and packaging and by increasing the longevity and durability of products (see above).

Almost all materials are capable of being recycled from a technologic point of view; the main blocks are generally economic — recycling is by no means inexpensive — and institutional. This section focuses on secondary metals, which, in contrast to many other recycled materials, are as useful to consumers as primary ones.

Advantages of recycling

The advantages of recycling are several. For one thing, recycling reduces the drain on resources caused by increasing consumption. (Even in a stable economy with complete recycling, some new inputs would be necessary to replace materials that are lost to dissipative uses, as with the cadmium and titanium consumed in paint, to erosion of scraping and grinding equipment, or to corrosion.) In addition, recycling reduces the need for land for waste disposal, which in urbanized areas is increasing in cost and decreasing in availability. In most cases recycling also reduces environmental impact (compared with primary production of equal quantities of metal) although some processes create their own environmental problems (detinning for example).

Secondary metals are already in the reduced form and generally require only re-refining to be equivalent to primary metals. The resulting energy saving (apart from collection costs) could theoretically be as high as 98%, but, be-

TABLE 9-V

Environmental impact comparison for 909 tonnes (1,000 tons) of steel product produced from primary vs. secondary materials (from U.S. Environmental Protection Agency, 1973; recalculated to metric basis)

Environmental effect	100% primary materials used	100% waste used*	Change from increased recycling (%)*
Primary materials	2,071 tonnes	227 tonnes	−90
Water use	75.4 × 10^6 litres	45.0 × 10^6 litres	−40
Energy consumption	24,626 × 10^5 kilojoules	6,370 × 10^5 kilojoules	−74
Air pollution	110 tonnes	15.5 tonnes	−86
Water pollution	61.4 tonnes	15.0 tonnes	−76
Consumer wastes generated	879 tonnes	−54.4 tonnes	−105
Mining wastes	2,591 tonnes	57.3 tonnes	−97

* Negative numbers represent a decrease in that category resulting from recycling.

cause of various inefficiencies 60—80% is more typical. The U.S. Environmental Protection Agency (1973) estimates that the substitution of scrap steel for primary material as an input to steel making results in a 74% saving in energy consumption, and 86% reduction in air pollution, and a 97% reduction in the mining waste (Table 9-V).

Despite all these advantages, the composite rate of scrap consumption for metals in the U.S. is not just low (about 25% excluding home scrap), but is declining; the rate is slightly higher in Europe. It has been cheaper, at least in accounting terms, for firms to use primary materials in most applications.

Industrial scrap

There are three types of metallic scrap. "Home scrap" or "run-around scrap", so-called because it returns to the melting furnaces without leaving the premises of the producer, is produced during mineral processing. It consists of spills, incorrect alloys and any other metal output not suitable for shipping, and is virtually all recycled. "Prompt industrial scrap" is produced during fabrication, as with cuttings from the machining of metals. Here too a high degree of recycling is achieved, partly because of the relative ease of gathering high concentrations of metal of known composition and partly because the costs of disposing of this waste, were it not to be recycled, would be borne by the industry itself. A specialized secondary materials industry has developed in most countries to provide the service of picking up scrap from the manufacturing industry, sorting and cleaning, and delivering the reusable metals to the smelter. "Obsolete scrap" is generated by both the manufacturing and consuming sectors. In manufacturing, this scrap comprises vehicles, machines or other equipment or materials which are worn out, damaged or obsolete. Most obsolete industrial scrap is recycled through the same channels as prompt industrial scrap. However, the process is less complete because the supply is erratic, and there is a need to dismantle and eliminate non-metallic components.

Consumer scrap
It is in the consumer sector that recycling has the greatest room for expansion. Consumers do not generally have to pay for the costs of disposing of their obsolete solid wastes and, until recently, little effort was made to organize a system of gathering or to develop a technology for processing regular consumer wastes. Thus, most metal-bearing goods from the consumer sector are still disposed of as solid waste in municipal dumps. U.S. urban wastes for one year have been estimated to contain 13.5 million tonnes of iron and steel, 12 million tonnes of glass and nearly million tonnes of non-ferrous metals. (These and other figures in this chapter appear in U.S. National Commission on Materials Policy, 1973, Chapter 4D.)

As Darney and Franklin (1972) show, the metallic composition of the municipal wastes in North American cities ranges from nearly 6% to more than 9%. Table 9-VI presents a U.S. estimate of the values of iron, aluminium, copper, glass and fuel in raw refuse. Similar calculations in other western

TABLE 9-VI

Value of products contained in 0.91 tonnes (1 ton) of raw municipal refuse in the U.S., 1973 (from Sullivan et al., 1973; recalculated to metric basis)

Product	Value of product (U.S.$/ weight)	kg/tonne of refuse	Percentage of total refuse	U.S.$/tonne of refuse
Iron	11.00/tonne	77	7.7	0.85
Aluminium	0.26/kg	4	0.4	1.06
Copper, lead, zinc	0.42/kg	3	0.3	1.22
Glass cullet, colour sorted	13.20/tonne	64	6.4	0.85
Combustibles (fuel)	1.35/tonne	814	31.4	1.10
Dirt	n.a.	17	1.7	—
Waste, fine glass	n.a.	21	2.1	—
Gross value	n.a.	n.a.	n.a.	5.10

n.a. = not applicable.
Note: the authors of this table assign values to paper and plastic as fuels rather than as recoverable items. Generally, they are more valuable in both dollar and conservation terms if they are recycled instead of being burned.

nations indicate potentials nearly as high. Evidently the typical refuse dump outside a modern city is comparable to a mine of not unprofitable grade. When one considers further that present solid waste management practices involve large costs to municipalities of gathering, transporting and burying these valuable resources, the attractiveness of recycling becomes clearer still. Costs for land fill disposal in North America are $2 to $3 per tonne exclusive of collection and assuming "waste land" is available. Unfortunately, because of their high capital costs, municipal recycling plants remain only marginally economic in accounting terms for cities with access to nearby land fill sites. One study indicated that only those communities facing waste disposal costs greater than about $8.50 per tonne would find recycling to be profitable on these terms (Abert et al., 1974), although other studies have been more optimistic (Sullivan et al., 1973). In many cases such projects might be socially desirable because of reduced external costs even if accounting losses are incurred.

Technology for recycling has been gradually improving over the past decade. However, obsolete scrap is spread over wide geographical areas, making the high cost of collection a major barrier to higher recycling rates. In addition, because shipments of scrap are often relatively small, processors have been unable to negotiate the preferred freight rates available to large primary mineral producers who can guarantee steady, large volume shipments. Another difficulty is that industries that exploit primary resources receive special tax benefits which, however justified on other grounds, do serve to give them a competitive advantage compared with the secondary materials industry. However, both freight rate and tax differences have tended to diminish in the last few years in many countries. Thus at the same time as changing social attitudes towards waste are promoting greater efforts at recycling in North America, the economics of recycling are also improving (U.S. National Academy of Sciences, 1973, pp. 57—58, 83—86).

EFFICIENT USE OF ASSOCIATED RENEWABLE RESOURCES

Land-use choices

The most difficult of all mining environmental conflicts is the question of whether mineral extraction will be permitted at all on specific sites or in certain regions. This is because, by their nature, such conflicts generally have to be resolved in an all-or-nothing-way; either mining will be permitted or it will not be permitted. In some instances, compromise positions can be found. A quarry or coal mine may be allowed to proceed as an underground but not as a surface operation (but at the expense of reducing the recovery ratio), or angle drilling may permit tapping a pool without occupying the immediately overlying surface. The problem then becomes one of how to operate in order to reduce environmental impacts (see below).

Most land-use decisions are, of course, made through the market process in which it is assumed that land will become dedicated to its "highest" use by letting it go to the highest acceptable bidder. It is not this process which is at issue here so much as the logically prior question of who will be considered as an acceptable bidder. Once it has been decided that mining is an acceptable land use in some region, it is logical to sell or lease that land to the highest bidder, but, if mining is not acceptable, even exploration should be prohibited.

For a variety of reasons the determination of acceptable land uses has become largely, if not entirely, a non-market, that is to say, a political process. Most land-use decisions have ramifications that go far beyond the immediate decision, and the capital gains from changed land use can be very high and concentrated while the losses tend to be high but widely dispersed. Moreover, decisions about land use are often emotionally charged as witness the bitter battles about the use of farm land for strip coal mining or about the introduction of mining into wilderness areas (large tracts of protected land in natural condition as defined in North America).

Not all conflicts are so complex. The single most common land-use conflict in mining involves the clash between operators of sand and gravel pits and nearby communities. In this case, the issues are clearer because both the extent and the distribution of benefits and costs are easier to identify and quantify (Bishko and Wallace, 1972). Moreover, the zoning system for defining appropriate land uses, which is all but universal in urbanized areas of North America, has a long history of finding accommodation between the needs of the construction industry and objections of nearby residents.

On most of the non-urban public lands of the U.S. and Canada, the legal presumption favours mining as opposed to other land uses. This is true, for example, in most national, state or provincial forests (but not parks), where the regulations generally permit mining unless a specific case can be made against it. From the point of view of both equity and efficiency, there is no more reason for a presumption favouring mining than one against it.

Perhaps the most useful guide to rational land-use decisions involving mining is the concept of sequential land use. Sequential land use is to mining what multiple land use is to forestry. It expresses the fact that, in most cases, a tract of land must be dedicated to mining as a single use while mining is in progress, whereas a forest can serve a variety of purposes even while the timber is growing and being harvested. Since timber is a renewable crop but mines are eventually exhausted, the different uses occur over time in the case of mining whereas they occur simultaneously in the case of forestry.

The main question suggested by sequential land use is: What will be lost irrevocably, or at least for a long time, if one permits (or does not permit) mining on this site? The greater this loss or the more irreversible the change, the stronger must be the rationale before one accepts the change in land use

(Krutilla et al., 1972; U.S. National Academy of Sciences, 1973, pp. 54—55, 91—93).

If it is proposed to build a housing subdivision on top of valuable sand and gravel deposits, the likelihood that these reserves could be mined in the future is small. It is hard to conceive of urban development being reversed in order to mine sand and gravel. Thus, there is a good a-priori case for mining the sand and gravel before the subdivision is built and conserving for the future reserves that are not so threatened. The social loss to the community in deferring development for a few years will typically be low in comparison with the loss of the sand and gravel reserves, although for the prospective developer the cost of deferral might be very high.

The opposite sort of case occurs when it is proposed to establish a mine in some region with a unique or uncommon biota. If it could be established that there would be irreversible damage to, say, some form of wildlife, the presumption would be strongly against mining. Few cases are this clear cut. Despite a growing body of research, it is difficult either to predict physical effects on wildlife or to quantify social losses. Moreover, these "costs" must be compared with those of *not* mining. (For an excellent example of an attempt to make this comparison, see Spore and Nephew, 1974.) For example, oil drilling in the Kenai Moose Range of Alaska (which was permitted only after a great deal of controversy) probably led to an increase in the size of the herd by opening pasture. On the other hand, while this may make drilling acceptable to hunters or photographers, it would be irrelevant to those interested in wilderness preservation for whom natural conditions, not biological productivity, is the goal.

The creation of three new national parks in Canada in 1972 provided an instructive set of case studies on mining-wilderness land-use conflicts. In one of them, Baffin Island, there was almost no possibility of valuable minerals being found and hence no conflict. In another, Nahani, there was Virginia Falls, three times the height of Niagara and both a splendid potential hydropower site and a magnificent unspoiled natural site. The conflict was as inevitable as it was irreconcilable: either there would be a wilderness park or there would be hydropower development. The choice, appropriate according to the criterion of avoiding irreversible change, was to not build the dam because other sources of power were available, whereas Virginia Falls was unique. The third park was Kluane. In one corner of this park, there was a significant potential for mineral discoveries, although no important deposits were known. Should the boundary of the proposed park be adjusted to keep potential mineral areas outside? The issue is not yet resolved, but, according to the sequential-use criterion the decision should depend upon whether adjustment of the boundary would cause an irreversible loss. If the original boundary line defines a particularly important part of the ecology, the presumption would be against mining; if it simply includes "typical" land of the region and an equal number of hectares could be included elsewhere, the presumption would be in favour of mining.

Land-use decisions will never be easy. Moreover, as personal incomes rise, non-market values of wildlife and scenery seem to increase relative to market values of minerals. Consequently, the mining industry must prepare for still more difficult decisions in the future than those of the past.

Environmental controls during mining

The major problems that arise during mining can be discussed under five categories: water pollution, air pollution and noise, land subsidence, coal fires, and radioactive wastes (see Brooks and Williams, 1973, for a review). The costs of control can be high, particularly if new procedures are added to mines and plants built when environmental protection was less of a concern. In most cases planning will not only reduce costs but also improve results. Except in special circumstances, water pollution will be the most serious problem during the exploration and mining stage, whereas air pollution will be most important at the smelting stage. Land subsidence, coal fires and radioactivity are less common but of critical long-term significance where they do occur. In a few cases, special efforts will be needed to protect particular cultural, historical, or ecological features, such as a native village or a fossil bed.

Reclamation after mining

If a mine has been planned and the mining carried out with an eye to later use of land after the mine closes, reclamation is the least difficult stage of environmental protection. The goal of reclamation can be stated quite simply: after completion of mining the land should be left in a condition that is no less pleasing to the eye and no less productive than it was before mining. While subject to interpretation, this requires either return to original site conditions or to new conditions that are stable and compatible with the surrounding area. It further requires that the mine site should not be a continuing source of pollution or of public danger. And, perhaps most important, to promote sequential land use neither ecological nor non-mineral economic productivity should be lost permanently because of mining. Land that was capable of producing marketable timber before mining should be capable of producing marketable timber afterwards; land that was habitat for prairie dogs before mining should support prairie dogs after mining. Of course, this goal has to be modified where the mine will create a permanent new land feature, as with deep pits and quarries, but these should be recognized as exceptions that are accepted only when there is no alternative.

Reclamation operations differ according to whether they are being undertaken at existing mines on an ongoing basis or at abandoned mines that have been left unreclaimed — "orphan" or "derelict" land of which more than 800,000 hectares had accumulated by 1965 in the U.S. from surface mining alone (U.S. Department of the Interior, 1967). The difference is less one of technique than of cost. Almost without exception, it is much more expensive to reclaim land after mining has been completed because the spoil was dumped helter-skelter and mixed with mining debris and because of the need to bring in equipment rather than using equipment already on the site.

The major determinants of reclamation costs are the level of reclamation desired and the slope of the land. The level of reclamation can vary from minimum regrading, just enough to ensure plant growth, to the creation of agricultural land, recreation facilities, or even home or industrial sites. Costs vary directly with the amount of work. Cost per tonne mined is sensitive to slope angle not only because steep slopes make flood and erosion control difficult but also because, with flat bedded deposits, they reduce the number of tonnes that can be mined.

Clearly, planning is the key to successful reclamation. With pits or deep mines that are expected to have a long life, the mine can be planned to minimize the need for later reclamation by considered location of access roads and careful placement of spoil or rock. Analysis of what is likely to be recoverable in the future will conserve resources by avoiding the dumping of waste or the location of structures on potential resources. Finally, attempts can be made to reduce the volume of waste that must be disposed. Certain waste materials can be used as road building material or as construction aggregate. In other cases, the waste can be used as fill underground, a technique which has the advantage of reducing subsidence and avoiding the need to leave ore in pillars.

With shallow or short-term surface mines, the eventual land use must be a consideration from the start, and it is difficult to distinguish mining from reclamation because the processes merge. All of the above considerations apply here too, and, in addition, there is much greater opportunity to mould the land surface during mining so as to prevent erosion, to bury unwanted material and to promote regrowth. In some cases it is possible to sell the reclaimed land, and in a few cases such sales may show a profit. However, true profits from reclamation are rare; they are generally found only where some form of urban development is close by the mine, as in the case of phosphate mines in Florida. (Statements to the effect that reclamation is profitable are common but are usually based on a comparison of the return from sales of reclaimed land with the direct costs of reclamation only; the investment costs represented by the equipment are ignored, as is the original cost of the land itself.) Regardless of private returns, in almost every case substantial reclamation will be profitable in an economic and social sense. Further details are provided in the review by Brooks and Williams (1973) and in the collection of papers prepared by a U.S. Senate Committee (1971).

CONCLUSIONS FOR THE FUTURE

If the distinctions between renewable and non-renewable resources are vague, so also are the distinctions between any principles applicable to conservation of mineral and of environmental resources. Indeed, in a technologically advanced society minerals may more closely approximate the classic concept of a renewable resource and the environment the concept of a non-renewable one (Brooks, 1973). That is, technology offers the promise of finding ways to renew the supply of minerals at about the same rate at which they are depleted through advances in exploration techniques, extraction processes and substitution. However, it can do so only by imposing certain largely irreversible changes on the environment, such as shifts in land use in order to make resources accessible or the production of long-lived radioactive wastes in order to supply the necessary energy. This makes it all the more important to search for some concepts that will help us to reach decisions about the use of our natural resources.

The need to conserve minerals stems most directly from the fact that mineral deposits are subject to exhaustion. Although in an economic sense world mineral potential may be nearly limitless, the quantity and quality of mineral resources available at a particular locale or region are fixed. With better conservation, exhaustion will be postponed and along with it such problems as declining communities, loss of export earnings for producers and higher mineral supply costs for consumers. Moreover, to the extent that minerals are conserved, renewable resources are also protected. Secondary recovery of metals in the form of scrap involves more intensive use of minerals and avoids land pollution from solid waste disposal. Efficient mining and processing increases recovery and, by prolonging mine life, avoids disturbance of the environment elsewhere.

Thus, it is not surprising that a great deal of confusion has resulted from loose use of the term "mineral conservation". Some writers are talking about the mineral resources, some about the associated environmental resources, and a few about both. Is it possible to bring ideas about these two dimensions into a single framework?

To retreat to some jargon adopted by economists from lawyers, air and water and sometimes land are common property resources in that they are owned by no one (or, better, by everyone). It is typical to regard such common property resources as free for any use without limit, and so they become ideal reservoirs for waste disposal; thus, we have the belching stack and raw sewage outfall. So long as the environment can handle the residuals without deterioration, there is no problem, but very quickly the impacts become perceptible. As emphasized by many observers, pollution problems will continue to occur until the common property nature of environmental resources is recognized and institutions are created to control demands placed upon them. This is in fact what new environmental laws around the world are beginning to do.

The problem of conserving minerals themselves can be viewed as the time counterpart of the spatial problem of conserving the environment. That is, pollution occurs because firms do not need to take account of their use of the spatial environment at some point of time, and as a result wastes spread too far laterally. In the case of the minerals, inappropriate entry or recovery decisions may be made by firms or government agencies because they try to optimize their operations over a shorter length of time than is best for the community. For example, a firm will close a mine when the returns are insufficient to cover out-of-pocket expenses, and it will not recover by-products, nor use scrap as an input unless such practices would yield a significant positive return on investment. The fact that such decisions would serve to lengthen the life of the mineral resource, and that many public costs arise because of shorter life, is irrelevant to that firm. However, the result is a socially inefficient rate of mineral production and consumption. Thus, the freedom to alter the time rate of non-renewable resource use has many analogies to excessive use of a common property resource.

Most of the suggestions for conserving minerals and the environment that have appeared in preceding pages represent ways to force the rate of mineral and environmental consumption towards a more efficient path over time. Unfortunately, no absolute criteria for dealing with the many explicit and implicit tradeoffs exist. It will always be necessary to resort to general guidelines pointing to, but not specifying, appropriate decisions, for the meaning of conservation is sensitive to changing economic conditions, technology and social values.

Therefore, what is needed above all is a change in outlook. We must begin to think of minerals in terms of management rather than exploitation. Most imperatively, we must adopt management concepts covering the full life cycle over which the mine operates and over which the metal or mineral can be used. If this is done, the integration of mineral development plans with plans for other natural resources in order to conserve both minerals and the environment will follow quite naturally.

ACKNOWLEDGEMENTS

The author would like to thank David R. Berry and Josef Lajzerowicz, both of the Mineral Development Sector, Department of Energy, Mines and Resources, Ottawa, Canada, who contributed significantly to several sections of the chapter. He also wishes to acknowledge the work of Alma Norman who assisted with the editing of this chapter.

REFERENCES CITED

Abert, J.G., Alter, H. and Bernheisel, J.F., 1974. The economics of resource recovery from municipal solid waste. *Science*, 183: 1052—1058.

Barnett, H.J. and Morse, C., 1963. *Scarcity and Growth.* Johns Hopkins, Resources for the Future, Baltimore, Md., 288 pp.

Beckerman, W., 1974. *In Defense of Economic Growth.* Cape, London, 287 pp.

Berg, C.A., 1974. Conservation in industry. *Science*, 184: 264—270.

Berry, R.S., Fels, M.F. and Makino, H., 1974. A thermodynamic valuation of resource use. In: M.S. Macrakis (Editor), *Energy: Demand, Conservation and Institutional Problems.* Massachusetts Institute of Technology, Cambridge, Mass., pp. 499—515.

Bishko, D. and Wallace, W.A., 1972. A planning model for construction minerals. *Manage. Science*, 18: B502—B517.

Brooks, D.B., 1973. Minerals: an expanding or a dwindling resource? *Miner. Resour. Bull.*, MR 134. Information Canada, Ottawa, Ont., 17 pp.

Brooks, D.B. and Andrews, P.W., 1974. Mineral resources, economic growth, and world population. *Science*, 185: 13—19.

Brooks, D.B. and Berry, D.R., 1972. Mine health and safety as an analytic problem. *Miner. Resour. Bull.*, MR 125. Information Canada, Ottawa, Ont., 33 pp.

Brooks, D.B. and Williams, R.L., 1973. Planning and designing for mining conservation. In: I.A. Given (Editor), *SME Mining Engineering Handbook.* AIME, New York, N.Y., 1—10; 19—23.

Commoner, B., 1971. *The Closing Circle.* Knopf, New York, N.Y., 326 pp.

Coomber, N.H. and Biswas, A.K., 1974. *Evaluation of Environmental Intangibles.* Ecological Systems Branch, Environment Canada, Ottawa, Ont., 74 pp.

Cranstone, D.A. and Martin, H.L., 1973. Are ore discovery costs increasing? *Can. Min. J.*, 94: 53—64.

Darney, A. and Franklin, W.E., 1972. *Salvage Markets for Materials in Solid Wastes.* U.S. Environmental Protection Agency, Washington, D.C., 187 pp.

Dawson, J., 1971. *Productivity Changes in Canadian Mining Industries.* Economic Council of Canada Staff Study No. 30. Information Canada, Ottawa, Ont., 63 pp.

Govett, M.H. and Govett, G.J.S., 1976. The problems of energy and mineral resources. In: F.R. Siegel (Editor), *Research in Problems in Modern Geochemistry.* UNESCO, Paris, in press.

Hammond, A.L. and Maugh, T.H., III, 1974. Stratospheric pollution. *Science*, 185: 335—338.

Heilbroner, R.L., 1974. *An Inquiry into the Human Prospect.* Norton, New York, N.Y., 150 pp.

Herendeen, R.A., 1974. Use of input-output analysis to determine the energy cost of goods and services. In: M.S. Macrakis (Editor), *Energy Demand, Conservation, and Institutional Problems.* Massachusetts Institute of Technology, Cambridge, Mass., pp. 141—158.

Herfindahl, O.C., 1961. The long-run cost of minerals. In: *Three Studies in Mineral Economics.* Resources for the Future, Washington, D.C., pp. 13—36.

Herfindahl, O.C. and Kneese, A.V., 1965. *Quality of the Environment: An Economic Approach to Some Problems in Using Land, Water and Air.* Resources for the Future, Washington, D.C., 96 pp.

Herfindahl, O.C. and Kneese, A.V., 1974. *An Introduction to the Economic Theory of Resources and Environment.* C.E. Merrill, Columbus, Ohio, 405 pp.

Kneese, A.V., 1966. Economics and resource engineering. *Eng. Educ.*, 57: 709—712. Also available as reprint No. 65 from Resources for the Future, Washington, D.C.

Krutilla, J.V., Cicchetti, C.J., Freeman, A.M., III and Russell, C.S., 1972. Observations on the economics of irreplaceable assets. In: A.V. Kneese and B.T. Bower (Editors), *Environmental Quality Analysis.* Johns Hopkins, Resources for the Future, Baltimore, Md., pp. 69—112.

Lajzerowicz, J. and MacKenzie, B.W., 1971. Planning the development of a mining district — an Eastern European approach. *Trans. Can. Inst. Min. Metall.*, 74: 213—223.

Malenbaum, W., Cichowski, C. and Mirzabagheri, F., 1973. *Material Requirements in the United States and Abroad in the Year 2000. A Reseach Project Prepared for the National Commission on Materials Policy.* University of Pennsylvania, University Park, Pa., 30 pp.

Meadows, D.H., Meadows, D.L., Randers, J. and Behrens, W.W., III, 1972. *The Limits to Growth.* Universe Books, New York, N.Y., 207 pp.

Nemetz, P.N., 1976. Mining and milling in British Columbia. In: J. Stephenson (Editor), *Economic Incentives for Pollution Control.* University of British Columbia, Vancouver, B.C., in press.

Newcomb, R.T., 1967. Measuring technical progress in the resource industries. *Proc. Counc. Econ. AIME*, 2: 53—68.

Roszak, T., 1968. *The Making of a Counter Culture.* Anchor, New York, N.Y., 303 pp.

Schumacher, E.F., 1973. *Small is Beautiful.* Blond and Briggs, London, 288 pp.

Scott, A., 1955. *Natural Resources: The Economics of Conservation.* University of Toronto, Toronto, Ont., 184 pp.

Spore, R.L. and Nephew, E.A., 1974. Opportunity cost of land use: the case of coal surface mining. In: M.S. Macrakis (Editor), *Energy: Demand, Conservation and Institutional Problems.* Massachusetts Institute of Technology, Cambridge, Mass., pp. 209—224.

Sprague, J.P., Elson, P.F. and Sanders, R.L., 1965. Sublethal copper-zinc pollution in a salmon river. *Int. J. Air Water Pollut.*, 9: 531—542.

Sullivan, P.M., Stanezyk, M.H. and Spendlove, M.J., 1973. Resource recovery from raw urban refuse. *U.S. Bur. Mines Rep. Invest.* No. 760, 28 pp.

Trudeau, P.E., 1974. *New Year's Day Message of the Prime Minister, Canada.* Press Release, Ottawa, Ont., 27 December, 1973, 3 pp.

United Nations, World Population Conference, 1974. *Report of the Symposium on Population, Resources and Environment.* New York, N.Y., 35 pp.

U.S. Department of the Interior, 1967. *Surface Mining and Our Environment.* U.S. Government Printing Office, Washington, D.C., 124 pp.

U.S. Environmental Protection Agency, 1973. *Report to Congress on Resource Recovery.* U.S. Government Printing Office, Washington, D.C., 61 pp.

U.S. National Academy of Sciences, National Academy of Engineering, 1973. *Man, Materials, and Environment.* Massachusetts Institute of Technology, Cambridge, Mass., 236 pp.

U.S. National Commission on Materials Policy, 1973. *Material Needs and the Environment, Today and Tomorrow.* U.S. Government Printing Office, Washington, D.C., paginated by chapters.

U.S. Senate, Committee on Interior and Insular Affairs, 1971. *The Issues Related to Surface Mining.* U.S. Government Printing Office, Washington, D.C., 255 pp.

Yasnowsky, P.N. and Colby, D.S., 1974. Impact of gasoline prices on the demand for minerals for the manufacture of automobiles. *Proc. Counc. Econ. AIME*, 9: 61—74.

Developments in mineral exploration and exploitation

Chapter 10

MINERAL EXPLORATION AND TECHNICAL COOPERATION IN THE
DEVELOPING COUNTRIES *

DANIEL A. HARKIN

INTRODUCTION

The radical changes which have steadily been intensifying in the global
mining and energy resource industries in recent years are the subject of extensive discussion by other contributors to this volume. The non-renewable nature of mineral resources, the limits of growth, the political, fiscal and mone-
tary uncertainties, the fluctuation of commodity prices, and nationalization
and expropriation are all matters which have attracted greatly increased attention and awareness, as well as affecting, often dramatically, the economies
of many countries, both developed and developing. While the economic near-
havoc wrought in 1973—1974 by production cut-backs and four-fold price increases by the petroleum exporting countries may not be anticipated in the
case of most other mineral commodities, there is a natural inclination, and in
some cases, an economic necessity, for the major producers of certain raw
materials entering into world trade to attempt to follow the example of the
oil-producing and -exporting countries (OPEC) — through production cut-
backs and "cartel" pricing arrangements — to impel price increases or at least
to stabilize prices in falling markets. The tin-producing countries of Southeast
Asia, the bauxite-rich countries of the Caribbean region, the phosphate producers of North Africa, and the large copper exporters (Zambia, Zaire, Chile,
and Peru) are cases in which production for world markets is concentrated in
a relatively few countries (see Chapter 4). There are, however, fundamental
differences between the fuel (hydrocarbon) and the non-fuel mineral sectors
which make it likely that free market forces and other economic factors will
continue to play a major part in market prices for the latter, although a
greater degree of market regulation could result, depending on the effectiveness of producer associations or producer-consumer associations, such as the
International Tin Agreement which is generally acknowledged to have been
effective in the past years.

Of the fundamental changes now far advanced, those concerning the operational arrangements of the international mining companies in the developing

---

* The views expressed in this chapter are those of the author and do not necessarily reflect
those of the United Nations Organization.

countries are the most pertinent in relation to the subject matter of this chapter. Transition from traditional long fixed-term agreement formulae permitting "ownership and exploitation of mineral deposits" to arrangements recognizing the concept of state sovereignty over national mineral resources has taken place in most developing countries. Mining companies now accept that they must work in partnership with governments and that there are advantages to be gained by such a course. Whereas the special nature and characteristics of the mineral industry offer some justification for apparently excessive profits in one country being necessary to offset risk capital lost in unsuccessful ventures in other countries, there is increased awareness that any agreement at the national level which provides for unduly favourable terms is unlikely to survive, even in the short-term. A number of examples — such as copper in Bougainville, petroleum in the North Sea — can be cited of actual or potential windfall profits, resulting from sharply rising commodity prices, which necessitated action by governments concerned to obtain greater benefits than were provided for in original development agreements. Neither is this concern of governments to exercise greater control over, and gain optimum benefits from, these non-renewable assets restricted to the developing countries. Canada, Australia, and Ireland are examples of developed countries in which nationalistic attitudes and fiscal decisions in recent years have been less favourable to private sector mining interests and have caused considerable reduction in risk capital investment for mineral exploration and development.

The need for participation of the international mining companies in many developing countries is scarcely lessened by current trends, for the technology, skilled manpower, access to markets and capital-raising capacity are in many cases largely concentrated in their hands and cannot be acquired by others rapidly. Although contractual services may be obtainable for individual components within the mineral development sequence, only large mining organizations have a comprehensive overall capacity in some cases with integration from production through to marketing of consumer products. Furthermore, some companies have established dominant global positions in respect of certain minerals such as nickel, platinum, and diamonds, particularly their processing and marketing, which may render new entries difficult without their cooperation.

In summary, therefore, the established mining industry still has much to offer the developing countries in their mineral resources development. As long as this situation exists, opportunities for participation may be open, but increasingly in the form of joint venture agreements with governments or in management services through which operational and marketing arrangements are provided to governments on a contractual basis. This may now be more in accordance with the inclinations and interests of the parties concerned — the mining companies obtaining the acceptable return for their technical services while still securing sources of supply for long-established industrial processing requirements, at the same time providing optimum benefits to the host country with no infringement of sovereign rights over national mineral resources. In discussing the changing scene from the viewpoint of the U.S. dependence on mineral imports from abroad, Ridge (1973) stressed the considerable advantages to American firms from adopting a system of service contracts, rather than mining concessions, for their overseas operations. His arguments apply in greater degree to the Western European countries (75% dependent) and Japan (90% dependent) which are much more dependent than the U.S. (15% dependent) on mineral imports.

While the foregoing mainly concerns the production aspects of mineral resources, it sets the stage for a discussion of the associated changes in the exploration sector. These changes have given rise to problems which must be overcome to ensure future supplies of minerals as a basis for industrial development in the developing countries, as well as meeting steadily rising global demand.

THE CHANGING SCENE IN MINERAL EXPLORATION

Exploration potential

Important mineral finds over the past decade or more in the Republic of Ireland, a country in which traditional geological mapping surveys have been taking place since the first half of the 19th century, point to the necessity for other well-mapped countries — particularly in Europe — to embark on the systematic surveys using the modern prospecting methods now necessary to locate sub-outcropping mineralization and to establish mineral potential. Many industrialized countries cannot be said to have obtained any full inventory of their mineral resources; new discoveries will be made in them as a result of systematic surveys and improved technology for subsurface exploration (see Chapters 11 and 12).

There is, nevertheless, a general consensus that the major new sources of minerals will increasingly be located in the developing countries which have not been well explored in the past; these countries may have some 40—45% of the world's non-fuel mineral resources. They also offer the prospect of better returns on exploration investment as they are less well covered by geological and mineral exploration than the more developed countries. The average cost of finding a major mine has been put in a recent World Bank study as U.S. $30 million in Canada, at an even higher figure in the U.S., and at U.S. $12 million in Australia, a relatively less-explored country but one still ahead of most developing countries in intensive exploration activities, particularly since World War II.

Fig. 10-1, which reproduces a diagram prepared for a United Nations report by Arce (1970) shows dramatically how far the developing countries have to go to catch up with the developed countries in terms of mineral pro-

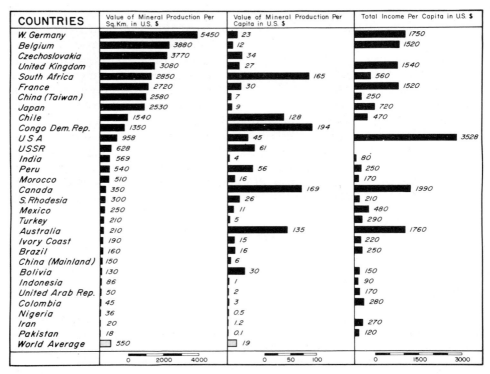

COUNTRIES	Value of Mineral Production Per Sq.Km. in U.S. $	Value of Mineral Production Per Capita in U.S. $	Total Income Per Capita in U.S. $
W. Germany	5450	23	1750
Belgium	3880	12	1520
Czechoslovakia	3770	34	
United Kingdom	3080	27	1540
South Africa	2850	165	560
France	2720	30	1520
China (Taiwan)	2580	7	250
Japan	2530	9	720
Chile	1540	128	470
Congo Dem. Rep.	1350	194	
USA	958	45	3528
USSR	628	61	
India	569	4	80
Peru	540	56	250
Morocco	510	16	170
Canada	350	169	1990
S. Rhodesia	300	26	210
Mexico	250	11	480
Turkey	210	5	290
Australia	210	135	1760
Ivory Coast	190	15	220
Brazil	160	16	250
China (Mainland)	150	6	
Bolivia	130	30	150
Indonesia	86	1	90
United Arab Rep.	50	2	170
Colombia	45	3	280
Nigeria	36	0.5	
Iran	20	1.2	270
Pakistan	18	0.1	120
World Average	550	19	

Fig. 10-1. Values of mineral production per km^2 and per capita compared to total per capita income for 1967 (based on international United Nations data compiled by the author).

duction. The imbalance shown cannot be regarded as due entirely or even principally to inherently greater resources in the present major producing countries; the exploitation of coal, iron, and construction materials needed for industrialization, rather than production of higher-valued minerals, accounts for the inclusion of a number of European countries in the upper part of the diagram. Given an adequate level of exploration investment the overall balance of known mineral reserves is expected to swing increasingly in favour of the developing countries as exploration progresses and the more easily accessible resources in the developed countries become depleted.

New concepts in metallogenesis and the emplacement of ore deposits, being developed as a result of a better understanding of the fundamental processes involved (for example, in plate tectonics and associated subduction zones) may give fresh impetus to exploration programmes and indicate regions or areas meriting priority attention. Exploration may also be encouraged by broad-scale studies of the kind undertaken by Petersen (1970) for South America;

from consideration of the distribution of certain ore deposit types, their relationships with major geotectonic features, and comparison with similar deposits in other continents, Petersen indicates several mineral exploration possibilities and concludes that a number of countries in the subcontinent could probably improve or diversify their mineral production significantly.

Historical review

Prior to and immediately after World War II, mineral exploration in the developing countries was principally in the hands of the private sector, generally carried out under exploration licences which place time and minimum expenditure limits (not always enforced) on the concession holder, while simultaneously giving the right to exploit any ore deposits found, in accordance with the provisions of the country's mining code. Field surveys were largely carried out by the traditional prospector — working either on an independent basis, for small national companies, for speculators who financed much of the activity, or for the larger mining companies. Except in a few cases, such as the "Bancroft Circus" in Northern Rhodesia (now Zambia) in the 1930s, there was little attempt at systematic coverage based on sound geological concepts. Geologists were more often engaged in basic geological mapping or in examination of prospects after they had been located, rather than playing any dominant role in the prospecting survey themselves.

In the post-World War II years increasing difficulty in making new discoveries on the basis of outcropping mineralization and the paucity of exposures in many countries — particularly those in the humid tropics — led to the development of modern exploration methods, especially photogeology, geochemical prospecting (Chapter 11), and exploration geophysics (Chapter 12) as fundamental aids to basic geology in exploration programmes. Exploration has thereby trended away from the traditional prospector and now largely depends on the integrated and systematic approach by geoscience professionals necessary to locate sub-outcropping mineralization. The technical capacity and financing needed for systematic surveys of this kind are largely beyond the resources of the small operator, so that the major mining companies have increasingly dominated the exploration scene in many countries.

These changes have been accompanied, particularly over the past decade, by a growing dissatisfaction on the part of some developing countries' governments at the terms under which their mineral resources were being exploited by the mining companies, particularly the international companies. The concept of proceeds from one successful exploration project having to pay for a large number of unsuccessful ventures in other countries is not easily accepted by a government concerned with maximizing its own country's benefits from development of national mineral resources. Neither is it always appreciated that the prospect of high profits is a factor that attracts risk capital for exploration ventures, and that political and economic instability are only offset by high profitability and a very short pay-back period. Relationships are scarcely improved if, on the company side, there has been repatriation of most income to the parent organization in the industrialized countries.

Whatever the reasons, the net result has increasingly been expropriation or nationalization of mining company assets in a number of developing countries, and in others renegotiation of agreements to provide the governments concerned with a proportion of the equity holding and a greater share in the benefits derived from the operations through increased taxation, royalties, or other means.

Exploration agreements and the exploration slowdown

Also of major importance in the exploration sector has been the tendency of some countries to exclude from exploration concession agreements the rights to develop any mineral deposits discovered by the concessionnaire. This departure from previous practice might still allow the discoverer a development option if his exploitation proposals are as favourable as those which other interested parties would negotiate with the government. However, it effectively discourages the entry of risk capital for large-scale regional ("grassroots") exploration surveys, companies naturally preferring to enter at the stage where development rights can be written into the contract — usually not before viable exploration targets have been located, and preferably when a mineral deposit showing some economic potential provides factual data on which to negotiate a mutually acceptable agreement.

It will seen from Table 10-I, in which an orthodox sequence of exploration activities is set out, that in these circumstances risk capital will tend to avoid the first two stages completely, will be reluctant to enter even at the third stage if preferential rights to any deposit found are not written into the contract, and may therefore largely concentrate their attention on fourth-stage activities, when risks are greatly reduced and a mineral deposit — though not necessarily an economic orebody — is known to exist. In fact, a number of international mining companies have, for many years, only been interested in participating at this stage, taking little or no part in the high-risk and high-cost systematic exploration surveys of large areas which are necessary to discover new deposits.

The net result has therefore been a further shift in the onus for primary systematic exploration being placed on national geological and mineral resource organizations. A number of developing countries have, for some time, taken the viewpoint that these basic surveys are, in any case, a national responsibility; many other countries have not taken this view and they may, moreover, lack the technical or financial capacity required to undertake large-scale surveys.

The location of specific economic deposits requires a more sophisticated and more costly sequence of exploration and evaluation activities as sub-surface deposits are sought. To this extent, therefore, a gap has appeared and widened in the exploration sequence of a number of developing countries; on a global basis there has been a reduction in the level of primary explora-

TABLE 10-I

Orthodox sequence of exploration activities

Stage and average duration	Operational activities	Operational objectives	Comments
(1) *Pre-project activities* (up to 1 year)	Review of existing maps and reports; limited air photo interpretation; field inspection of known occurrences, etc., to determine whether a viable exploration project can be developed	Project formulation in the case of positive assessment	Project areas of up to or over 10,000 km^2 are acceptable at this stage
(2) *Regional reconnaissance surveys* (up to 2 years)	Systematic screening of project area by geological/geochemical sampling and possibly airborne geophysical surveys (usually after photo-geological maps have been prepared) in order to discard — as rapidly, safely and economically as possibly — reduced potential and low priority areas and locate exploration targets	Exploration targets located for detailed exploration	Growing reluctance of private sector to venture risk capital at this stage in many countries. In virtually all cases can be expected to locate exploration targets for next stage, if project area soundly selected
(3) *Detailed exploration surveys* (up to 2 years)	Follow-up investigations on exploration targets utilizing geological, geochemical, and geophysical methods — leading to pitting, trenching and limited diamond drilling where results warrant	Mineral deposits identified and possible economic importance demonstrated	A number of targets can be written off quickly at low cost; others may require considerable inputs. Risk capital easier to attract at this stage

TABLE 10-I (continued)

Stage and average duration	Operational activities	Operational objectives	Comments
(4) *Mineral deposit definition* (up to 3 years)	Pre-development exploration by grid drilling and tunnelling to determine orebody configuration, tonnage, grade, beneficiation characteristics; other studies (infrastructure, mining legislation, etc.) necessary to establish economic feasibility and provide data for design of mine	Ore deposits proved and mine development data obtained	Exploration risk greatly reduced and costs often greatly increased. Scope for joint venture or service contracts with private sector

tion activity which might otherwise have been reached. Without an increasing degree of external cooperation in pursuing systematic exploration programmes, many developing countries, and particularly the least developed, will have great difficulty in making any real progress in this sector.

This overall situation is occurring at a time when steadily increasing demand for minerals requires the expansion of world-wide exploration activities to meet anticipated future needs. The long-term danger of underinvestment in prospecting and mining is recognized by the European Economic Community (see *Economist*, 1974), by the international mining industry, and no doubt by other bodies in the consuming countries which appreciate that a decade or more passes between the start of an exploration survey and production from any deposits which may be discovered. A greater degree of external cooperation in mineral exploration and development ventures in the developing countries, at least until such time as independent national capacity for this work is established, would help to alleviate this situation. Cooperation of the developing countries in pursuing expanded exploration programmes will be encouraged if there are expectations that optimum benefits will accrue, both in terms of obtaining raw materials required for their own national development and in equitable returns from mineral exports.

TECHNICAL COOPERATION IN MINERAL EXPLORATION

Types of assistance

Those countries lacking the necessary national capacity — in expertise or finance — may arrange for technical cooperation in mineral exploration from a number of sources. Technical services are offered by private consultant and contracting companies in a number of the developed — and some developing — countries. Those countries which have been most active in pursuing mineral exploration programmes and in developing modern methods and techniques (Canada, for example) offer a comprehensive range of services for both airborne and ground surveys, but the high costs of such services are often a limiting factor for many developing countries.

Official geological and mineral resource organizations of a number of industrialized countries have established separate divisions catering specifically for overseas operations, mainly in technical assistance to developing countries. Most of the organizations count on a comprehensive technical capacity, but may incline to or emphasize different objectives in their operational activities; the orientation of some is towards basic geological mapping, others concentrate on fundamental metallogenetic studies, certain countries endeavour to sustain national geological services established by them in former dependencies, and others assist in the investigation of mineral occurrences. Similarly, there is variation in the exploration philosophy and methodology of the various countries concerned.

Bilateral assistance is provided under a variety of arrangements, ranging from outright grants, through various types of loan agreements, to what are, in effect, service contracts with payments often being made in commodities. A number of the centrally planned countries provide the latter type of assistance.

Some indication of the nature and extent of assistance by a number of major donor countries — Canada, France, the U.K., the U.S., and U.S.S.R. — is given by Tremblay (1972), Lespine (1972), Pallister (1972), Reinemund et al. (1972), and Gorbunov (1972), respectively, in papers presented at a symposium on Earth Science Aid to Developing Countries held as part of the 24th International Geological Congress. While the overall philosophy of all of these programmes is broadly similar and most place emphasis on the training of national personnel and the transfer of technology, the orientation of the individual programmes has differed, particularly in respect of whether fundamental mapping and other basic surveys or direct ore-finding objectives are emphasized. Tremblay (1972), for example, noted that at least three-quarters of Canadian aid (which amounted to some C.$60 million between 1953 and 1971) was for single-phase airborne reconnaissance surveys which produced raw data but made little provision for ground follow-up surveys; he recommended that greater use be made of Canada's comprehensive and high-level capability in multiphase mineral resource programmes. The U.S.S.R. stresses the search for minerals and uses a wide range of geological and exploration methods. The British programme is comprehensive in nature but is at a modest level of size and expenditure; most emphasis is placed on training in the simpler, well-tried techniques of mapping and prospecting. The U.S. has undertaken about 1,000 assignments in cooperation programmes in more than 70 developing countries and accepted 1,200 personnel for training in the U.S. over the past three decades; strengthening of national institutions, geoscience training, transfer of technology and quick publication of maps, and reports (over 1,200 prepared) have been basic concepts of the programme.

Among other countries, West Germany has given aid to a number of countries, which often takes the form of small teams assigned to regional geological mapping and inventory of known mineral occurrences. Fundamental surveys of this type provide a very good basis for later systematic mineral exploration programmes.

Another important source of bilateral assistance is given by geology and mineral resource departments of universities. The technical capacity of these institutions ranges from basic geological mapping and other fundamental studies through to, in a few cases, ore-oriented exploration methodology. The universities can offer advantages in programmes involving applied research, and their services have been utilized by a number of developing countries, often under bilateral assistance arrangements or in the form of private consulting or contractual services.

Multilateral assistance has been provided mainly through the United Nations and by other intergovernmental organizations such as OECD. The United Nations system is responsible for the largest of the multilateral technical cooperation programmes. Within the system, UNESCO is responsible for the basic scientific and research aspects for training, and for regional and global compilation and correlation programmes, which it conducts in association with scientific bodies such as the International Union of Geological Sciences. Activities in the developing countries are principally in cooperation with national universities or other academic and research institutions. More direct development-oriented programmes for the discovery and evaluation of mineral resources are carried out by the Department of Economic and Social Affairs (ESA) within which the Centre for Natural Resources, Energy and Transport is technically responsible for all geological and mineral resource development activities; these involve pre-investment projects executed in cooperation with national geological survey and mineral resource institutions, with emphasis placed on the location of economic mineral deposits and on applied and practical on-the-job training given in the course of multi-mineral exploration and institute-strengthening programmes. Radioactive minerals are dealt with separately, but in a similar fashion, by the International Atomic Energy Agency (IAEA). See Brand (1972), Cameron (1972), Berger (1972), and Katz (1972) for details of these programmes.

The projects undertaken with United Nations cooperation are financed principally through the United Nations Development Programme (UNDP), on a free and unfettered basis. However, limited resources are available, and the many development fields other than minerals which require assistance are greatly limiting factors.

A United Nations Revolving Fund for Natural Resources Exploration (discussed below) has recently been established. It differs from other UNDP assistance in that recipient countries incur a certain repayment obligation based on production obtained from any deposit discovered as a result of Revolving Fund operational activities.

A number of other intergovernmental and regional organizations are active in providing technical assistance. The services provided, however, are usually short-term and advisory in nature, rather than involving large-scale operational activities. OECD and certain other regional cooperation and defence pact groupings such as the Colombo Plan, Central Treaty Organization (CENTO), and NATO are examples of organizations so involved.

Analysis of technical cooperation programmes

The International Geological Congress Symposium (referred to above) provided a forum for frank and critical discussion of earth science assistance afforded to developing countries by bilateral and multilateral agencies. Not unexpectedly non-technical comment — conditions of service of technical assistance personnel, their personal characteristics and life styles, and other similar subjects — figured in a number of contributions. Overall substantive comment was not uniform; Bonis (1972), for example, felt that United Nations exploration projects included too little effective training, while Tamale-Ssali (1972), a recipient country representative, considered the fact that national personnel are regarded as a necessary component of the programmes a drawback since there are often in short supply. Nevertheless, concensus is discerned in a number of important aspects of technical assistance which should be given due attention in future programmes. These are summarized below.

Project preparation
Technical assistance projects are often formulated too hurriedly and after only short preliminary visits by donor organization representatives to the recipient country. Too little time may therefore be spent in adequate field examination, resulting in ill-prepared projects, and in some cases in projects which are not fully relevant to the country's needs. The time necessary to assess needs and formulate a sound and viable project naturally depends on the extent and reliability of existing geological and mineral resource data. Pre-project activities by an experienced geologist (the first stage of Table 10-I) over a period of several months would be necessary in many cases. A criticism made of earlier United Nations assistance under the former United Na-

tions Special Fund — that only relatively large-scale costly projects could be considered — no longer applies; since the amalgamation in 1965 of the Special Fund and the Technical Assistance Agency, to form the UNDP, there has been no lower limit on expenditure.

Training of national personnel

The importance of training was emphasized by a number of contributors. Special attention was, however, drawn to the desirability of training being carried out in the national environment rather than in foreign countries where the physical, geological and metallogenetic settings and work problems bear little relationship to those encountered in the trainee's home country. In this respect, the importance of associating universities in the developing countries with geological and mineral exploration projects being executed jointly by the national geological surveys and technical cooperation agencies was stressed. The involvement in mineral exploration projects of universities in developed countries which possess high-level expertise and specialization in this field was also considered to be very desirable. The lines along which training of this kind could best be accomplished are discussed by Govett and Govett (1972, 1975) who stress the need for sound formulation of a country's requirements for trained earth scientists, prior to embarking on programmes for this task; increased emphasis on the training of the sub-professional and technical staff required for routine and supporting services is another important requirement.

Major project objectives

As mentioned above, some differences in viewpoint as to where emphasis should be placed in earth science aid to developing countries is apparent in the symposium contributions — specifically as to whether fundamental training or the direct search for mineral deposits should take preference. Institute strengthening and training of national personnel has probably predominated in bilateral programmes, whereas the direct search for mineral deposits, at the same time strengthening national institute capability in the methodology employed, has been emphasized in United Nations activities; this is in keeping with the pre-investment development objectives of the overall UNDP programme. The search for mineral deposits is usually more costly and, mineral exploration being a high-risk venture, more often than not results in no important discoveries; at the same time, training may receive less specific attention as pressures build up towards project completion. Where important mineral finds are made, however, the eventual benefits to the country can be very considerable and disproportionately large in relation to the input made.

General conclusions

One would conclude from the foregoing that, in mineral exploration programmes at least, greater care should be taken to ensure (1) sound project selection after adequate preparatory work has provided the necessary data for assessment and formulation within an integrated development plan; (2) more suitable and timely inputs to meet project objectives in staffing, equipment, and other components; (3) close attention to training of national personnel to meet, but not exceed, a country's needs, and, where possible and appropriate, training within the trainee's home country or similar environment; (4) provision for follow-up investigation of positive exploration results which may lead to identification of ore deposits; and (5) close cooperation and coordination of efforts among all agencies and institutions extending earth science aid to the developing countries.

UNITED NATIONS MINERAL SURVEY PROGRAMMES

Review of operations

Donor country contributions to the UNDP are voluntary; they amounted to U.S. $363 million for 1974, with U.S. $410 million anticipated for 1975. Expenditures are in accordance with UNDP country programmes, which are formulated in cooperation with recipient governments, and, based on their requests, with priorities in funding specific projects assigned within the limited overall UNDP financial allocation to the country concerned. The pre-investment nature of the overall programme is stressed.

As a rule, the UNDP contribution covers all external costs in staff, equipment, and sub-contracts; recipient governments meet all local expenditures of the project which vary in accordance with the nature of the operation, but may approach or even exceed the UNDP contribution. To the end of 1974 expenditures and commitments for larger-scale projects in geology and mineral resources totalled about U.S. $106 million for UNDP and a roughly equivalent contribution by recipient governments. Taking smaller-scale assistance into account, average combined annual expenditure over the past 15 years is about U.S. $15 million. The exploration budgets of many major international mining corporations are roughly similar to, or may exceed, this order of expenditure.

Virtually all aspects of mineral resources development — geological survey and mineral resource department organization, geological mapping, photogeology, mineralogy-petrology, mineral exploration using all modern methods (geochemical prospecting, exploration geophysics), drilling, analytical chemistry and assaying, economic feasibility studies, mining and mineral processing, mining legislation — have received attention in the programme. Operations have, however, largely been concentrated on the first three exploration stages outlined in Table 10-I.

An indication of the extent of the programme is shown by the listing of large-scale projects in the Appendix at the back of this volume. The projects may be considered as falling within four main categories discussed below.

Institute organization and strengthening

In essence, virtually all projects executed assist in the organization and strengthening of government institutions, and training of national personnel at all levels is an important part of them all. However, a number of projects have this as their primary objective, while concurrently carrying out the routine geological, mineral exploration, or other field surveys required. Positive exploration results have encouraged mineral resources development; for example, a Geological Survey Institute project in Iran identified previously known copper mineralization as being of porphyry type and hence of major development importance.

Mineral exploration

This is the largest single operational sector, since the direct search for ore deposits, while simultaneously training national personnel in the modern methods employed, has been the prime motivation for the bulk of the UNDP-assisted projects executed — in keeping with the pre-investment objectives of the overall UNDP programme. As a general rule, investigations are carried to the stage where investment is attracted for detailed exploration and assessment of any mineral deposits discovered, and include the obtaining of adequate technical data to place government in a satisfactory position to negotiate with interested principals for further work. The operations have involved virtually all methods — in-ground, airborne, and offshore. Within this category a number of combined mineral and groundwater projects have been executed as, for example, in Cyprus and Somalia.

Mining and mineral processing

Mineral processing components have been included in certain of the exploration projects, either as semi-independent activities or in cases where mineral deposits located cannot be evaluated without establishing beneficiation characteristics. It is not unusual for mineral deposits of otherwise satisfactory tonnage and grade to be rendered unworkable through mineral processing difficulties.

There are also a number of projects in which mining or mineral processing investigations form the major part of operations — in other words, projects not necessarily tied to exploration activities, but designed to meet the needs of an existing mining industry. Projects of this kind have been carried out at the Bawdwin lead-zinc mines in Burma and in Bolivia, where assistance was given in establishing a mining and metallurgical research institute.

University schools in applied geology

The emphasis placed by certain developing countries on applied rather than purely scientific or basic university training has resulted in a few projects which were executed by the United Nations (ESA) jointly with national universities, usually in close cooperation with national geological or mines organizations which will be employing the specialists trained, and in association with UNESCO. Training at the post-graduate level combines academic course work with concentrated practical field training in the student's home country. Positive exploration results are by no means ruled out in these programmes, as has been demonstrated by the results of the Institute of Applied Geology project in the Philippines (Hale and Govett, 1968).

Other technical cooperation

Assistance on a more modest scale has been given in the form of the services of individual experts assigned to advise governments or to meet specific needs, usually without the operational responsibility assumed in full-scale projects. During the eight-year period 1966—1973 a total of 340 man-years of services were provided in the fields of general exploration and economic geology (127), mining and mineral processing (85), geological and mining institute organization (30), analytical chemistry, mineralogy and other laboratory work (23), mineral economics (22), drilling (22), mining legislation (14), geophysics (8), photogeology (4), geochemistry (3) and marine geology (2).

Assessment of United Nations mineral survey programmes

In a review of United Nations exploration activities covering the first decade of the programme, Carman (1971) estimated that for expenditures of U.S. $78 million by the UNDP and U.S. $72 million equivalent in government contributions, capital investment of nearly U.S. $650 million had been attracted, with gross value of minerals discovered estimated at around U.S. $13 billion. In respect of the capital investment, these figures can now be more than doubled, and the gross value of minerals discovered has certainly increased considerably.

Review of the projects listed in the Appendix demonstrates that, overall, the recipient countries have contributed to the projects executed in amounts roughly equalling the UNDP contribution. As regards geographical distribution, UNDP project expenditures of nearly U.S. $50 million in Africa are approximately equal to those in Latin America (U.S. $24 million) and Asia (U.S. $25 million) combined. Long unsettled conditions in parts of Southeast Asia are reflected in the absence of a number of countries from the list.

Positive exploration results of potential or actual major importance are mainly located in the orogenic zones, and in particular the circum-Pacific belt, where discovery of porphyry copper-type mineralization in virgin areas has been much facilitated by geochemical prospecting methods. There have been fewer discoveries reported by projects exploring in cratonic areas; this can probably be attributed to a number of factors, including generally lesser prospects of economic mineral concentrations, more subdued physical relief resulting in fewer exposures and less secondary dispersion lending itself to prominent geochemical expression. As discussed by Brand (1972), reconnaissance geochemical drainage sediment sampling surveys have been the most rewarding screening method utilized in metallic mineral exploration, offering the incomparable advantage of allowing simultaneous geological observation and a more confident approach in detailed follow-up investigation than has been the United Nations experience with airborne geophysical screening methods. Regional geochemical surveys are now considered to be necessary adjuncts to the geological map; the data obtained are basic and of fundamental importance in geological and metallogenetic interpretation, as well as in a number of other important fields such as agriculture, veterinary science, and health.

While by no means all UNDP-assisted projects can be regarded as well formulated or executed, and the nature of the programme and variable conditions in the developing countries have made inevitable unevenness in performance, a number of mineral exploration projects demonstrate that effective programmes can lay the basis for accelerated development, both at national and regional levels. The United Nations Mineral Surveys in Panama (1965—1971) are a major success story from all points of view. Prior to the projects the metallic mineral potential of Panama was not considered promising. Geochemical reconnaissance sampling revealed a considerable number of anomalies, principally for copper. Follow-up work led to the discovery of the Cerro Petaquilla porphyry copper prospects. These attracted $4 million commercial investment for detailed proving work which so far is reported to have indicated about 3×10^8 tonnes of 0.65% copper. This discovery also caused a number of international mining companies to start exploration in Panama and in neighboring countries; in Panama itself the private sector has subsequently discovered the Cerro Colorado porphyry copper deposit, one of the world's largest (Hargreaves, 1974); its present proved reserves are stated to be 2×10^9 tonnes with 0.61% copper, 0.015% molybdenum, 0.002 oz/tonne gold, and 0.14 oz/tonne silver. A further 1×10^9 tonnes are estimated in each of the "probable" and "possible" categories. The cost estimate for this mining project exceeds $600 million. The Panama projects have filled a large gap in the known distribution of porphyry copper-type mineralization within the circum-Pacific orogenic belt in the Americas; the stimulus provided for exploration in the region has since resulted in the discovery of porphyry copper deposits at Pantanos-Pedagorcito in Colombia, and at Chaucha in Ecuador by a UNDP-assisted mineral survey project (present reserves are 7.4×10^7 tonnes of 0.7% copper). (For details of the Panama project see United Nations, 1970, 1972.)

Other UNDP successes in Latin America include the delineation in the Argentine Cordillera of a new porphyry copper belt and identification of specific deposits; some of these have attracted commercial interest (the Campaña Mahuida deposit has 2×10^7 tonnes of 1.2% copper). In Chile drilling and aditing have outlined 2.5×10^8 tonnes porphyry copper ore of 0.9% copper (including molybdenum equivalent) at the Los Pelambres deposit located 6 km from the Pachon deposit in Argentina, and the two governments are studying the possibility of a joint development; commercial production should commence around 1980. Another UNDP project in Chile located 6.3 × 10^7 tonnes of 60% iron at Boqueron Chañar. In Honduras a recently completed project has delineated some 5.0 × 10^6 tons of 1.0 g/ton gold ore and 5.0 × 10^6 tons of 0.5% copper ore suitable for inexpensive surface working; an additional 5 × 10^6 tons of each are considered probable. In Mexico UNDP projects have led to the discovery of the La Caridad porphyry copper deposit which will commence operations in 1977 (at a capital cost of $500 million) with an annual production of 150,000 tonnes of copper. An additional 50 × 10^6 tonnes of 60% iron at the Las Truchas iron ore deposit have been delineated.

In Asia follow-up work on geochemical anomalies in Sabah revealed the Mamut porphyry copper deposit which is reported to have reserves of 85 × 10^6 tons of 0.7% copper; it is expected to be brought into production in 1975 at a capital cost of $120 million. In Indonesia exploration delineated 12,000 tons of fine tin in offshore placers. Geochemical exploration by the Institute of Applied Geology in the Philippines outlined additional exploration targets for copper porphyry deposits at Santo Niño and Boneng, both of which are now being prepared for production. The recognition by the United Nations Geological Institute Survey project of the Sar Chesmeh copper mineralization in Iran as a porphyry type stimulated interest in these long-known deposits; annual production of 145,000 tons of copper is due to commence in 1975 from the deposit.

In Africa UNDP projects have contributed to the discovery of a variety of deposits. In Morocco 3 × 10^9 tons of better than 98% pure rock salt will be produced from a medium-sized underground mine costing about $5 million and will supply domestic needs as well as provide raw materials for a $85 million complex to produce polyvinyl chloride, chlorine, and soda ash. In Somalia reserves of 5,000 tons of uranium have been established; in Upper Volta 10 × 10^6 tons of 52% manganese ore (some of it battery-grade) is awaiting a rail link for development; in Burundi geochemical surveys have located more than 100 million tons of nickel laterite deposits grading 1.8% nickel and 4 g/ton platinoid metals; in Guinea 6 × 10^8 tons of iron ore (65% iron) has been delineated; in Togo UNDP work has led to development of limestone for a cement industry, clays for a ceramic industry, and marble for local finished products and for export to Italy.

In spite of these successful projects, any global programme of mineral exploration, no matter how favourable the geological and metallogenetic settings, must count on a greater number of projects ending without discovery of economic mineral deposits. Even where no discoveries have been made, however, positive benefits have been derived where national institutions have been strengthened, their technical staff upgraded, and a soundly based inventory of mineral resources (or lack of them) established or initiated to assist in development planning. The benefits are naturally greatest where worthwhile finds are made and especially where, as in Panama, the positive results have regional exploration significance.

Identification of porphyry copper-type mineralization in Iran was, therefore, also important in demonstrating potential in this sector of the Alpine orogenic belt, which takes in the neighbouring countries of Pakistan, Afghanistan, and Turkey. Mineralization of this type has, in fact, since been identified in Turkey and Pakistan and is also likely to be present in Afghanistan. Another pertinent example is the recent discovery of potentially important and extensive nickel laterite deposits developed on ultrabasic bodies intruding the Karagwe-Ankolean formation in Burundi; the same formations — and similar intrusions within them — extending into Tanzania thus became attractive exploration targets which are now receiving attention. The Burundi find is the first record of important lateritic nickel deposits in this part of Africa, and its significance is therefore considerable.

While there have been a fair number of positive exploration results in the United Nations projects, relatively few of the mineral discoveries have yet reached the production stage. In some cases this is not entirely the result of the normal passage of time between discovery and production which is usually between five and ten years for large deposits. Excessive delays in negotiating mutually satisfactory development agreements have been a contributory factor. In the market economy countries, this often reflects need for clarification or modification of national mining policy and legislation to provide reasonable terms for capital investment, while ensuring optimum benefits to the country from development of its mineral resources. The United Nations provides technical advice in these matters, but the subject is complex, with many non-technical considerations and national pressures.

On balance, the exploration record of the United Nations-assisted programmes can be considered satisfactory, while the subsequent stages leading to mine development have been subject to individual and varying national attitudes and influences. In discussing the limiting factors in external assistance for mineral exploration, Carman (1972) has drawn attention to the mixed results of training programmes for national staff in developing countries. With some exceptions, external fellowship training in institutions abroad may be judged as generally less effective than practical on-the-job training during project operations in the trainee's national environment; adequate professional staffing in government geological and mineral resources departments will pose problems in many developing countries for some years to come. Where there is a limited supply of trainees the geosciences generally do not compete well for good-quality students against medicine, law, and other professions which may confer greater social prestige. Furthermore, the most able geologists and engineers in some developing countries are often advanced to higher executive posts after only a short period of work in their profession;

others are attracted to better-paid positions in the private sector, not uncommonly in foreign countries. The net result is that in only a few of the more developed of the developing countries is the professional capacity fully adequate to sustain efficient independent programmes in mineral exploration. If this situation continues, there is little expectation that overall technical cooperation needs will diminish rapidly in this development sector.

FUTURE PROBLEMS AND DEVELOPMENT

The problems now to be faced in the global mineral resources field are complex and embrace all sectors — exploration, production, and marketing; comment on them at any one time is apt to be quickly overtaken by new developments. Some of these problems were considered at the 6th Special Session of the United Nations General Assembly (1974) when a call was made for a new international economic order "... based on equity, sovereign equality, interdependence, common interest and cooperation among all States, which shall correct inequalities and redress existing injustices, make it possible to correct the widening gap between the developed and developing countries and ensure steadily accelerating economic and social development for present and future generations".

The Special Session focussed particular attention on the plight of those developing countries poorly endowed with natural resources, and thus placed in critical situations through the greatly increased costs of food, fertilizers, oil, other raw materials, and manufactured imports. As noted by Barbara Ward (1974) at the time, the raw materials situation has created at least four different types of developing states: the oil-rich Arabian peninsula states (Saudi Arabia, Kuwait, Abu Dhabi, and Qatar) whose problem is to make rational use of their tremendous wealth; the more populous oil producers, such as Iran, Nigeria, Venezuela and Indonesia who can well use their increased revenues in economic development; countries such as China, Columbia, Mexico, Bolivia, Morocco, Malaysia, Zambia, and Zaire who are either more or less self-sufficient in oil or have major export earnings from other raw materials such as phosphates, copper, rubber, etc.; and, finally, the poorest developing countries which are importers of a large proportion of their needs. The Indian subcontinent, tropical Africa, the Caribbean and parts of Latin America are well represented in the last group.

The situation is not static; the major copper-exporting countries, to mention a notable example, have been hard hit as the market price of copper dropped more sharply in the middle six months of 1974 than it had risen in previous months. Some form of stabilization of commodity markets for minerals such as copper, lead, and zinc has long been considered necessary and in the interests of both producers and consumers. The difficulties in formulating a scheme that would afford consumers stability of supply and producers stability of revenue are many and no resolution of the problem is yet in sight.

The International Tin Agreement, involving both producer and consumer interests, has certainly dampened price swings for most of the post-World War II period and offers encouragement to those who advocate solutions based on a greater degree of international cooperation between producing and consuming interests. The non-fuel mineral sector is, however, heterogeneous, and a stabilization system for one mineral commodity may well not be applicable to another. In particular, the problem of developing country mineral producers will not be resolved without recognition of their need to make progress in industrial development based on their indigenous resources and to obtain equitable benefits in terms of real purchasing power, to take account of inflation and the high prices to be paid for essential inputs such as food, fertilizers, oil, other commodities, and manufactured goods (see also Chapter 4).

In the mineral exploration sector of most developing countries, the major problem is considered to be that of providing for a level of activity which will discover the new mines necessary to stimulate national economic development, taking this to include lower-valued minerals such as construction and other non-metallic minerals for local use as well as the higher-valued metallic minerals with industrial development and export possibilities. National development needs, as well as future global requirements, are unlikely to be met without considerable expansion in exploration effort.

Of the external sources of technical or financial support for mineral exploration in the developing countries, the capacity of the established mining industry and other private sector organizations greatly exceeds that of the official bilateral and multilateral agencies. While no doubt the latter could be expanded if financial resources were made available, the level of exploration activities necessary to make an impact in terms of future needs may be difficult to achieve without greater utilization of the former, under mutually acceptable agreements. In a number of developing countries in which mining companies are operating, it has been shown that such cooperation need not be at variance with the increasing insistence of virtually all nations — developed and developing — to obtain a greater measure of control over their natural resources. Risk capital from developed countries for mineral exploration will, however, still prefer those countries in which national mineral policy and legislation are clearly defined and where there is reasonable expectation that the rules of the game will not be subject to radical change.

Those developing countries with adequate financial resources will find little difficulty in obtaining contractual services for their mineral exploration and development needs. Given their large and increasing currency reserves, it may be that certain of the petroleum exporting nations would be inclined to provide financial assistance for mineral exploration to other developing countries, possibly channelled through development banks or other agencies; a number of the least developed countries have ethnic ties which might encourage such assistance. As a whole, the OPEC countries, in terms of their gross national products, have in fact been relatively generous in committing and

Lespine, J., 1972. Les sciences de la terre et leur action dans les pays en voie de développement — le cas de la France. *Proc., 24th Int. Geol. Congr., Symp. 2, Earth Science Aid to Developing Countries*, pp. 41—51.

Pallister, J.W., 1972. British overseas aid in the field of earth sciences. *Proc., 24th Int. Geol. Congr., Symp. 2, Earth Science Aid to Developing Countries*, pp. 27—36.

Petersen, U., 1970. Metallogenic provinces in South America. *Geol. Rundsch.*, 59: 834 — 897.

Reinemund, J.A., Taylor, G.C. and Schoechle, G.L., 1972. Scope and concepts of United States Geological Survey cooperative assistance to developing countries. *Proc. 24th Int. Geol. Congr., Symp. 2, Earth Science Aid to Developing Countries*, pp. 13—26.

Ridge, J.D., 1973. Minerals from abroad; the changing scene. In: E.N. Cameron (Editor), *The Mineral Position of the United States, 1975—2000*. University of Wisconsin, Madison, Wisc., pp. 127—152.

Tamale-Ssali, C.E., 1972. Earth science aid experience in Uganda. *Proc., 24th Int. Geol. Congr., Symp. 2, Earth Science Aid to Developing Countries*, pp. 67—71.

Tremblay, T.M., 1972. Canadian earth science aid. *Proc., 24th Int. Geol. Congr., Symp. 2, Earth Science Aid to Developing Countries*, pp. 3—12.

UNDP, 1975. Report of the Administrator on General Review of Programmes and Policies of UNDP — The Future Role of UNDP in World Development in the context of the Preparations for the Seventh Special Session of the General Assembly. UNDP Document DP/114 of 24 March 1975.

United Nations, 1970. *Technical Reports of the Mineral Survey of the Axuero Area, Panama*. New York, N.Y., 7 volumes.

United Nations, 1972. *Mineral Survey (Phase II) Panama Technical Reports*. New York, N.Y., 4 volumes.

Ward, B., 1974. First, second, third and fourth worlds. *Economist*, May 18, pp. 65—73.

Chapter 11

THE DEVELOPMENT OF GEOCHEMICAL EXPLORATION METHODS AND TECHNIQUES

G.J.S. GOVETT

INTRODUCTION

It can plausibly be argued that, in the long-term, dependence on conventional sources of minerals to meet world requirements will be reduced by advances in mineral processing which will allow the use of very low-grade ores, or by the development of techniques to recover resources from the sea and the ocean floors. In the short- to medium-term, however, demand will have to be met from mineral deposits similar to those being exploited today; therefore, fundamental to the question of whether adequate mineral supplies are available to maintain world consumption and the rate of growth of consumption is the ability of exploration geologists to continue to find workable mineral deposits.

The increasing difficulty and alarming escalation in the cost of finding a new economic metallic deposit is graphically illustrated by data from Canada for the years 1951—1969 in Fig. 11-1 (Derry, 1968; Roscoe, 1971). Exploration expenditures increased four-fold in real terms during this period, while the probability of finding a new orebody (based on an arbitrary average cost of $30,000 for each exploration try) decreased ten times. During this period, Canadian ore reserves increased by only 50%, and there is considerable doubt that enough conventional new reserves are being identified to replenish depleting stock in the long-term.

The average cost of each new discovery in Canada in the period 1951—1953 was $2.7 million (1968 dollars); the average cost of each new discovery in 1967—1969 was $27 million (1968 dollars). Morgan (1970) has suggested that 3—12% of gross operating profit in the mining industry must be spent on exploration by the large companies, and that an exploration expenditure of 9 to 30 million dollars over a six- to ten-year period should be expected to yield only one medium to large deposit (exploration costs are also discussed in Chapter 2).

The underlying cause of the increasing cost and difficulty of finding new economic deposits is that most of the deposits which have a surface expression in accessible areas of the globe have already been found; this is reflected in the changing pattern of discovery techniques. According to Derry (1968, 1970) 85% of Canadian mines in production in 1955 were found by surface

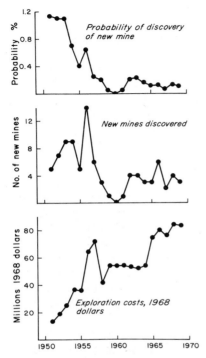

Fig. 11-1. Exploration costs, numbers of new mines discovered, and probability of discovery of a new mine calculated on the basis of an average cost of $30,000 for each exploration try in Canada. All dollars standardized to 1968 dollars. (Data from Roscoe, 1971.)

prospecting. During the period 1956—1968 only 22% of new discoveries were made by conventional prospecting; almost one-quarter of the new deposits were found by using geological methods, 52% were found by geophysical methods, and one new deposit was credited to geochemistry. These data illustrate the fact that techniques dependent on an obvious surface expression of a mineral deposit are becoming progressively less important as techniques which are capable of detecting hidden and buried deposits are taking their place. The discovery of adequate new mineral deposits must obviously depend to an increasing extent upon the potential of geophysical and geochemical methods to locate mineral deposits at greater depths in the earth's crust.

The value and effectiveness of geochemistry in mineral exploration is now unquestioned, but there is considerable difficulty in assigning primary credit for a new discovery to a particular technique in modern exploration, since there is an increasing tendency to use integrated geochemical-geophysical-geological surveys. It is probable that geochemical techniques have had their greatest successes outside of North America; for example, most of the new discoveries in Ireland have been attributed primarily to geochemistry (Schultz, 1971), and the number of new deposits which have been found by geochemistry in the more tropical climates of the less developed countries is impressive (see Chapter 10).

The use of geochemistry in mineral exploration is a relatively new development, originating in the U.S.S.R. in the 1930s (Fersman, 1939) and in Scandinavia in the late 1930s and early 1940s (Rankama, 1940). The first tentative steps to investigate the technique in North America were taken by H.V. Warren at the University of British Columbia in 1944, by H.E. Hawkes and others at the U.S. Geological Survey in 1946, and by R.W. Boyle at the Canadian Geological Survey in 1949. In 1953 the Geochemical Prospecting Research Centre (now the Applied Geochemistry Research Group) was established by J.S. Webb at Imperial College (London). The first major geochemical survey outside of the U.S.S.R. was conducted in 1954 in New Brunswick, Canada (Hawkes et al., 1960); since then the use of geochemistry has expanded rapidly until today more than $5 million are expended annually on geochemical exploration and some 7 million samples are collected annually outside of the U.S.S.R. (For historical surveys see Boyle, 1967; Boyle and Smith, 1968; Boyle and Garrett, 1970; Bloom, 1971; Warren, 1972.)

There can be no attempt to describe fully the application of geochemical exploration in this chapter. The reader is referred to texts on the subject (Hawkes and Webb, 1962; Levinson, 1974) and to surveys of developments in the proceedings of the biannual exploration geochemistry symposia which started in 1966 (Cameron, 1967; Canney et al., 1969; Boyle and McGerrigle, 1971; Jones, 1973; Elliott and Fletcher, 1975).

In this chapter the general principles of exploration geochemistry are illustrated and two main problems for future mineral exploration are identified for discussion in the context of geochemical contributions to their solution. These problems are:

(1) The areas of the world which have been less heavily prospected are generally remote or inhospitable; costs of exploration in such regions are high, and exploration techniques are required which can be used on a regional reconnaissance basis to reliably identify broad regions favourable for the occurrence of economic mineral deposits.

(2) In the well-prospected parts of the world most of the near-surface mineral deposits which give a response to well-established geochemical techniques have already been found; it is, therefore, essential that new techniques be developed which are capable of detecting deposits which have no near-surface expression and which may be buried by thick overburden or lie beneath barren cap rock.

In the latter part of the chapter some specific topics which are believed to be potentially important in the future use of geochemistry in exploration are discussed and assessed.

GENERAL PRINCIPLES OF GEOCHEMICAL EXPLORATION

A mineral deposit is an abnormal and rare geological event, and as such is the focus of abnormal physical and chemical conditions in the earth's crust. Mineral exploration by *geological* techniques relies on the recognition of locally abnormal geological features such as alteration zones and particular structural, petrological, and stratigraphic features; *geophysical* techniques depend on the measurement of various physical characteristics, such as electrical or magnetic parameters and the recognition of abnormal values or patterns which may be due to the presence of a mineral deposit; *geochemical* techniques are based on the premise that materials of the earth around a mineral deposit — rocks, soils, stream and lake waters and sediments, vegetation, and even the air — may be expected to differ in chemical composition from similar materials where there is no mineral deposit present.

As with geophysical methods, the attraction of using geochemical tech-

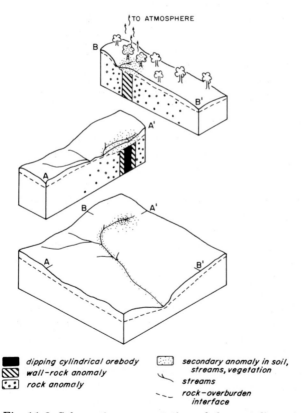

dipping cylindrical orebody
wall-rock anomaly
rock anomaly
secondary anomaly in soil, streams, vegetation
streams
rock-overburden interface

Fig. 11-2. Schematic representation of element dispersion around an orebody.

niques is that abnormal conditions can generally be detected far beyond the limits of economic ore, and thus a large target is provided for exploration, and the possibility of detecting ore deposits which are hidden by surface over-burden or which lie beneath barren cap-rock is enhanced. While geochemical methods, at least until quite recently, have suffered from the disadvantage of not having as great a depth penetration as geophysical techniques, they have an advantage in being direct, inasmuch as chemical differences directly re-lated to mineralized conditions are measured and, in many cases, the concen-tration level of the actual ore elements being sought are determined.

The basic principles of geochemical exploration are simply stated in Fig. 11-2. Changes in the chemistry of the host rock, due either to processes asso-ciated with the mineralizing event in the past, or with secondary processes which post-date mineralization, will give rise to *primary halos* in the rock adjacent to an ore deposit which can be recognized by abnormal concen-trations of elements (wall-rock anomalies); in some cases the host rock it-self may show an abnormal background content of elements on a regional scale. *Secondary halos* of abnormal element concentrations will occur in the soil and vegetation overlying a mineralized occurrence and, through dispersion in solution in groundwater and through physical surface transport, will also occur in the waters and sediments of the drainage system. Abnormal concentrations of some elements may also be found in the atmo-sphere above a mineralized occurrence.

The operational procedures of exploration geochemistry require that one or more types of earth material — rock, soil, stream sediment or water, vege-tation, air — be sampled and chemically analyzed for one or more elements in an attempt to recognize anomalous element distribution patterns which may be related to a mineral deposit. The choice of the sampling material will be dictated to some extent by its local availability: rock geochemical tech-niques have little applicability (except for drilling control) in areas covered by tens of metres of soil; similarly, geochemical surveys based on analyses of drainage systems are not practical in arid areas devoid of streams. There is normally little advantage in sampling vegetation in areas of moderately thick residual soil; however, if the soil cover is extremely thick, or the parent ma-terial is transported overburden having no relation to the underlying bedrock (e.g., glacial till), geochemical anomalies can, in many cases, be detected in organs of deep-rooted trees.

Given a choice of sample material, the sampling medium will be determined by the objective of the particular phase of exploration being undertaken. Successful exploration by any technique requires that a logical sequence of steps be followed to progressively reduce the size of the target until an ore deposit is found. Thus, sampling of a drainage system with a sparse sample density (and hence at relatively low cost) may be used as a *reconnaissance* technique to delimit broad areas of interest wherein detailed drainage sam-pling can be subsequently undertaken to define a target; soil sampling may

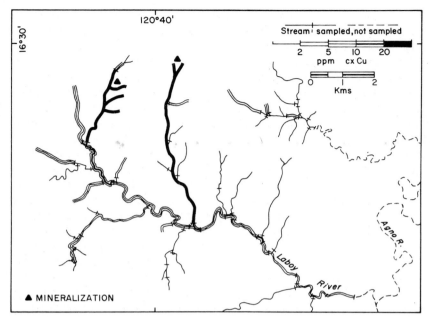

Fig. 11-3. Distribution of cold-extractable copper (cxCu) in wet, unsieved stream sediments around the St. Nino copper porphyry orebodies, Philippines. (After Govett and Brown, 1968, with permission of Philippine Bureau of Mines.)

be used to locate and define a potentially mineralized zone as a drilling target.

In the context of the schematic illustration of element dispersion shown in Fig. 11-2, some actual examples of dispersion in drainage, soils, vegetation, and rock are shown in Figs. 11-3, 11-4 and 11-5. The distribution of copper in stream sediments collected from a drainage basin of 100 km² which contains a copper porphyry deposit is shown in Fig. 11-3. Stream geochemistry is ideally suited to this situation; extremely rugged mountainous topography makes access difficult except along water courses, and rock outcrop rarely occurs except along the stream bed. The copper deposit is clearly reflected 16 km away by anomalous concentrations of copper in stream sediments; the concentration of copper increases towards the mineralized source, thus indicating the areas of greatest interest for more detailed exploration.

The type of response obtained in soils and vegetation is illustrated in Fig. 11-4 by the distribution of nickel in the vicinity of nickel-iron mineralization in the Pioneer area in Western Australia. The area is semi-arid and, although the soil cover is thin (15 cm), there is almost no rock outcrop. The soil is partly transported, but has enough residual content for its chemistry to reflect the underlying bedrock; strongly anomalous values of nickel and copper are coincident with mineralization. Analyses of leaves, twigs, bark, and roots of the

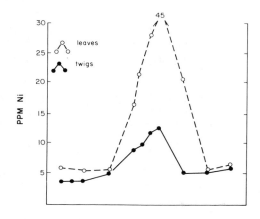

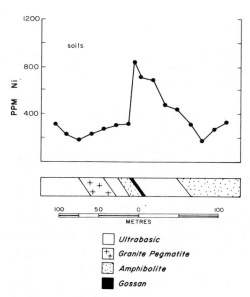

Fig. 11-4. Distribution of total nickel in soil, and leaves and twigs of the tea tree *(Melaleuca sheathiana)* over gossanized nickel mineralization, Pioneer area, Western Australia. (Redrawn with permission of Hall et al., 1973, *Proceedings of the Australasian Institute of Mining and Metallurgy*, No. 247, 1973, pp. 18 and 20.)

tea tree *(Melaleuca sheathiana)* show anomalous nickel and copper contents in the vicinity of mineralization; leaves give the best contrast between background and anomalous values and, although the anomaly is not as extensive as that in soils, it is more intense. Both soil and vegetation geochemistry (biogeochemistry) in this example are clearly of value in locating hidden

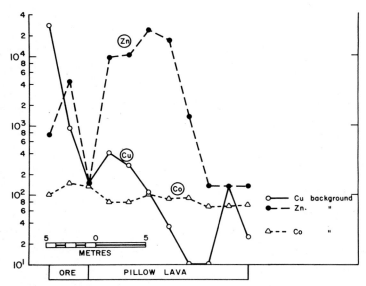

Fig. 11-5. Distribution of copper, zinc and cobalt in the centre of basaltic pillow lavas adjacent to Skouriotissa cupriferous pyrite orebody, Cyprus. (Modified with permission after Constantinou and Govett, 1973, *Economic Geology*, Vol. 68, 1973, fig. 7, p. 852.)

mineralization. While there is no apparent relation between distribution of vegetation type and bedrock in this example, in many parts of the world plant assemblages characteristic of mineralized conditions and variations in underlying bedrock type have been described (e.g., Cole et al., 1968).

A typical example of a wall-rock anomaly of copper, zinc, and cobalt adjacent to a cupriferous pyrite orebody in Cyprus is shown in Fig. 11-5; apart from the normal positive anomaly displayed by zinc and cobalt, it shows a negative anomaly for copper (which is below background levels) adjacent to the ore deposit. While this type of wall-rock response is of little value for finding new deposits because the extent of the anomaly is so small, it is extremely useful for tracing extensions of known deposits and for guiding drilling operations. The identification and use of *regional* rock anomalies and atmospheric anomalies, which are not so securely based as the methods described above, are considered in more detail below.

ANALYTICAL TECHNIQUES AND STATISTICAL METHODS

Successful development of regional reconnaissance and deep prospecting methods demands that considerably more information be derived from each sample than just a measurement of the ore elements being sought; regional anomalies indicating a mineral province and surface anomalies indicating a deep deposit are generally very small and are only recognizable through the combined variation of many elements. There are, therefore, two essential prerequisites for the development of new geochemical methods: an analytical capability for rapid determination of very many elements, and improved techniques for interpreting the resultant increase in analytical data.

The number of elements which can be cheaply and routinely analyzed with a high degree of sensitivity, accuracy, and precision has increased enormously since the advent of the commercial atomic absorption spectrophotometer (AAS) in the early 1960s. Continued development of the AAS, and the introduction of flameless techniques, has allowed measurement of many rare elements in the parts-per-billion range; the commercial introduction of plasma torch emission spectrometers in 1974 promises to provide even greater speed, sensitivity, and expansion of the number of elements which can be analyzed. Other instrumental techniques — such as X-ray fluorescence, neutron activation, specific ion — will increase the range of analytical capability; new instrumentation capable of measuring gaseous and volatile constituents is of particular importance (the relevance of gaseous geochemistry is discussed in more detail below). The improved analytical capability has also increased the use of major elements (e.g., iron, sodium, potassium, calcium, magnesium) in addition to trace elements in exploration; this is especially important in the application of rock geochemistry (Boyle, 1974; Goodfellow, 1975a; Govett and Goodfellow, 1975).

The handling and interpretation of the greatly increased flow of data generated by improved analytical capability would not be possible without the use of the high-speed digital computer; it facilitates storage and retrieval of vast amounts of data and can be used for automatic data plotting and display. Since all geochemical techniques ultimately depend upon recognition of spatial patterns of element distribution which can be related to mineralized conditions and distinguished from normal — background — conditions, the application of computerized statistical analysis has allowed important advances in geochemical interpretation. (Govett et al., 1975).

Considerable progress has been made in the use of univariate statistical techniques such as trend surface analysis (Nichol et al., 1969) and moving averages (Gleeson and Martin, 1969). Multivariate cluster analysis (Obial, 1970) and discriminant analysis (Howarth, 1971; Govett, 1972; Whitehead and Govett, 1974) permits the classification of samples into groups on the basis of two or more variables and assists in the identification of subtle anomalies which cannot easily be recognized on the basis of one or two elements alone; factor analysis is used to simplify a complex system of multi-element geochemical determinations into a smaller number of factors which may be related to natural geological associations (Garrett and Nichol, 1969); regression analysis may be used for prediction purposes, e.g., to predict the mineral potential of an area based upon the relation of a series of measured variables (Sinclair and Woodsworth, 1970). (Useful reviews of these and other statistical

352

techniques are given in Nichol et al., 1969; Agterberg and Kelley, 1971; and Nichol, 1973.)

REGIONAL RECONNAISSANCE METHODS

The development of new analytical capabilities and the use of computerized statistical analysis have contributed to attempts to characterize a particular stratigraphic unit or a particular granite intrusion as being potentially ore-bearing as distinct from apparently similar rock units which are barren. In the past geologists prospected in particular areas by looking for a particular rock unit, perhaps in a particular structural setting, perhaps with a particular and characteristic alteration. Now two geochemical techniques show promise for the identification of regions or geological units favourable for the occurrence of mineralization: these are regional drainage surveys (both streams and lakes), and rock geochemical surveys. With both types of surveys the objective is not to identify individual ore deposits as such, but to provide broad targets for subsequent more detailed exploration.

Stream sediment regional reconnaissance

Intuitively it may be expected that general background levels of elements in stream sediments should reflect characteristics of regional mineralization;

TABLE 11-I

Variation in background of copper, lead, and molybdenum in stream sediments, and dominant element assemblages in mineral deposits in different geological zones across the Canadian Cordillera, British Columbia (derived from Brown, 1974)

	Geological zones				
	Insular	Coast	Intermediate	Omineca	East marginal
Dominant element assemblage in deposit in order of importance	Fe Cu Mo	Fe, Cu Zn, Mo Ni	Cu, Mo Ag Hg	Pb, Zn, Ag W, Cu Mo, Sn, U	Pb, Zn Ag Th, Cu
Background in stream sediment (ppm)					
Cu	63	45	38	26	25
Pb	6.5	4.0	2.5	22	18
Mo	2.5	2.1	2.6	<1.0	1.0

that this may be so generally is illustrated by data from the Canadian Cordillera shown in Table 11-I. There is more than a two-fold variation in background levels of copper and molybdenum and almost a ten-fold variation in the background of lead across the five main geological units; there is a general (but not precise) correlation between background levels and dominant element assemblages in mineral deposits in the five zones.

Clearly stream sediment surveys should be increasingly useful to reveal these patterns and could provide a relatively rapid and cheap method of assessing some of the more remote areas of the world. However, while stream sediment surveys designed to locate targets of less than 50 km^2 — using varying sample densities from a minimum of about one sample per 6 km^2 for reconnaissance purposes, down to extremely detailed sampling (perhaps one sample every 30 m) along a particular stream — are well established, the use of more widely spaced samples to locate and identify geochemical provinces is much less well developed. One of the earliest demonstrations of the value of regional stream sediment surveys was in an area of about 8,000 km^2 in Zambia, where Webb et al. (1964a, b) showed that mineralized rock units were generally reflected by high metal contents in stream sediments, and that the major geological units were associated with characteristic variations in the range and average contents of some elements in stream sediments. A more comprehensive study over an area of 207,000 km^2 of central Zambia was undertaken with a sample density of about one stream sediment per 200 km^2 (Armour-Brown and Nichol, 1970); each sample normally represented a catchment area of about 26 km^2, although some samples representing a catchment area of as much as 260 km^2 were also collected; the 26-km^2 catchment area samples showed a more consistent relation to the geochemistry in the upstream area. A similar study over the Basement Complex in Sierra Leone (Garrett and Nichol, 1967) used drainage samples with a catchment area of up to 39 km^2.

Interpretation of drainage data is especially difficult on a regional basis because of the problem of discriminating between the effects of general background fluctuations (due to inherent local variations, sampling, and analytical variability) and the effects due to surface environment, bedrock, and mineralization. Notwithstanding these problems, Armour-Brown and Nichol (1970) showed that, after smoothing the analytical data by a computerized moving average technique, variations in background metal contents of stream sediments (anomalous samples were removed from the data set) were associated with the main metallogenic zones; the results for the distribution of copper are shown in Fig. 11-6 (the high copper value in the southwest is due to basalts). The controlling role of bedrock geochemistry in determining the trace element distribution in stream sediments was confirmed through analysis of rock and soil in selected areas.

The validity of low-sample-density drainage surveys is further illustrated in Fig. 11-7 by comparison between the distribution of copper in stream sedi-

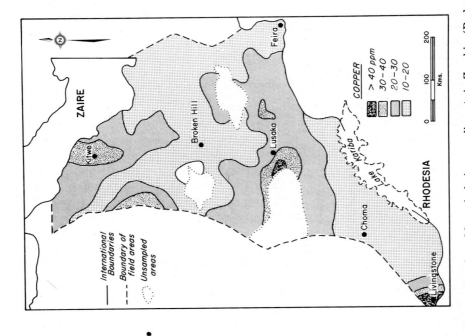

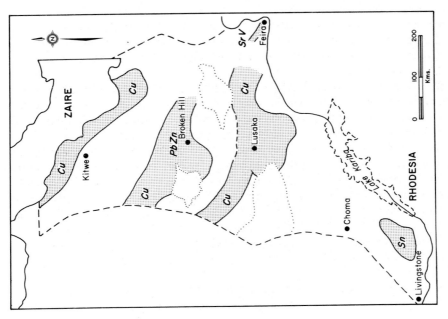

Fig. 11-6. Metallogenic zones and moving average distribution of copper in minus 80-mesh drainage sediments in Zambia. (Redrawn with permission from Armour-Brown and Nichol, 1970, *Economic Geology*, Vol. 65, 1970, fig. 2B (p. 314) and fig. 3A (p. 319).)

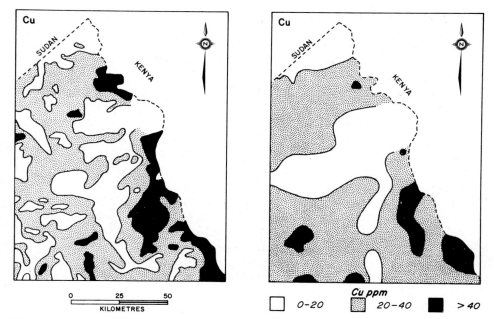

Fig. 11-7. Moving average distributions of copper in stream sediments from northern Karamoja, Uganda, based on 10,000 samples (left) and on 100 samples (right). (Redrawn with permission from Reedman and Gould, 1970, *Transactions Institution of Mining and Metallurgy, Section B*, Vol. 79, 1970, fig. 1b and fig. 1e, p. B247.)

ments at a density of one sample per 1.8 km^2 and a simulated sample density of one sample per 180 km^2 over 18,000 km^2 of basement rocks in Uganda (Reedman and Gould, 1970); these results are also presented as moving average contours to eliminate background "noise". It is evident that the broad pattern of regional variation is reflected in the low-density sample pattern. The practical significance of this is that the collection and analysis of 10,000 samples for the high-density survey took the equivalent of ten man-years, while it is estimated that the one hundred samples of the simulated low-density survey would have taken six man-months to collect and analyze.

In the above example, a simple univariate data treatment of data smoothing was used; similar results can be obtained from more complex polynomial trend surface fitting (Nichol et al., 1969), and additional information can usually be derived from consideration of multi-element data. Visual interpretation of the distribution of more than several elements normally reveals only the obvious features, whereas, especially in the case of low-density sample surveys in areas of complex geology, subtle but possibly important features may be revealed through statistical interpretation of many elements. For example, anomalous contents of arsenic in regional stream sediments from a mineralized schist belt in Sierra Leone were found to be related to disseminated arsenic in ultrabasic rocks and to gold mineralization; the two populations could not be distinguished on the basis of arsenic data alone, but could be separated by R-mode factor analysis of multi-element analyses of the samples (Garrett and Nichol, 1969).

The use of regional reconnaissance level drainage surveys is not possible in all environments: on the Canadian Shield disorganized and impeded drainage militates against the use of stream geochemistry on almost any scale. On the other hand, it has been estimated (Allan, 1971) that the Shield has enough lakes to provide samples at a density of at least one per 25 km^2. The general viability of lake sediment sampling as an exploration tool on both a regional and a local scale has been convincingly demonstrated by a number of recent studies (Allan, 1971; Allan et al., 1973). A comparison between a lake sediment survey with sample sites at 5-km intervals and an airborne (gamma-ray) survey over 47,000 km^2 of the Canadian Shield showed that both techniques detected major crustal variations in potassium and uranium (Allan and Richardson, 1974). The time involved and cost of both surveys was similar, but the lake survey technique is potentially the more useful since analysis of the samples could be extended to include many more elements than the two actually determined.

Regional rock geochemical techniques

The success or failure of any geochemical technique depends ultimately upon there being geochemical differences in bedrock. The regional reconnaissance techniques described above detect broad geochemical variations in bedrock; the advantage of using stream or lake sediments rather than the bedrock itself as a sampling medium is that the sediments provide a natural composite of a large area and thus the background noise is reduced, essentially, by averaging. To achieve the same results with rocks as a sampling medium would, in most cases, require that a far greater number of samples be collected and analyzed.

Exploration geochemists (at least outside of the Soviet Union) have apparently been loath to use rock as a sample medium, although stream sediments are an average of the *weathered* residue and are only a distant reflection of bedrock conditions. The need to detect hidden mineralization (which may nevertheless be sub-outcropping) through thick overburden — whether glacial till or tropical soils — and the requirement for rapid and cheap reconnaissance techniques to assess very large regions led to the use of soil and stream geochemical techniques to identify secondary anomalies which tended to be extensive geographically and generally intensive. Also, much of the early routine analytical facility was able to detect only numerically large chemical differences; anomalies in bedrock tend to be numerically very small and hence were undetected or regarded as having too limited an extent to be of much value in exploration except in very detailed studies.

Rock geochemistry is now being used in an attempt to recognize potentially mineralized granitic rocks (granitoids) — a problem which has long been a major preoccupation of economic geologists and geochemists, since these rocks are host to many important mineral deposits, including copper porphyries and many of the tin, tungsten, and molybdenum deposits in the world. The intuitive expectation that granitoids associated with mineralization should be chemically distinct from barren granitoids, at least in terms of a significant enrichment or depletion of ore elements, generally has not been substantiated; despite an enormous amount of work which has resulted in a large literature, there are no satisfactory criteria which appear to have universal acceptance. Many writers make the point that there seems to be no simple correlation between ore elements in granitoids and apparently genetically related ore deposits within them (Tauson, 1966); for example, Barsukov (1966) has stated that for deposits of tungsten, lead, zinc, molybdenum, and tin in granitoids in the U.S.S.R., only tin shows a marked enrichment in granitoids associated with deposits of that element. Similarly, Sheraton and Black (1973) showed that there is no systematic enrichment or depletion of copper, lead, or zinc in granitic rocks associated with the corresponding mineralization in northeast Queensland (Australia), but again granites associated with tin deposits are significantly enriched in this element (as well as in boron, beryllium, lithium, fluorine). The enrichment of tin has been remarked in tin districts in many parts of the world, but it is by no means an accepted reliable criterion.

The concentration of ore elements in particular minerals in granitoids appears to offer greater promise for recognition of element concentrations related to ore deposits. Thus, Bradshaw (1967) showed that tin, lead, and zinc are distinctly enriched in feldspars (one of the dominant minerals in granite, along with quartz and mica) for mineralized granites in the west of England compared with non-mineralized granitic stocks elsewhere in the U.K. Granitoid intrusions in the Basin and Range structural province of the U.S. have associated copper, lead, and zinc ore deposits, and it has been found that the lead content of feldspars is higher in granitoids associated with mineralization than those which are not (Slawson and Nackowski, 1959), and also that there is more copper in biotite mica in mineralized than in barren granitoids (Parry and Nackowski, 1963). On the other hand, Blaxland (1971), on the basis of only 44 samples from 37 granitoids in the U.S. and the U.K. claimed to find no significant difference in the amount of zinc in biotite mica from granitic rocks associated with mineralization compared with barren granitoids.

Another approach to the problem of recognizing potentially productive granitic rocks is based on the conclusion that most hydrothermal ore solutions are chloride-rich brines; fluid inclusions in ore minerals have a high chloride content, and in some cases also have a high fluoride content. Thus, if the rock is crushed, most of the material in the fluid inclusions is released, and

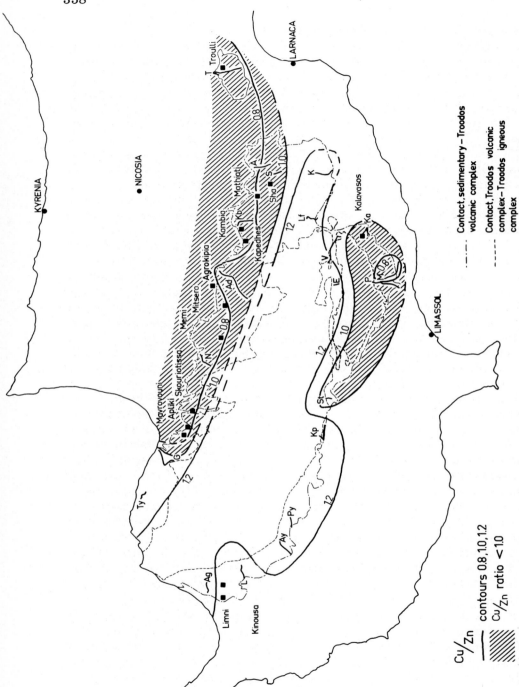

Fig. 11-8. Distribution of hot nitric acid-soluble copper/zinc ratios in centres of basaltic pillow lavas, Troodos Volcanic Series, Cyprus in relation to cupriferous pyritic sulphide deposits.

halides can be readily extracted in water. Kesler et al. (1973) measured the water-soluble chloride and fluoride contents of crushed samples from nine granitic intrusions from the Greater Antilles and Central America (two of the nine intrusions have porphyry-type copper mineralization and two are associated with contact-type copper mineralization); they found that there is no distinction in the chloride abundance between the mineralized and barren intrusions, although the average fluoride content and the range of individual values is greater in the mineralized than in the barren intrusions. Stollery et al. (1971) showed that both biotite mica and whole rock of the Provindencia granodiorite stock in Mexico which has associated lead-zinc-silver mineralization have chloride enrichment of two to three times normal values; they suggested that chlorine may be of use in regional exploration, but pointed out the many intrusions which are not associated with any mineralization and are also abnormally rich in chloride.

Although some very simple situations do exist, such as that shown in Fig. 11-8 where most of the cupriferous pyrite deposits in Cyprus are shown to be spatially associated with regionally low concentrations of copper and regionally high concentrations of zinc, regional geochemical indices of the ore-bearing potential of rock units are expected to be difficult to recognize. The difference in the regional mean value of individual elements between mineralized and non-mineralized rock units is likely to be very small, and a number of workers have suggested that a more realistic assessment of differences can be expected by consideration of the frequency distributions of elements (Cameron and Baragar, 1971; Garrett, 1971; Govett and Pantazis, 1971). This concept is illustrated in Fig. 11-9 for the frequency distribution of cobalt in basic lavas from Cyprus; there is an almost total overlap of the frequency distributions of the background and the anomalous populations, but the modal value of the anomalous group of samples is some 10 ppm higher than the background samples and, equally important, the anomalous distribution has a positive tail of higher values and hence a greater variance.

A comparison of the distribution of copper in basaltic lavas of the Coppermine River Group and the Yellowknife Group in the Northwest Territories, Canada (Cameron and Baragar, 1971) also illustrates this approach. The Coppermine River Group contains many copper occurrences, including one of economic size and grade (not less than 3.6×10^6 tons of 3.4% copper); the Yellowknife Group has no known copper mineralization in the southern part of the Mackenzie District where the investigation was conducted. In Fig. 11-10 the cumulative frequency distribution of logarithmic copper values are plotted for the two series; the Yellowknife Group of samples shows a curved distribution (which approaches an arithmetic normal distribution), whereas the Coppermine Group approximates a log-normal distribution (i.e., has a high positive skew). The barren Yellowknife Group has a higher modal value than the mineralized Coppermine Group and also appears to be inherently richer in copper. The very important distinction between the two groups is

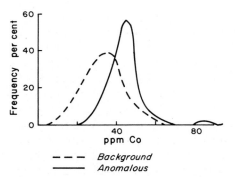

Fig. 11-9. Frequency distributions of cobalt in centres of basaltic pillow lavas from back-ground areas and an anomalous area near a cupriferous pyritic sulphide deposit, Cyprus.

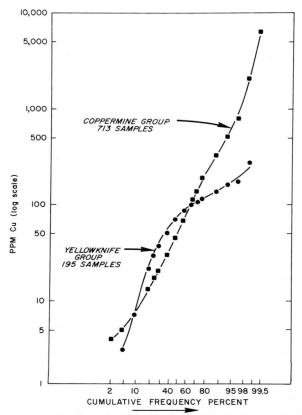

Fig. 11-10. Cumulative frequency distributions of copper in basaltic lavas of the Copper-mine and Yellowknife Group, Northwest Territories, Canada. (Redrawn with permission from Cameron and Baragar, 1971, *Canadian Institute of Mining and Metallurgy, Special Volume*, 11, 1971, p. 575.)

TABLE 11-II

Arithmetic mean ($\bar{x}$) and geometric mean (G) values of ascorbic acid/hydrogen peroxide-soluble copper, nickel, and cobalt and sulphur in ultrabasic rocks from various locations in the Canadian Shield and Eastern Townships of Quebec (Cameron et al., 1971)

	Cu (ppm)		Ni (ppm)		Co (ppm)		S (%)	
	$\bar{x}$	G	$\bar{x}$	G	$\bar{x}$	G	$\bar{x}$	G
Ore-bearing ultrabasic (16 locations, 372 samples)	439	68	1875	715	84	57	1.37	5.15
Barren ultrabasic (40 locations, 616 samples)	26	7	579	354	44	31	0.059	0.031

in the samples above the 70th percentile which in the Coppermine Group display a long tail of high values indicative of sulphide segregation and hence potentially mineralized conditions.

It has been recognized that the differences in element concentration between ore-bearing and barren rocks are generally extremely small; however, some of the trace element variation can be attributed simply to major element variation reflecting petrological changes. A more powerful technique than the single-element studies described above is to use the combined variation of many elements. Principal component analysis and multiple regression of major and minor element data for Cretaceous acid plutonic rocks of the northern Canadian Cordillera enabled Garrett (1973) to delineate certain intrusions as being of interest for potential mineralization; the raw data were not nearly selective enough, and in some cases, failed to reveal nearby deposits. Discriminant analysis was used by Cameron et al. (1971) to distinguish nickel-copper bearing from barren ultrabasic rock intrusions from 56 locations across the Canadian Shield and the Eastern Townships of Quebec; in this example, the differences in element concentration between barren intrusions and those containing significant mineralization is large, as shown in Table 11-II.

METHODS FOR DEEPLY BURIED DEPOSITS

Conventional soil geochemistry has proved successful for detecting deposits covered by tens of metres of soil but, until quite recently, it has not been

362

considered probable that geochemical techniques would prove effective in locating deposits covered by hundreds of metres of barren rock or very thick deposits of transported overburden. However, local-scale rock geochemical techniques based on measurements of the usual trace elements (copper, lead, zinc) and major elements (iron, calcium, magnesium, sodium, potassium, etc.), and the measurement of volatile and gaseous constituents have extended the depth penetration capability of geochemistry to the hundreds of metres range.

Rock geochemical methods

Rock geochemical techniques offer great promise for detecting sulphide deposits which are covered by considerable thicknesses of non-mineralized and, in some cases, post-mineralization cap rock. Dispersion halos of trace and major elements are now being regularly reported beneath, lateral to, and above sulphide deposits in various host rocks at distances far beyond those of recognizable mineralogical alteration. (In the ensuing discussion dispersion halos of tens of metres, such as those illustrated in Fig. 11-5, are excluded; only dispersion of hundreds of metres or more is considered).

Considerably more work appears to have been done in the U.S.S.R. than elsewhere on primary halos associated with specific ore deposits. The Russian work is characterized by the determination of a very large number of elements. For example, Ovchinnikov and Baranov (1972) state that pyritic ores in Caucasus, Urals and Rudny Altay, regardless of province or age of the ore, all have halos of some 28 elements; halos are said to extend for several kilometres along strike, hundreds of metres to a few kilometres downwards, and 200—300 m over orebodies. Another feature of the work in the U.S.S.R. is the establishment of a comprehensive zoning pattern of the elements in the halos which allows the determination of the level of the mineralization, i.e., whether the surface intersection is far above the ore or nearby, and also allows discrimination between hanging-wall halos related to ore beneath the ground and footwall halos related to ore that has been removed by erosion. It is claimed that there is correspondence of the zoning sequences in halos around hydrothermal ore deposits of many types and composition, and a generalized sequence from those elements most common above the ore to those most common below the ore has been proposed (Grigoryan, 1974).

Studies outside the U.S.S.R. have been far less comprehensive, but the work that has been done shows that ore deposits may be detected at considerable distances through anomalous halos in rocks. An interesting feature of this work has been to show that even volcanic-sedimentary massive sulphide deposits have extensive halos. A common characteristic is enrichment of magnesium and a tendency for a loss of alkali, e.g., Archaean massive sulphides at Timmins-Matagami-Noranda on the Canadian Shield (Bennett and Rose, 1973); the copper-zinc deposits of the Abitibi volcanic belt in Superior province of the Canadian Shield have positive magnesium halos of

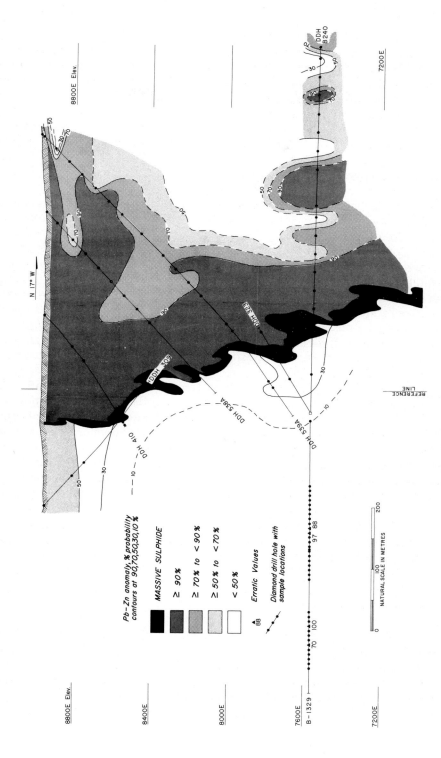

Fig. 11-11. Contours of probability of samples of quartz feldspar porphyry containing anomalous concentrations of lead and zinc around the Heath Steele B-zone Zn-Pb-Cu orebody, New Brunswick, Canada. (Compiled from Whitehead and Govett, 1974.)

about 200 m and negative sodium halos of about 600 m in extent (Descarreaux, 1973). At the Brunswick No. 12 zinc-lead-copper massive sulphide deposit in New Brunswick a broad halo extending 450 m laterally away from the deposit and 450 m below the deposit has been defined by an increase in potassium/sodium and magnesium/calcium ratios towards the deposit (Goodfellow, 1975a); a hemispherical halo of 450—600 m radius has been similarly defined in hanging-wall rocks at Brunswick No. 12 deposit. It is of interest that a marked enrichment of magnesium has also been recorded in biotite mica around copper porphyry deposits (Moore and Czamanske, 1973; Patton et al., 1973).

Less work appears to have been done on trace elements which, in general, seem to give weaker but, in some cases, more extensive anomalies. Computer processing by discriminant analysis of trace element data from basic volcanic rocks in Cyprus has shown that anomalous halos of copper-zinc-cobalt can be indentified several kilometres laterally from cupriferous sulphide deposits and for several hundreds of metres over deposits in rocks which post-date mineralization (Govett, 1972). A similar technique applied to lead-zinc data from acid volcanic rocks around the Heath Steele zinc-lead deposit in New Brunswick defined an anomalous halo zone extending 370 m stratigraphically above the deposit in hanging-wall rocks (which are presumed to post-date mineralization) and 1,200 m laterally from the deposit, also in hanging-wall rocks (Fig. 11-11; Whitehead and Govett, 1974).

An extensive study of the distribution of trace elements in rocks (and soils) in the lead-zinc-silver district of the Coeur d'Alene district of northern Idaho revealed extensive halos and pronounced zoning of metals related to ore zones (Gott and Botbol, 1973). The ore metals lead, zinc, copper, and arsenic are not themselves as good indicators of mineralization as ratios of these elements to silver and copper. Shallow- to intermediate-depth deposits are best shown by silver (and antimony) anomalies; cadmium is enriched relative to zinc at greater distances from the deposits, while tellurium and manganese appear to have been dispersed furthest and are most useful for detection of deeply buried deposits. Gott and Botbol (1973) refer to a tellurium halo in soils over a deposit lying at a depth of more than 600 m below the surface; they also make the interesting point that the areal dispersion patterns in soils are as large or larger than in rocks.

Volatile constituents

One of the most exciting new developments which holds great promise for the detection of deeply buried mineral deposits is the measurement of volatile constituents in both the solid phases of the earth and in the atmosphere; measurements in the atmosphere offer the possibility of direct airborne geochemical exploration (see next section). The fundamental premise upon which this work is based is that if volatile constituents in abnormal amounts are associated with mineralized conditions they will, naturally, migrate considerably further from their source than the elements which are conventionally measured in exploration.

The continuous escape of various gases into the atmosphere is pronounced in areas of deep tectonic fracturing and jointing; the composition and concentration of gaseous emanations from the earth also varies according to local geothermal conditions. Not all high concentrations of volatile constituents are related to ore deposits, and because of the strong control of the physical character of the rocks, interpretation of results is often difficult. These geological difficulties are compounded in the interpretation of the distribution of volatile components in the air by the influence of climate and weather (temperature, wind, atmospheric pressures) and sampling level. The strong structural control is indicated by variations caused by seismic activity; a carbon dioxide anomaly over a mercury deposit in the Caucasus showed a threefold increase during an earthquake, and during the Tashkent earthquake in 1966 the helium content in water in a borehole rose twelve times and radon increased three times (Ovchinnikov et al., 1973).

The possibility of measuring volatile constituents in surface materials as a guide to metallic deposits probably dates from the work of Saukov (1946) who commented on the association of mercury with many sulphide minerals, an observation which is now firmly established. The use of mercury dispersion halos in exploration work has been slow, largely because of the need for extremely sensitive instrumentation to differentiate anomalous concentrations from very low background levels (normally less than 100 ppb) and also because of problems caused by interference (these problems also arise with other volatile constituents). Today laboratory and portable instruments are available that allow mercury to be rapidly and reliably measured in rocks, soils, soil air, and the atmosphere; there is a considerable body of case history data which firmly establishes mercury as a potentially extremely useful indicator element for many types of sulphide deposits.

The measurement of mercury in bedrock is having an increasing application (Ozerova, 1959; Bradshaw and Koksoy, 1968; Smith and Webber, 1973). The distribution of cold-extractable mercury in rocks along a drill hole around the Brunswick No. 12 zinc-lead-copper deposit in New Brunswick is shown in Fig. 11-12; ore shoots at depth below the drill holes are clearly indicated by strong peaks in the mercury content. Measurement of mercury is widely used in areas of thick overburden. Friedrich and Hawkes (1966) report mercury anomalies of 150—600 ppb (compared with background values of 50 ppb) in soils where there is no known mineralization closer than 120 m to the surface in the Pachuca—Real del Monte silver-producing district of Mexico. Warren et al. (1966) discussed the use of mercury in soil and vegetation for exploration purposes in British Columbia (Canada), while Fursov (1970) gives comparative data for mercury in both soils and soil gas (a review of recent work is given in McCarthy, 1972). Comparative results for mercury in soils

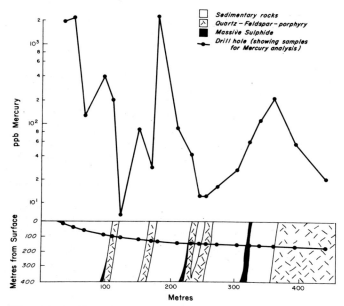

Fig. 11-12. Distribution of mercury in rocks around the Brunswick No. 12 Zn-Pb-Cu orebody, New Brunswick, Canada. (Compiled from data in Goodfellow, 1975b.)

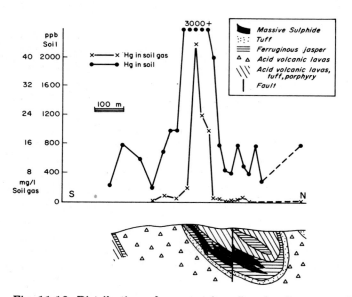

Fig. 11-13. Distribution of mercury in soil and soil gas over Almagrera massive sulphide deposit, Spain. (Data courtesy J. Robbins, Scintrex Ltd., Tharsis Co., and Union Explosives-Rio Tinto; geological information based on Strauss and Madel, 1974.)

and soil gas over the Almagrera massive sulphide deposit (1.5% zinc and 0.6—0.8% lead) in Spain are shown in Fig. 11-13 (J. Robbins, Scintrex Ltd., Canada, personal communication, 1975). Both soils and soil gas show strongly anomalous values associated with the location of the ore deposit, and it is of interest that there are no particularly high concentrations associated with the vertical post-mineralization fault through the deposit.

Many other gaseous constituents are being investigated as possible useful pathfinders for buried mineralization, although their utility is not nearly as well established as that of mercury. These include sulphur dioxide (Rouse and Stevens, 1971), fluorine, and iodine (Frederickson et al., 1971). Recent reviews of the use of these and other gases are given by Bristow and Jonasson (1972) and McCarthy (1972).

SOME OTHER APPLICATIONS AND NEW TECHNIQUES

Theoretical models

Since the actual measured concentration of an element or a chemical component of the materials of the earth's crust is the result of many complex independent and interdependent processes, exploration geochemists are now giving considerably more attention to theoretical studies to provide models of element distribution relevant to exploration. Underlying bedrock (which may or may not be host of a mineral deposit), concentration of elements in the soil, the soil horizon, the climatic regime, local physiographic conditions, and local biological and microbiological processes will all affect secondary dispersion patterns; the genetic origin of an ore deposit and its subsequent geological history will intimately affect the primary dispersion halo in the rocks around a deposit. Obviously, interpretation of geochemical patterns would be much more reliable if the processes of element distribution were better understood.

Landscape geochemistry studies explain the movement, distribution, and interaction of all elements between bedrock, soil, drainage systems, biosphere, and atmosphere in terms of local geological, geomorphological, and climatic conditions (Glazovskaya, 1963; Fortescue, 1975); conceptual models of element distribution for type situations have been proposed by Bradshaw (1975). The experimental work of Govett (1973, 1975) and Bölviken and Logn (1975) seeks to provide some explanations of element distribution based on electrochemical processes for secondary movement of elements around sulphides. Conceptual models for palaeoenvironmental conditions and deposition of volcanic-sedimentary massive sulphides are being used to improve interpretation of primary halos in the rocks around them (Whitehead, 1973; Goodfellow, 1975b).

Techniques in glaciated terrain

In the case of two of the richest mineralized areas of the world — Canada, and the U.S.S.R. and Scandinavia — there are special problems, inasmuch as these regions were heavily glaciated during the Quaternary; large areas (89% of Canada; Nichol and Bjorklund, 1973) were blanketed by thick deposits of transported glacial debris which have little relation to the underlying rock in many places. Stream drainage in such situations is generally poorly developed, and conventional geochemical responses in soils developed upon glacial material is often weak or non-existent. A further problem is that about 50% of the area of Canada and of the U.S.S.R. is within the permafrost zone. The problems of such glaciated and permafrost regions are being solved by a variety of techniques.

Contrary to earlier belief, it has been convincingly demonstrated (Allan and Hornbrook, 1970; Shvartsev, 1972; Ridler and Shilts, 1974) that permafrost areas are geochemically active, and that measurable dispersion can be found in soils and summer drainage. Sampling of sediments in the numerous lakes, which are one of the apparent disadvantages of glacially disrupted drainage, has been shown to be an effective means of geochemical exploration (see above). To overcome problems caused by thick glacial cover, a number of workers have used shallow drilling to obtain samples of the basal till (Gleeson and Cormier, 1971; Wennervirta, 1973), while others have had considerable success with vegetation sampling (Warren et al., 1964; Fortescue, 1967).

The problems of glaciated terrains have spawned some quite novel ideals, e.g., the successful use of dogs, especially in Scandinavia, to trace glacial boulders containing sulphides (Hyvarinen et al., 1973), and the sampling of snow (Kolotov et al., 1965; Jonasson and Allan, 1973). An electrochemical approach to the problem of detecting mineralization beneath thick glacial (and other) cover suggested by Govett (1973, 1975) relies on the measurement of total water-soluble ions by measuring the conductivity of a water slurry or on direct measurement of individual water-soluble cations and anions.

Petroleum and marine geochemistry

Geochemical techniques for hydrocarbons have had as long a history as those for metals — dating from Russian and German work in the early 1930s and American work in the late 1930s (Pirson, 1940; Rosaire, 1940). While work in this field has not prospered in the past two decades, half a dozen papers were presented at the 3rd International Geochemical Symposium (Boyle and McGerrigle, 1971), and an interesting new development is the possibility of airborne detection of iodine which is associated with oilfield brines and oil source rocks (Barringer, 1970).

Recent interest in the oceans as a source of mineral raw materials has promoted work in marine geochemistry (Siegel, 1971; Webb, 1971). A specific case history is reported by Holmes and Tooms (1973), who showed that element distribution in particulate matter and filtered seawater could be related to the metalliferous deposits in the Red Sea.

Airborne geochemical techniques

Airborne geochemical methods are in a very early stage of development but deserve mention here because of their potential application for rapid reconnaissance mineral exploration. There are essentially two approaches to airborne geochemistry: one is the detection and measurement of trace gases by adsorption spectra present in reflected solar radiation with an aircraft-monitered correlation spectrometer (Barringer, 1970); the other is actual sampling of air and its included particulate and aerosol fraction and analysis of the physical samples. The former technique has been successfully applied to the pollution problem, but, as pointed out by Barringer (1970), it has obvious application in detection of gases such as sulphur dioxide and other sulphur gases liberated from oxidizing ore zones, and of iodine and hydrocarbon gases associated with oil fields.

Air sampling techniques appear to offer a broader scope for development. As with ground methods based on gaseous components, mercury seems to have been the first gas to be shown to give an atmospheric anomaly associated with mineral deposits on the ground (Williston, 1964, 1968); Barringer (1971) has published results of an airborne traverse at an altitude of 60 m obtained by continuous profiling optical spectrometer measurements of mercury around Cobalt, Ontario (Canada), which shows a distinct anomaly corresponding with the silver mining district.

A potentially more significant and useful technique for rapid airborne geochemical exploration, particularly in heavily forested regions, relies upon measurement of metals in the organic fraction of the atmosphere (A.R. Barringer, personal communication, 1975). The organic matter is generated at the surface of the earth and rises into the atmosphere, where it absorbs and concentrates volatile constituents which have also travelled upwards from the earth's surface in gaseous form; this process appears to enhance geochemical contrast. Some metals may become incorporated into the organic fraction on the ground before it rises into the atmosphere (Curtin et al., 1974).

The AIRTRACE® * system is an airborne technique which measures elements in particulate compounds in the atmosphere. Most of the principal ore elements and some major elements are determined by an ultra-sensitive analytical technique. It is not possible to weigh each individual sample because of the small amount of material collected; however, the concentration of major

_____

* This is a registered name of a system developed by Barringer Resaerch Ltd., Toronto, Ontario, Canada.

elements is relatively constant, and the ore elements can therefore be expressed as ratios of a major element.

Other airborne techniques are more indirect in their use, and include airborne gamma-ray spectrophotometer measurements for radioactive mineralization (Barringer, 1970; Bennett, 1971) and infrared and colour photography. The latter techniques have been used to show that colour infrared indicates changes in vegetation growth associated with unusual trace element concentrations in soil, as well as general vegetation changes due to broad geological differences (Lyon and Lee, 1970; Press and Norman, 1972; McCarthy, 1973). Remote-sensing techniques are discussed in more detail in Chapter 12.

CONCLUSIONS

Geochemical techniques for locating near-surface ore deposits by stream sediment sampling and detailed soil sampling in areas of residual soil are now well-established. These methods will continue to have wide application for the search for sub-outcropping mineralization in the less intensively prospected areas of the world, especially in tropical zones with thick residual overburden; in such areas they may be expected to make significant contributions to the search for new deposits. However, in the intensively prospected areas of the world now — and in other areas later — the search must inevitably penetrate to greater depths to locate deposits beneath hundreds of metres of barren rock.

Extensive halos around blind, deeply buried mineral deposits have been revealed by multi-element rock geochemical surveys, and further development of these techniques will provide one of the major methods for locating deep deposits. The use of gaseous constituents is expected to be particularly important because of the ability of gases to migrate considerable distances from a deposit. Gaseous anomalies are similarly expected to be of increasing utility in detecting mineral deposits beneath thick transported overburden (including glacial debris and desert sand) and in regions of permafrost.

Widely spaced drainage sediment sampling and statistical interpretation of multi-element analyses may be expected to delineate regional geochemical provinces; similarly, multi-element analysis of reconnaissance-scale rock samples should assist in discriminating between potentially ore-bearing and barren geological units. The compilation of regional geochemical maps — based on stream sediment, soil, and rock analyses — are likely to become as standard an adjunct to mineral exploration as the geological map, and, in fact, to provide an invaluable aid to geological mapping itself. Such maps will also have wide-ranging applications in the fields of agriculture, medicine, and environmental control.

Progressive improvement in exploration geochemical methods followed improved analytical and interpretative capabilities. Whereas considerable advances are being made in empirical interpretation through computer-assisted multivariate statistical analysis, major improvements now require a better understanding of the fundamental processes which control element concentration in the earth's crust. Some progress is being made on this aspect of geochemistry; the long-term significance of the work is not merely a higher rate of success in finding individual ore deposits, but the possibility of forecasting the actual quantity of ore in an area; suggestions for this have already been made by Russian geochemists using stream geochemical data (Polikarpochkin, 1971) and rock geochemical data (Solovov et al., 1971).

Airborne techniques, based upon direct measurement of gaseous, organometallic, and possibly aerosol constituents of the atmosphere are in the very early stages of development. Nevertheless, results are sufficiently encouraging to allow a prediction that airborne geochemistry will become a viable reconnaissance exploration method of particular importance in areas of the world where ground access is difficult.

Mineral exploration cannot depend on geochemical methods alone, any more than it can depend on geophysical or geological methods alone. Each has a contribution to make in the search for more mineral deposits; their relative contributions will vary from case to case. However, it is probably reasonable to expect that geochemistry, as the newest of the methods for mineral exploration, will play an increasingly important role in the future. There is every reason to expect, from present trends, that geochemical methods can be developed to significantly increase the efficiency of mineral exploration and to detect deeply buried deposits. Whether these techniques are developed in time to assist in avoiding a mineral supply shortage depends less upon the innate ability of geochemists and the inherent potential of geochemistry than on the requisite pressure and financial resources being brought to bear on the problem.

REFERENCES CITED

Agterberg, F.P. and Kelly, A.M., 1971. Geomathematical methods for use in prospecting. *Can. Min. J.*, 92: 37—58.

Allan, R.J., 1971. Lake sediment: a medium for regional exploration of the Canadian Shield. *Can. Inst. Min. Metall. Bull.*, 64(715): 43—59.

Allan, R.J. and Hornbrook, E.H.W., 1970. Development of geochemical techniques in permafrost Coppermine River region. *Can. Min. J.*, 91: 45—49.

Allan, R.J. and Richardson, K.A., 1974. Uranium and potassium distribution by lake-sediment geochemistry and airborne gamma-ray spectrometry; a comparison of reconnaissance techniques. *Can. Inst. Min. Metall. Bull.*, 67(746): 109—120.

Allan, R.J., Cameron, E.M. and Durham, C.C., 1973. Lake geochemistry — a low sample density technique for reconnaissance geochemical exploration and mapping of the Canadian Shield. In: M.J. Jones (Editor), *Geochemical Exploration 1972*. Institution of Mining and Metallurgy, London, pp. 131—160.

Armour-Brown, A. and Nichol, I., 1970. Regional geochemical reconnaissance and the location of metallogenic provinces. *Econ. Geol.*, 65: 312—330.

372

Barringer, A.R., 1970. Remote-sensing techniques for mineral discovery. In: M.J. Jones (Editor), *Proceedings of the Ninth Commonwealth Mining and Metallurgical Congress, 1969*. Institution of Mining and Metallurgy, London, 2: 649—690.

Barringer, A.R., 1971. Airborne exploration. *Min. Mag.*, 124: 182—189.

Barsukov, V.L., 1966. Metallogenic specialization of granitoid intrusions. In: A.P. Vinogradov (Editor), *Chemistry of the Earth's Crust*. Israel Program for Scientific Translations, Jerusalem, 2: 211—231.

Bennett, R.A., 1971. Exploration for hydrothermal mineralization with airborne geochemistry. *Trans. AIME*, 250: 109—113.

Bennett, R.A. and Rose, W.I., 1973. Some compositional changes in Archean felsic volcanic rocks related to massive sulfide mineralization. *Econ. Geol.*, 68: 886—891.

Blaxland, A.B., 1971. Occurrence of zinc in granitic biotites. *Miner. Deposita*, 6: 313—320.

Bloom, H., 1971. Exploration geochemistry — past, present and future. *Mines Mag.*, May, pp. 12—14.

Bölviken, B. and Logn, Ö., 1975. An electrochemical model for element distribution around sulphide bodies. In: I.L. Elliott and W.K. Fletcher (Editors), *Geochemical Exploration 1974*. Elsevier, Amsterdam, pp. 631—648.

Boyle, R.W., 1967. Geochemical prospecting — retrospect and prospect. *Can. Inst. Min. Metall. Bull.*, 60(657): 44—49.

Boyle, R.W., 1974. The use of major elemental ratios in detailed geochemical prospecting utilizing primary halos. *J. Geochem. Explor.*, 3: 345—369.

Boyle, R.W. and Garrett, R.G., 1970. Geochemical prospecting — a review of the status and future. *Earth-Sci. Rev.*, 6: 51—75.

Boyle, R.W. and McGerrigle, J.I. (Editors), 1971. *Geochemical Exploration. Can. Inst. Min. Metall., Spec. Vol.*, 11, 594 pp.

Boyle, R.W. and Smith, A.Y., 1968. The evolution of techniques and concepts in geochemical prospecting. In: E.R.W. Neale (Editor), *The Earth Sciences in Canada*. University of Toronto Press, Toronto, Ont., pp. 117—128.

Bradshaw, P.M.D., 1967. Distribution of selected elements in feldspar, biotite, and muscovite from British granites in relation to mineralization. *Trans. Inst. Min. Metall.*, 76: B137—B148.

Bradshaw, P.M.D. (Editor and Compiler), 1975. Conceptual models in geochemical exploration — Canadian Cordillera and Canadian Shield. *J. Geochem. Explor.*, 4: 1—213.

Bradshaw, P.M.D. and Koksoy, M., 1968. Primary dispersion of mercury from cinnabar and stibnite deposits, W. Turkey. *Rep., 23rd Int. Geol. Congr., Prague*, 7: 341—355.

Bristow, O. and Jonasson, I.R., 1972. Vapour sensing for mineral exploration. *Can. Min. J.*, 93: 39—44; 85.

Brown, A. Sutherland, 1974. Aspects of metal abundances and mineral deposits in the Canadian Cordillera. *Can. Inst. Min. Metall. Bull.*, 67(741): 48—55.

Cameron, E.M. (Editor), 1967. *Proceedings, Symposium on Geochemical Prospecting, Ottawa. Geol. Surv. Can. Paper* 66-54, 282 pp.

Cameron, E.M. and Baragar, W.R.A., 1971. Distribution of ore elements in rocks for evaluating ore potential; frequency distribution of copper in the Coppermine River Group and Yellowknife Group volcanic rocks, N.W.T., Canada. In: R.W. Boyle and J.I. McGerrigle (Editors), *Geochemical Exploration. Can. Inst. Min. Metall., Spec. Vol.*, 11: 570—576.

Cameron, E.M., Siddeley, G. and Durham, C.C., 1971. Distribution of ore elements in rocks for evaluating ore potential; nickel, copper, cobalt, and sulphur in ultramafic rocks of the Canadian Shield. In: R.W. Boyle and J.I. McGerrigle (Editors), *Geochemical Exploration. Can. Inst. Min. Metall., Spec. Vol.*, 11: 298—313.

Canney, F.C., Bloom, H. and Hansuld, J.A. (Editors), 1969. *Proceedings, International Geochemical Exploration Symposium, Golden, Colorado, 1968. Q. Colo. Sch. Mines*, 64, 520 pp.

373

Cole, M.M., Provan, D.M.J. and Tooms, J.S., 1968. Geobotany, biogeochemistry and geochemistry in mineral exploration in the Bulman-Waimuna Springs area, Northern Territory, Australia. *Trans. Inst. Min. Metall.*, 77: B81—B104.

Constantinou, G. and Govett, G.J.S., 1973. Geology, geochemistry and genesis of Cyprus sulfide deposits. *Econ. Geol.*, 68: 843—858.

Curtin, G.C., King, H.D. and Mosier, E.L., 1974. Movement of elements into the atmosphere from coniferous trees in subalpine forests of Colorado and Idaho. *J. Geochem. Explor.*, 3: 245—263.

Derry, D.R., 1968. Man and minerals . . . exploration. *Can. Inst. Min. Metall. Bull.*, 61(670): 200—205.

Derry, D.R., 1970. Exploration expenditure, discovery rate and methods. *Can. Inst. Min. Metall., Bull.*, 63(695): 362—366.

Descarreaux, J., 1973. A petrochemical study of the Abitibi volcanic belt and its bearing on the occurrence of massive sulphide ores. *Can. Inst. Min. Metall. Bull.*, 66(730): 61—69.

Elliott, I.L. and Fletcher, W.K. (Editors), 1975. *Geochemical Exploration 1974*. Elsevier, Amsterdam, 720 pp.

Fersman, A.E., 1939. *Mineralogical Methods of Prospecting for Mineral Resources*. Academy of Sciences of the U.S.S.R., Moscow, 446 pp.

Fortescue, J.A.C., 1967. Research program for examination of scope of geobotanical and biogeochemical prospecting in Canada. *Geol. Surv. Can. Paper* 66-61, pp. 32—35.

Fortescue, J.A.C., 1975. The use of landscape geochemistry to process exploration geochemical data. *J. Geochem. Explor.*, 4: 3—7.

Frederickson, A.F., Lehnertz, C.A. and Kellog, H.E., 1971. Mobility, flexibility highlight a mass spectrometer-computer technique for regional exploration. *Eng. Min. J.*, 172(6): 116—118.

Friedrich, G.H. and Hawkes, H.E., 1966. Mercury as an ore guide in the Pachuca—Real del Monte district, Hildego, Mexico. *Econ. Geol.*, 61: 744—753.

Fursov, V.Z., 1970. Mercury emanations over mercury deposits. *Dokl. Akad. Nauk S.S.S.R.*, 194(6): 1421—1423.

Garrett, R.G., 1971. Molybdenum, tungsten and uranium in acid plutonic rocks as a guide to regional exploration, southeast Yukon. *Can. Min. J.*, April, 37—40.

Garrett, R.G., 1973. Regional geochemical study of Cretaceous acidic rocks in the northern Canadian Cordillera as a tool for broad mineral exploration. In: M.J. Jones (Editor), *Geochemical Exploration 1972*. Institution of Mining and Metallurgy, London, pp. 203—219.

Garrett, R.G. and Nichol, I., 1967. Regional geochemical reconnaissance in eastern Sierra Leone. *Trans. Inst. Min. Metall.*, 76: B97—B112.

Garrett, R.G. and Nichol, I., 1969. Factor analysis as an aid in the interpretation of regional geochemical stream sediment data. In: F.C. Canney, H. Bloom and J.A. Hansuld (Editors), *Proceedings, International Geochemical Exploration Symposium, Golden, Colorado, 1968*. Q. Colo. Sch. Mines, 64; 245—264.

Glazovskaya, M.A., 1963. On geochemical principles of the classification of natural landscapes. *Int. Geol. Rev.*, 5: 1403—1431.

Gleeson, C.F. and Cormier, R., 1971. Evaluation by geochemistry of geophysical anomalies and geological targets using overburden sampling at depth. In: R.W. Boyle and J.I. McGerrigle (Editors), *Geochemical Exploration. Can. Inst. Min. Metall., Spec. Vol.*, 11: 159—165.

Gleeson, C.F. and Martin, L., 1969. Application of reconnaissance geochemistry in Gaspé. *Can. Inst. Min. Metall. Bull.*, 62(688): 819—823.

Goodfellow, W.D., 1975a. Major and minor element halos in volcanic rocks at Brunswick

No. 12 sulphide deposit, N.B. Canada. In: I.L. Elliott and W.K. Fletcher (Editors), *Geochemical Exploration 1974*. Elsevier, Amsterdam, pp. 279—295.

Goodfellow, W.D., 1975b. Rock geochemical exploration and ore genesis at Brunswick No. 12 massive sulphide deposit, N.B. Unpubl. Ph. D. Thesis, Univ. of New Brunswick, Fredericton, N.B.

Gott, G.B. and Botbol, J.M., 1973. Zoning of major and minor metals in the Coeur d'Alene mining district, Idaho, U.S.A. In: M.J. Jones (Editor), *Geochemical Exploration 1972*. Institution of Mining and Metallurgy, London, pp. 1—12.

Govett, G.J.S., 1972. Interpretation of a rock geochemical exploration survey in Cyprus — statistical and graphical techniques. *J. Geochem. Explor.*, 1: 77—102.

Govett, G.J.S., 1973. Differential secondary dispersion in transported soils and post-mineralization rocks: an electrochemical interpretation. In: M.J. Jones (Editor), *Geochemical Exploration 1972*. Institution of Mining and Metallurgy, London, pp. 81—91.

Govett, G.J.S., 1975. Soil conductivities: assessment of an electrogeochemical technique. In: I.L. Elliott and W.K. Fletcher (Editors), *Geochemical Exploration 1974*. Elsevier, Amsterdam, pp. 101—118.

Govett, G.J.S. and Brown, W.W., 1968. Geochemical prospecting. In: M.H. Tupas, W.E. Hale and G.J.S. Govett (Editors), Exploration for Disseminated Copper Deposits in Humid, Mountainous, Tropical Terrain. *Philippine Bur. Mines, Manila, Rep. Invest.* No. 66, pp. 48—92.

Govett, G.J.S. and Goodfellow, W.D., 1975. The use of rock geochemistry in detecting blind sulphide deposits — a discussion. *Trans. Inst. Min. Metall.*, 84: B134—B140.

Govett, G.J.S. and Pantazis, Th. M., 1971. Distribution of Cu, Zn, Ni and Co in the Troodos Pillow Lava Series, Cyprus. *Trans. Inst. Min. Metall.*, 80: B27—B46.

Govett, G.J.S., Goodfellow, W.D., Chapman, R.P. and Chork, C.Y., 1975. Exploration geochemistry — distribution of elements and recognition of anomalies. *Math. Geol.*, 7: 415—446.

Grigoryan, S.V., 1974. Primary geochemical halos in prospecting and exploration of hydrothermal deposits. *Int. Geol. Rev.*, 16: 12—25.

Hall, J.S., Both, R.A. and Smith, F.A., 1973. A comparative study of rock, soil and plant chemistry in relation to nickel mineralization in the Pioneer area, Western Australia. *Proc. Australas. Inst. Min. Metall.*, 247: 11—22.

Hawkes, H.E. and Webb, J.S., 1962. *Geochemistry in Mineral Exploration*. Harper and Row, New York, N.Y., 415 pp.

Hawkes, H.E., Bloom, H., Riddell, J.E. and Webb, J.S., 1960. Geochemical reconnaissance in eastern Canada. *Proc. Int. Geol. Congr., 20th Sess., Mexico*, 3: 607—621.

Holmes, R. and Tooms, J.S., 1973. Dispersion from a submarine exhalative orebody. In: M.J. Jones (Editor), *Geochemical Exploration 1972*. Institution of Mining and Metallurgy, London, pp. 193—202.

Howarth, R.J., 1971. Empirical discriminant classification of regional stream-sediment geochemistry in Devon and East Cornwall. *Trans. Inst. Min. Metall.*, 80: B142—B149.

Hyvarinen, L., Kauranne, K. and Yletyinen, V., 1973. Modern boulder tracing in prospecting. In: M.J. Jones (Editor), *Prospecting in Areas of Glacial Terrain*. Institution of Mining and Metallurgy, London, pp. 87—95.

Jonasson, I.R. and Allan, R.J., 1973. Snow: a sampling medium in hydrogeochemical prospecting in temperate and permafrost regions. In: M.J. Jones (Editor), *Geochemical Exploration 1972*. Institution of Mining and Metallurgy, London, pp. 161—176.

Jones, M.J. (Editor), 1973. *Geochemical Exploration 1972*. Institution of Mining and Metallurgy, London, 458 pp.

Kesler, S.E., Van Loon, J.S. and Moore, C.M., 1973. Evaluation of ore potential of granodioritic rocks using water-extractable chloride and fluoride. *Can. Inst. Min. Metall. Bull.*, 66(730): 56—60.

Kolotov, B.A., Kiselera, Ye.A. and Ruboykin, V.A., 1965. On the secondary dispersion aureoles in the vicinity of ore deposits. *Geochem. Int.*, 2(4): 675—678.

Levinson, A.A., 1974. *Introduction to Exploration Geochemistry.* Applied Publishing Ltd., Calgary, Alta., 612 pp.

Lyon, R.J.P. and Lee, K., 1970. Remote sensing in exploration for mineral deposits. *Econ. Geol.*, 65: 785—800.

McCarthy, J.H., Jr., 1972. Mercury vapour and other volatile components in the air as guides to ore deposits. *J. Geochem. Explor.*, 1: 143—162.

McCarthy, J.H., Jr., 1973. Annual review 1972. Geochemistry. *Min. Eng.*, 25(2): 61—63.

Moore, W.T. and Czamanske, G.K., 1973. Composition of biotites from unaltered and altered monzonitic rocks in the Bingham mining district, Utah. *Econ. Geol.*, 68: 269—274.

Morgan, D.A.O., 1970. A look at the economics of mineral exploration. In: M.J. Jones (Editor), *Proceedings of the Ninth Commonwealth Mining and Metallurgical Congress, 1969.* Institution of Mining and Metallurgy, London, 2: 305—324.

Nichol, I., 1973. The role of computerized data systems in geochemical exploration. *Can. Inst. Min. Metall. Bull.*, 66(729): 59—68.

Nichol, I. and Bjorklund, A., 1973. Glacial geology as a key to geochemical exploration in areas of glacial overburden with particular reference to Canada. *J. Geochem. Explor.*, 2: 133—170.

Nichol, I., Garrett, R.G. and Webb, J.S., 1969. The role of some statistical and mathematical methods in the interpretation of regional geochemical data. *Econ. Geol.*, 64: 204—220.

Obial, R.C., 1970. Cluster analysis as an aid in the interpretation of multi-element geochemical data. *Trans. Inst. Min. Metall.*, 79: B175—B180.

Ovchinnikov, L.N. and Baranov, E.N., 1972. Endogenic geochemical halos of pyritic ore deposits. *Int. Geol. Rev.*, 14: 419—429.

Ovchinnikov, L.N., Sokolov, V.A., Fridman, A.I. and Yanitskii, I.N., 1973. Gaseous geochemical methods in structural mapping and prospecting for ore deposits. In: M.J. Jones (Editor), *Geochemical Exploration 1972.* Institution of Mining and Metallurgy, London, pp. 177—182.

Ozerova, N.A., 1959. The use of primary dispersion halos of mercury in the search for lead-zinc deposits. *Geochemistry*, 7: 793—802.

Parry, W.T. and Nackowski, M.P., 1963. Copper, lead and zinc in biotite from Basin and Range quartz monzonites. *Econ. Geol.*, 58: 1126—1144.

Patton, T.C., Grant, A.R. and Cheney, E.S., 1973. Hydrothermal alteration at the Middle Fork copper prospect, Central Cascades, Washington. *Econ. Geol.*, 68: 816—830.

Pirson, S.J., 1940. Critical survey of recent developments in geochemical prospecting. *Bull. Am. Assoc. Pet. Geol.*, 24(8): 1464—1474.

Polikarpochkin, V.V., 1971. The quantitative estimation of ore-bearing areas from sample data of the drainage system. In: R.W. Boyle and J.I. McGerrigle (Editors), *Geochemical Exploration. Can. Inst. Min. Metall., Spec. Vol.*, 11: 585—586.

Press, N.P. and Norman, J.W., 1972. Detection of orebodies by remote-sensing the effects of metals on vegetation. *Trans. Inst. Min. Metall.*, 81: B166—B168.

Rankama, K., 1940. On the use of trace elements in some problems of practical geology. *Bull. Comm. Geol. Finlande*, 126: 90—106.

Reedman, A.J. and Gould, D., 1970. Low sample density stream sediment surveys in geochemical prospecting: an example from northeast Uganda. *Trans. Inst. Min. Metall.*, 79: B246—B248.

Ridler, R.H. and Shilts, W.W., 1974. Mineral potential of the Rankin Inlet. *Can. Min. J.*, 7: 32—42.

Rosaire, E.E., 1940. Geochemical prospecting for petroleum. *Bull. Am. Assoc. Pet. Geol.*, 24(8): 1400—1433.

Roscoe, W.E., 1971. Probability of an exploration discovery in Canada. *Can. Inst. Min. Metall. Bull.*, 64(707): 134—137.

Rouse, G.E. and Stevens, D.H., 1971. The use of sulphur dioxide gas geochemistry in the detection of sulphide deposits. Paper presented at *AIME Annu. Meet., 1971,* mimeogr.

Saukov, A.A., 1946. Geochemistry of mercury. *Akad. Nauk S.S.S.R., Inst. Geol. Nauk,* No. 78 (Mineralogy-Geochemistry Series, No. 17), 129 pp.

Schultz, R.W., 1971. Mineral exploration practice in Ireland. *Trans. Inst. Min. Metall.,* 80: B238—B258.

Sheraton, J.W. and Black, L.P., 1973. Geochemistry of mineralized granitic rocks of northwest Queensland. *J. Geochem. Explor.,* 2: 331—348.

Shvartsev, S.L., 1972. Hydrogeochemical prospecting for blind ores in permafrost. *Int. Geol. Rev.,* 14: 1037—1043.

Siegel, F.R., 1971. Marine geochemical prospecting — present and future. In: R.W. Boyle and J.I. McGerrigle (Editors), *Geochemical Exploration. Can. Inst. Min. Metall., Spec. Vol.,* 11: 251—257.

Sinclair, A.J. and Woodsworth, G.J., 1970. Multiple regression as a method of estimating exploration potential in an area near Terrace, B.C. *Econ. Geol.,* 65: 998—1003.

Slawson, W.F. and Nackowski, M.P., 1959. Trace lead in potash feldspars associated with ore deposits. *Econ. Geol.,* 54: 1543—1555.

Smith, E.C. and Webber, G.R., 1973. Nature of mercury anomalies at the New Calumet Mines area, Quebec, Canada. In: M.J. Jones (Editor), *Geochemical Exploration 1972.* Institution of Mining and Metallurgy, London, pp. 71—80.

Solovov, A.P., Dosanova, B.A. and Kosheleva, I.A., 1971. Analysis of results of lithochemical exploration in metallogenic prospecting. *Int. Geol. Rev.,* 13: 954—965.

Stollery, G., Borcsik, M., and Holland, H.D., 1971. Chlorine in intrusives: a possible prospecting tool. *Econ. Geol.,* 66: 361—367.

Strauss, G.K. and Madel, J., 1974. Geology of massive sulphide deposits in the Spanish-Portuguese pyrite belt. *Geol. Rundsch.,* 63: 191—211.

Tauson, L.V., 1966. Geochemistry of rare elements in igneous rocks and metallogenic specialization of magmas. In: A.P. Vinogradov, (Editor), *Chemistry of the Earth's Crust.* Israel Program for Scientific Translations, Jerusalem, 2: 248—259.

Warren, H.V., 1972. Biogeochemistry in Canada. *Endeavour,* 31: 46—49.

Warren, H.V., Delavault, R.E. and Barakso, J., 1964. The role of arsenic as a pathfinder in biogeochemical prospecting. *Econ. Geol.,* 59: 1381—1385.

Warren, H.V., Delavault, R.E. and Barakso, J., 1966. Some observations on the geochemistry of mercury as applied to prospecting. *Econ. Geol.,* 61: 1010—1029.

Webb, J.S., 1971. Research in applied geochemistry at Imperial College, London (abstract). In: R.W. Boyle and J.I. McGerrigle (Editors), *Geochemical Exploration. Can. Inst. Min. Metall., Spec. Vol.,* 11: 45.

Webb, J.S., Fortescue, J.A.C., Nichol, I. and Tooms, J.S., 1964a. Regional geochemical reconnaissance in the Namwala Concession Area, Zambia. *Tech. Comm., Geochem. Prospect. Res. Centre,* No. 47, 42 pp.

Webb, J.S., Fortescue, J.A.C., Nichol, I. and Tooms, J.S., 1964b. *Regional Geochemical Maps of the Namwala Concession Area, Zambia, Nos. 1—8.* Lusaka Geological Survey, Lusaka.

Wennervirta, H., 1973. Sampling of the bedrock-till interface in geochemical exploration. In: M.J. Jones (Editor), *Prospecting in Areas of Glacial Terrain.* Institution of Mining and Metallurgy, London, pp. 67—71.

Whitehead, R.E.S., 1973. Environment of stratiform sulphide deposition: variation in Mn/Fe ratio in host rocks at Heath Steele Mine, New Brunswick, Canada. *Miner. Deposita,* 8: 148—160.

Whitehead, R.E.S. and Govett, G.J.S., 1974. Exploration rock geochemistry — detection of trace element halos at Heath Steele Mines (N.B., Canada) by discriminant analysis. *J. Geochem. Explor.,* 3: 371—386.

Williston, S.H., 1964. The mercury halo method of exploration. *Eng. Min. J.,* 165: 98.

Williston, S.H., 1968. Mercury in the atmosphere. *J. Geophys. Res.,* 73: 7051—7055.

Chapter 12

FUTURE TRENDS IN GEOPHYSICAL MINERAL EXPLORATION

L.S. COLLETT

INTRODUCTION

Geophysics is concerned with the application of the principles of physics to the study of the earth and includes such diverse topics as geodesy (size and form of the earth and its gravitational field), geomagnetism, seismology, geothermometry, and radioactivity. In this chapter we are concerned with the utilization of the concepts, instruments, and techniques of geophysics in mineral exploration.

Geophysical methods were used to locate concealed magnetic iron ore deposits in Sweden more than a century ago; an oil well was drilled in the U.S. in 1926 following the delineation of concealed salt domes; geophysical testing of oil drill holes was used in the 1930s; and the airborne magnetometer was developed and used during World War II. Nevertheless, the period from 1850 to 1950 can be considered as the century of the prospector in mineral exploration. Methods in use were largely the conventional ones of surface prospecting, geological mapping, sampling, pitting and trenching, followed by drilling of favourable showings.

The period since 1950 has been the one in which geophysical exploration techniques experienced steady growth. The airborne magnetometer was modified after World War II for conducting aeromagnetic surveys in Canada with a further development of the direct-reading nuclear precession magnetometer about 1959 (Collett and Sawatzky, 1966; Serson, 1961). Airborne electromagnetic (AEM) surveys were first performed in 1950 after the development of the International Nickel Company (INCO) system. During the period 1950—1970 a number of airborne (Pemberton, 1962) and various ground electromagnetic (EM) systems were developed. A group at Newmont Exploration Ltd. also developed the induced polarization (IP) time domain method for detecting disseminated sulphides (Wait, 1959), and the frequency domain IP method was developed about 1956 (Hallof, 1957). Between 1956 and 1967 the very low frequency (VLF, 15—24 KHz) method was developed by Vaino Ronka, Geonics Ltd., followed by airborne Radiophase and E-Phase (Barringer and McNeill, 1968; McNeill, 1971) and ground Radiohm (Collett and Becker, 1968), and the airborne and ground gamma-ray spectrometer developed at the Geological Survey of Canada (Darnley, 1970).

Both geochemical techniques for mineral exploration (see Chapter 11) and geological concepts of mineral deposits were developed during the period after 1950. Prior to this most exploration geologists had adhered to the hydrothermal theory of ore deposition; the present widely held theories of volcanic origin for massive sulphide ores and of syngenetic origin for base metal deposits in sedimentary rocks were earlier regarded as heretical by most geologists, and the recent concepts relating mineral deposits to the plate tectonic history of a region were unknown (Hutchinson, 1974).

It is fair to say that the last 25 years have seen major developments in all phases of mineral exploration and geological concepts. Major developments are also expected in the next 25 years, although the exact nature of these changes is as difficult to foresee as it would have been for exploration scientists in 1950 to visualize the exploration industry today. The purpose of this chapter is to take a look "over the horizon" at future trends and developments in geophysics and remote-sensing techniques. It can be assumed that the quantitative methods that are already well established will continue to be used and refined, and it is quite unlikely that any new physical properties will be utilized. The need for quantification of lithological identification of rocks in geology is obvious. On the other hand, the product of the geophysicist must be what the geologist needs and can use. In the coming decades the integrated approach will necessitate more team work between the exploration geologist, the geochemist, and the geophysicist.

GEOPHYSICAL METHODS AND MINERAL EXPLORATION

It is doubtful that any major new deposit within 30—60 m of the surface will be found, at least in Canada, by the present exploration methods since most favourable geological areas have been covered at least once by geophysical methods (Hallof, 1974a). However, present methods are adequate for prospecting in areas of the world where they have not been used previously. As the demand for minerals increases, the need to search for deeper ore bodies without surface expression is obvious (see Chapter 11). This need will motivate geophysicists and engineers to design systems and techniques that can delineate and differentiate anomalies at depth.

Physical parameters of geophysical methods

Before proceeding further, it is relevant to review some fundamental considerations that apply to geophysical exploration methods which are based on the detection of a contrast in a significant physical property between an orebody and its enclosing rocks. Such contrasts, expressed as anomalies, may yield valuable information as to the presence, location, attitude, and nature of the orebody.

According to Brant (1967) any geophysical anomaly may be expressed by:

$$A = \sum \Delta p \cdot F \cdot \frac{V}{r^n}$$

where A = measure of the anomaly; Δp = physical property difference; F = acting force, natural or artifical; V = active volume including the shape factor of the causative body; r = distance of the body from the observation point; n = some integral member depending on the form of the body, method used, and measurement made; and Σ = the sum of the effects from the various causative bodies, formations, and surface materials, as several may be contributing to the measured anomaly.

The type of anomaly and method varies, of course, with the physical property being measured. The magnetic method relies on the detection of variations in the geomagnetic field which results from the differing intensity of magnetism in the underlying rocks. The magnetic properties of rocks may be *induced* and/or be remanent. Induced magnetism is developed in the rocks as a result of its presence in the magnetic field of the earth; the intensity depends on the *magnetic susceptibility (k)* of the rock such that:

$$k \text{ (magnetic susceptibility)} = \frac{I \text{ (induced magnetism)}}{H \text{ (magnetic field)}}$$

The magnetic susceptibility of a rock is almost completely dependent upon the amount and nature of its ferromagnetic mineral constituents. By definition, induced magnetism is proportional to the geomagnetic field and is parallel to it. Remanent magnetism, on the other hand, is that part of the magnetic properties of rock which cannot be accounted for by induction. Its origin dates back to the formation of the rock; it may be acquired during the cooling of an igneous rock or during the deposition of detrital grains in a sedimentary rock. Its magnitude and direction bear no relation to the present geomagnetic field.

Gravity prospecting depends on the differences in density of different rock types — where one "rock" type could, in favourable circumstances, be a sulphide ore deposit. Seismic methods are based on the fact that acoustic waves travel at different velocities in rocks of different elasticities; the acoustic waves are generated by detonating an explosive charge at the surface and measuring the time of arrival of these waves at measuring stations at different distances from the point of detonation. Measurement of gamma-ray emissions from the spontaneous decay of radioactive minerals forms the basis of radioactivity geophysical methods; this technique is useful for the search for any mineralization which has some radioactive minerals associated with it (in addition to the search for uranium). Its drawback is that gamma rays are adsorbed by less than a metre of soil and 30 cm of water.

Electrical geophysical methods include: self potential (SP), earth resistivity (ER), and induced polarization (IP). The SP method measures and maps electrical potential differences at the surface which have been generated by electrochemical reactions at depth (commonly due to the presence of a sulphide body). On the other hand, the ER method takes advantage of differences in electrical conductivity (or its reciprocal, resistivity) among different rock types; current (direct current or low frequency alternating current) is fed into the ground at one point and the resultant potential differences are measured at other surface stations. IP is a relatively new method which is generally best-suited to exploration for disseminated sulphides; the method depends upon the charging effects of metallic conductors compared to effects on other materials and measurement of the resultant impedance changes.

Electromagnetic (EM) methods depend upon the imposition of an electromagnetic field at the surface; this causes an electric current to be generated in any subsurface geological conductor. Magnetic or electrical parameters associated with the current are measured. The method is dependent upon the area normal to the inducing field, i.e., on the induction parameter which is related to the dimensions of the conducting body and the effective depth of penetration (skin depth). For the audio frequencies, the skin depth (δ) is given by:

$$\delta = 500 \sqrt{\rho/f}$$

where δ = skin depth in metres; ρ = resistivity in ohm metres, f = frequency in hertz (cps).

Ground-cable induction systems, as well as methods relying upon the natural electromagnetic field such as audio frequency magnetics (AFMAG), and audio magnetotellurics (AMT) depend on both conductivity differences (as in the resistivity methods) and the induction parameter (as in the electromagnetic methods). The very low frequency method (VLF) makes use of radio waves in the 15—24 KHz range (transmitted by some naval stations for communication with submerged submarines) as a source of energy and depends on differences in conductivity and induction.

Among the remote-sensing techniques radar and microwave frequency methods depend mainly on the relative dielectric constant differences where porosity and moisture content are important. For most applications, frequencies above 10^9 Hz penetrate only a few centimetres of most earth materials. Infrared emission is a function of body temperature and the emissivity and reflectance of the soil and vegetation.

In all of these geophysical methods the limitation in detecting causative bodies at depth arises from two main considerations: the decrease of the anomaly response with depth; and the increase of noise with depth. The volume of a body and its distance from the observation point (the factor V/r^n given in the generalized formula above) is critical for deeper exploration. In general, the response for tabular bodies falls off linearly, for cylindrical bodies as the square, and for lenses or spheres as the cube. Lenticular and pipe-like bodies are always more difficult to detect than tabular or dike-like zones because the $1/r^n$ factor involves a higher order of n for the former. Airborne EM and ground methods are limited to between 30 and 120 m of penetration, IP and resistivity methods to about twice the minimum dimension of the body sought, gravity methods to about 120 m for mining work, and magnetic methods to some five times the minimum dimension.

Noise, encountered in geophysical surveys, can be classed into four types (Ward and Rogers, 1967): instrument noise, geologic noise, errors in location and orientation, and disturbance field noise. With present geophysical instrumentation, instrument noise can be made lower than the threshold of other noise sources so that it is not a real problem. The geologic noise consists of many influences depending on the method used; geologic noise is exhibited by physical inhomogeneities, bedrock relief, contacts, shear zones, graphitic conducting zones, and serpentinites (see Ward and Rogers, 1967). Error noise, due to terrain effects and the resultant precise location and orientation of the sensors can seriously affect the quality of the measured field data. Disturbance field noise can be identified for each of the methods used as follows:

(a) gravity — earth tides;

(b) magnetics — diurnal variations, magnetic storms, sudden impulses and micropulsations, and power line interference;

(c) resistivity and IP — telluric currents (natural, low-frequency earth currents), power line interference, and cultural structures (pipelines, fences, etc.);

(d) electromagnetics — sferics (electromagnetic fluctuations taking place in the atmosphere), power line field, and cultural structures.

For gravity and magnetic surveys, the noise effects can be eliminated by repeating a given station at regular time intervals, or by continuously recording at a base station; for resistivity, IP, and EM surveys, natural electric and magnetic field noise often limits applications of these methods, although they can be reduced by proper choice of frequencies and filtering techniques.

Geologic information

One of the most aggravating problems in mineral exploration is the difficulty of obtaining consistently accurate geologic data (Cook, 1969). Geologic maps too often combine fact and interpretation with no indication of where one begins and the other ends. As a consequence, maps of the same area by two competent geologists commonly show remarkable differences. Part of the difficulty lies in the failure to standardize mapping techniques and the general lack of standards. Geologic maps should convey historical, physical, and chemical information on the rocks. Probably the greatest single need in routine field description and identification of rocks is to be able to quantify geologic data.

Geologic guidance is necessary for the choice of physical characteristics to be measured by geophysical techniques. Too little use is made of geology by geophysicists in the final geophysical interpretation. Frequently, geophysical techniques are chosen without an appreciation of the mineralogical components in the material being measured. Interpretation of the exploration data must be based on the characteristics of the orebodies being sought and the variables of their environments. The selection of the appropriate techniques in exploration work is one of the most critical problems faced by the explorationist. The future places a heavy responsibility on the geologist who is now being asked to quantify his data; there is no doubt that great advances will be made in this area in the future with inputs from the geochemist and the geophysicist.

PROBLEM-ORIENTED RESEARCH AND DEVELOPMENT

The advance of geophysics during the next decade depends to a considerable extent on the necessity to find new mineral deposits in the world to meet the demand of mineral consumption. During the immediate future it is expected that technical developments will take place in the following areas:

(a) differentiating between non-economic (geologic noise etc.) and economic geophysical anomalies;

(b) deep penetrating geophysical techniques, including borehole exploration;

(c) multi-frequency and broad-band EM systems, airborne and ground;

(d) multi-frequency and broad-band induced polarization (IP) apparatus and removal of EM coupling;

(e) increased use of multi-sensor airborne systems (EM, gamma-ray spectrometer, high-sensitivity magnetometer, and gradiometer) which will require installed mini-computers for controlling sensors and handling data for digital data recording systems;

(f) electronic navigation systems for digital flight path recording and recovery;

(g) marine geophysical technique development for detection of minerals on and below the sea floor;

(h) mineral prospecting in difficult regions (permafrost, conductive overburden, and rugged topography);

(i) improvements in interpretation and mathematical scale modelling;

(j) exploration application of remote sensing technology (image-forming systems);

(k) exploration strategy.

An excellent review of trends and development in mineral exploration has been made each year since 1964 by Hood (1965—1975) who summarizes the state-of-the-art in geophysical exploration techniques. Since the thrust in the future will be directed to the development of geophysical methods and systems that can detect mineral deposits at depth and, at the same time, aid in mapping of geologic structures, the main emphasis in the following sections will be on methods which are particularly useful for deep penetration.

GROUND GEOPHYSICAL TECHNIQUES

Electromagnetic techniques have been used in mineral exploration for more than fifty years; the principles of EM methods may be found in a few good texts, e.g., Keller and Frischknecht (1966), Grant and West (1965), and Parasnis (1973). The technique is constantly being refined to deal with specific problems; these include:

(1) Difficulties with highly conductive overburden (Paterson, 1973a, b). An instrument developed by Geonics (the EM-25) to deal with this problem also shows promise for detecting large, massive sulphide bodies in the presence of disseminated sulphides and electrolytic conductors.

(2) An experimental multi-frequency EM system (Ward, 1972), which may be the forerunner of EM equipment in the future, is designed to separate signals due to a massive sulphide deposit from noise caused by halo, host rocks, weathered layers, carbonaceous sedimentary rocks, overburden, and topography.

(3) An extremely low frequency (ELF) transmitter (below 100 Hz) is being tested by the U.S. Navy to communicate with deeply submerged submarines (Bernstein et al., 1974; Sturrock et al., 1974).Propagation at these frequencies takes place in the "waveguide" formed between the earth and the ionosphere; low propagation losses allow nearly worldwide communication from a single transmitter. When this station finally commences transmitting, the EM fields will be useful to prospect for mineral deposits. Techniques similar to AMT and magnetotelluric sounding can be used. At frequencies below 100 Hz the depth of investigation should be considerable.

There are many other newly developed EM systems of various types. These include MaxMin II (Apex Parametrics); Pulse EM (Crone Geophysics), a lightweight version of the Newmont equipment that was designed for use on deeply buried sulphides in Cyprus (Dolan, 1969); the Time Domain Ground EM, an experimental system utilizing a primary magnetic field (Lamontagne and West, 1973); Geoprobe EMR-14 (Geoprobe Ltd.), a multispectral system which uses artificial magnetic dipole sources (Ghosh and Hallof, 1974) which is presently used in delineating geothermal and permafrost zones, but which may have use in exploration for deep conductors; and Audio Frequency Magnetotellurics (AMT), a multi-frequency, narrow-band analogue system which employs natural lightning transients as the source field (Strangway et al., 1973; Koziar et al., 1973) and which is useful for penetration of a resistivity cover.

Since Newmont Exploration Ltd. developed the IP method after World War II, the method has been universally accepted for prospecting for disseminated sulphide deposits. During the post-war period many IP anomalies have been drilled, only to find them non-economic and due to clay, graphite, and pyrite; since 1970 studies of the IP phenomena has been concentrated on the complex behaviour of economic and non-economic responses. Collett and

Katsube (1973) and Katsube and Collett (1973) found by laboratory measurements over a frequency range from 0.3 to 100 Hz that it was possible to distinguish iron-rich sulphides and graphite from copper-rich sulphides. Zonge Engineering and Research Organization have developed a relatively sophisticated computerized system for measuring phase angles between 0.01 and 110 Hz (Zonge and Wynn, 1975). Field work to date indicates that this technique offers hope of discriminating IP reponses between barren pyrite and chalcopyrite. In addition, the complex resistivity spectra can be used to facilitate the removal of electromagnetic coupling and to identify pipeline, fence, and other cultural coupling effects (see Dey and Morrison, 1973; Hallof, 1974b).

The magnetic induced polarization (MIP) method has been developed recently by Scintrex (Seigel, 1974). In this technique current is caused to flow in the ground by the usual IP technique for time and frequency domain, but the magnetic fields associated with galvanic current flow are detected by an alternating current magnetometer. In general it can be said that the falloff of the magnetic field is less than the electric field and therefore deeper penetration capability can be achieved. Tests have shown that the MIP method can detect sulphides beneath very conductive overburden where EM and the normal IP techniques are unsuitable. For the first time the MIP method can be made airborne, and tests are in progress.

An electrical prospecting method, known as Magnetometric Resistivity (MMR) measures the horizontal component of the magnetic field perpendicular to the line joining the two current electrodes. The MMR anomaly is the difference between the measured values and the normal values expected at the surface of a uniform earth (Edwards, 1974). Uniform, horizontal layers produce no MMR anomaly. The MMR method appears to have potential for mapping geological structures and for detecting deep conductive orebodies.

AIRBORNE GEOPHYSICAL TECHNIQUES

The most comprehensive review of airborne geophysical methods is by Hood and Ward (1969). Most airborne surveys used today in the search for mineral deposits employ a combination of three main methods: airborne electromagnetic, magnetometer and gamma-ray spectrometer. Most aircraft are equipped with data acquisition systems so that the vast amount of data can be conveniently processed by computer methods on the ground. The one remaining bottle-neck is an adequate electronic navigational system that can recover the flight path in a digital format.

Paterson (1971, 1973b) has summarized the airborne electromagnetic methods as applied to the search for mineral deposits. The Barringer VLF E-Phase system has been developed recently, but it is not generally used for detecting mineral deposits because of the higher frequencies. Research is presently being done on the materials in boom construction in order to improve on the rigidity of the structure for increasing the signal-to-noise ratio. Cross-correlation techniques are being considered to improve on the detection of signals buried in noise which should enhance the depth of detection.

Two recent development should be mentioned — a multi-frequency system (Tridem), and research in cryogenic coils for EM prospecting. The Tridem is an AEM system using three widely spaced frequencies: 500, 2,000, and 8,000 Hz (Bosschart and Seigel, 1974). The six channels permit the determination of the overburden properties (thickness and conductivity) as well as detection of underlying conductors. The low frequency will pick up the good conductors, and the high frequency will respond more to the poorer conductors and the overburden; many mineral conductors fall in the medium conductivity range and these will be detected by the medium frequencies. The Tridem system will observe a different conductivity for each frequency.

Existing AEM systems generally consist of a transmitter which induces current flow in the ground, and a remotely located receiver measures the secondary induced fields. Since the secondary induced field must be measured in the presence of the much larger primary field of the transmitter, complex nulling and orientation procedures are required which limit the sensitivity. If a single coil is used for transmitting and receiving, the change in inductance due to a nearby conductor is measurable only when the distance is approximately the same dimensions as the transmitting loop. However, this constraint can be overcome by using a cryogenic coil cooled with liquid helium; such a unicoil device has been constructed and is being tested (Morrison and Dolan, 1973; Morrison, 1974). The advantages of the superconducting single coil are that it provides a means of enhancing geologic signal-to-noise ratios, there is an increased magnetic dipole field strength at low frequencies (even below 100 Hz), there is complete freedom from noise due to relative coil positioning experienced with two coils systems, the conducting bodies can be detected at distances comparable to the size of such bodies, and multiple frequencies can be employed. This development is a forerunner of what can be expected in the way of improved technology for deeper penetration for mineral deposits.

BOREHOLE EXPLORATION TECHNIQUES

Drilling is one of the most expensive phases of exploration, while borehole geophysics is one of the least-used methods of exploration. Borehole logging has been well developed in petroleum exploration for making in-situ measurements of the physical properties of rocks in the immediate vicinity of the hole; however, the diameter of the holes drilled for oil and gas structures is generally much greater than those drilled for other minerals. Therefore, the equipment for logging in oil and gas wells is not usually compatible for diamond drill holes. Besides, the approach in mineral exploration is usually to sample as large a radius from the hole as possible. As surface and airborne geophysical techniques are designed for deeper penetration, more use will be made of borehole exploration techniques to minimize the cost of drilling deeper holes.

The principle geophysical methods used in boreholes today are SP and resistivity, EM, IP, and magnetics. With the interest in uranium, small diameter gamma-ray spectrometers will see further development; the future will also see development of the use of pulse EM and VLF instrumentation for drill holes. One of the major advances will be in the use of neutron activation techniques for directly detecting and analyzing for nickel, copper, and other metals. Scintrex Ltd. has been testing the technique in 4-inch (10 cm) diameter holes on lateritic nickel deposits; the accuracy achievable to date is ± 5% at the 1% nickel level and ± 15% at the 0.5% copper level.

MARINE GEOPHYSICS

The mining of ocean-floor nodules containing manganese, nickel, and cobalt will have a profound impact on these mineral supplies and possibly will result in a decreased effort in land exploration for these minerals (Mero, 1972a, b; Archer, 1974). However, as interest in offshore mining increases, geophysicists will be turning their exploration expertise to developing geophysical techniques for locating mineral deposits in offshore areas.

Several workers have written on the use of geophysics at sea (Terekhin, 1962; Sarkisov and Andreyev, 1964; Paterson, 1967; Kermabon et al., 1969; Henderson, 1970; Coggon and Morrison, 1970; Morat, 1974; Whiteley, 1974; Srivastava and Folinsbee, 1975). The most useful methods for mineral exploration are marine seismic, resistivity and electromagnetics, magnetics, and magnetotellurics; some work has also been done in underwater measurements of radioactivity. Darnley (1975) reports on a technique which has been developed in the U.K. whereby a radiation borehole probe has been mounted inside an armoured hose and this assemblage has been towed for thousands of kilometres along the North Sea and Irish Sea floors. Progress in underwater radiometry in Lake Athabasca areas has been reported by Goldak (1975).

REMOTE SENSING

In the fast growing field of remote sensing — acquiring information through the use of cameras and related devices (such as radar and thermal and colour infrared sensors) operated from aircraft and spacecraft — several useful developments are taking place that are applicable to the mineral resource industry. Remote sensing is opening up new techniques for map-making, geology, and mineral exploration. It is too early in the development of remote sensing to assess its usefulness for locating mineral deposits, but as the expertise develops in the interpretation of remote sensing imagery and comparisons are made to areas of known ore deposits, this new information will help to direct geophysical exploration programmes to regions where it is likely that economic mineral deposits exist.

Earth Resource Technology Satellite (ERTS-1)

On July 23, 1972 the U.S. National Aeronautics and Space Administration (NASA) launched ERTS-1 into a near-polar orbit at a height of 920 km (406 nautical miles). Two spectral bands (0.5—0.6 and 0.6—0.7 μm), a near-infrared band (0.7—0.8 μm), and a far infrared band (0.8—1.1 μm) are scanned by the Multispectral Scanner (MSS). The satellite images cover an area of 185 km × 185 km on the ground with a resolution of 79 m. ERTS signals are received by three ground stations in the U.S.: at Fairbanks, Alaska; at Goldstone, California; and at Goddard, Maryland. Canada and Brazil have built their own receiving stations; Italy is building a station; Iran and Zaire are each planning one; and many other countries, including West Germany and Japan are interested in the idea.

Weather permitting, ERTS-1 collects MSS data every 18 days for pre-selected areas over which the satellite passes in its near polar, sun-synchronous orbit. Each class of objects on earth possess its own distinctive fingerprint, or spectral signature, determined by its atomic and molecular structure. Vegetation, for example, stands out especially well in infrared, while rocks and soils are more visible in other spectral regions. Most interpretation that is now being done for mineral exploration is either pattern recognition using various types of remote-sensing and photogeologic techniques, or spectral discrimination with selected wavelengths and specific objectives (Gregory and Moore, 1974). Contrary to early assumptions, repetitive ERTS data is proving extremely useful for mineral exploration, primarily because of various seasonal enhancements of geologic structure. During seasonal changes geologic features are enhanced by the pseudo-radar effect resulting from low inclination of the sun, emphasis of minor topography by a uniform thin cover of snow or sand, emphasis of lineaments by residual snow and ice, emphasis of drainage and lineaments by meltwater, moisture patterns in residual soils, and preferential growth of vegetation on favourable rocks and soils. Several anomalies in spectral reflectance have been interpreted as rock alterations and gossans; other anomalies have been attributed to stress in vegetation as a result of poisoning by metals. Spectral discrimination remains an interesting possibility that requires more research; the amount of relevant research in progress is difficult to assess.

Bylinsky (1975) reports several examples of the use of ERTS imagery; a geologist, after seeing a promising new linear feature not far from La Paz, Bolivia, went into the area and returned with rock containing copper mineralization that was found directly over the fault; in the southwestern U.S. ERTS images revealed a large number of circular structures believed to be remnants of old volcanoes, some of them containing copper mineralization. Shallow ore deposits sometimes stunt the growth of vegetation in a way that can be detected by the satellite; ERTS images of vegetation on islands off Indonesia gave U.S. Steel geologists ideas about where to look for chromium.

Using a computer technique called "band ratioing", ERTS images can bring out geologic details invisible to the eye; Goetz and Roman of the U.S. Geological Survey used this band ratioing technique on an image of the famous Goldfield mining district of central Nevada — a green colour on the image indicated that highly hydrated iron oxides and clays, which are often associated with deposits of silver, gold, and copper were lying on the surface. After detecting twenty other spots in Nevada marked by that colour, the geologists found that every single place turned out to have hydrated iron oxides; only about one-half of these areas had been previously explored.

Baker (1974) reports the use of ERTS imagery in conjunction with seismic data to investigate the tectonic evolution of Alaska; two intersecting sets of regional lineaments trended nearly northeast and northwest in Alaska. Similar trending lineament sets in western Canadà are known to be associated with tin, mercury, and tungsten deposits. Instead of the classic concept of arcuate mineralized belts paralleling the western Cordillera geosyncline, Lathram and Gryc (1972) suggested that the fault and fracture intersections might also provide likely sites of mineral localization; this suggestion has been partly verified by the recent discovery of copper porphyry deposits in Alaska, associated with these lineament intersections.

Results of the use of ERTS imagery to unravel the tectonic history of a region are also being used for petroleum exploration in Alaska, Oklahoma, Texas, and in Africa in Kenya (Bylinsky, 1975). The implications are far-reaching; ERTS imagery cannot alone discover an ore deposit, but it can direct attention to a likely site.

Thermal infrared applied to mineral exploration

Above absolute zero objects radiate energy in the infrared band from 0.7 to 14 μm and higher wavelengths; the intensity of radiation is proportional to the temperature of the object. Most infrared radiation is attenuated by the atmosphere due mainly to moisture content; however, there are two wave bands, or windows, within which the atmosphere does not cause attenuation to any great extent, 3.5—5.0 and 8.0—14.0 μm, although these bands are seriously affected by cloud cover. The longer wavelength band is the one commonly used for geological purposes.

Various types of geologic conditions may be associated with infrared anomalies (Rumsey, 1972). Shallow salt domes surrounded by sedimentary rocks of different thermal conductivities are likely to have an associated thermal anomaly which persists to the surface (Poley and Van Steveninck, 1970); detection of such anomalies would be applicable to hydrocarbon exploration. In mineral exploration a massive orebody would have a thermal anomaly persisting at surface for the same reason as a salt dome; some work in the early 1960s was carried out in the search for ore deposits by Kennecott Copper with rather inconclusive results (Strangway and Holmer, 1966).

In a similar way, faults that are filled with gouge material are usually higher in moisture content than the surrounding rocks and exhibit thermal anomalies. Faults and soils with high moisture content have lower temperatures than the surrounding materials and hence exhibit less thermal radiation. Howard (1973) describes some experimental work done in Australia on metamorphic rock, granodiorite, shale, dry clay, and dry sand using infrared surface reflectivity in the 1.0—1.9 μm band.

The most significant development is taking place in the 8—14 μm wavelength region where compositional information about silicate rocks can be determined by infrared multispectral scanners (Vincent, 1975). Emittance minima in this spectra region, caused by interatomic oscillations, occur at different wavelengths depending on silicate rock type. A ratio signal from two thermal scanner channels can be used to map chemical and mineral variations in silicates, while suppressing temperature variations across a scanned scene. Theoretical studies indicate that future infrared scanners with eight to twelve channels in the 8—14 μm region might be used to produce an image for mapping silicate rocks according to traditional rock classification charts. Sensor technology is still the limiting factor; Watson (1975) expands on this study of geothermal and thermal inertia mapping for differentiating rock types. These applications for rock-type discrimination will only be applicable in areas of good geologic exposure, such as those found in arid or semi-arid regions or in the shield areas devoid of overburden and vegetation cover; however, this phenomena will be an aid to assist in quantification of geologic materials in the future.

Colour infrared applied to mineral exploration

Colour infrared film is sensitive to reflectance in the near-infrared region from 0.7 to 1.0 μm and the entire visible spectrum; when film is exposed, an adequate filter is used to exclude all blue light. On the film infrared reflectance is represented by red, red reflectance by green, and green reflectance by blue, with appropriate components in between. This film is often referred to as "false colour" film. This portion of the infrared spectrum is *reflected* and not *emitted* energy and has nothing to do with heat; therefore, the near-infrared energy is reflected energy from the sun. Nearly all green vegetation reflects near-infrared more strongly than visible wave energy (Rumsey, 1972).

Healthy trees appear on infrared film as red or magenta in colour; trees under stress, that is, poisoned by trace elements, do not show the normal colour. Yost and Wenderoth (1971) made in-situ measurements of incident and reflected radiation on red spruce and balsam fir which grew upon and adjacent to a copper-molybdenum anomaly at Catheart Mountain in Maine; soil mineralization directly affects the colour reflectance of these two species of trees both in the visible and the near-infrared spectral region. Howard et al. (1971) made measurements of the visible and near-infrared spectra of foliage of *Pinus ponderosa* growing in a copper-rich area and those growing in an area of lower copper content; differences occurred in the near-infrared spectrum at 0.800 μm and in the red and infrared end of the visible spectrum (0.675—0.775 μm). A number of workers have suggested that the concentration of trace elements in trees is at its highest at the time of maximum growth. There are many parameters that can affect false colour imagery, and much more work must be done before it becomes a reliable method for application in mineral exploration.

Side-looking radar applied to mineral exploration

Side-looking radar (SLAR) is an active system in which an impulse of microwave energy is transmitted at a low angle of incidence (between 16° and 45° below the horizon), and the return signal is representative of the ground that reflects the energy. A hill returns a considerable amount of energy, and at the same time will create a shadow on the far side. Radar imagery has been likened to aerial photography taken with a low sun angle (Rumsey, 1972).

Three main parameters affect the back scatter signal: topography, surface roughness, and dielectric constant of the material. The ability of radar to de-emphasize the vegetation contrast, which is a dominant feature of photography, leads to the "removal" of the jungle cover in the case of the tropics; another advantage of SLAR is that it can "see" through cloud cover, haze, and light rain. When the first SLAR survey of the continuously cloud-covered Darien Province in Panama was done in 1967, no one had any idea of the topography because it had not been possible to use photographic techniques; this opened up an entirely new mapping concept, and by 1971 vast areas in several Latin American countries were being mapped. The most concentrated and one of the largest efforts ever made to provide basic data for planning and development of the Amazon Region was made by the Government of Brazil beginning in 1970 (De Azevedo, 1971); known as Project RADAM (for *Rad*ar *Am*azon) the survey employed the Goodyear SLAR system, a wide-angle Zeiss camera with colour infrared film and a multiband camera at an altitude of 11,000—12,000 m above the terrain. Maps were produced from uncontrolled and semi-controlled mosaics in quadrangles of 1° in latitude and 1.5° in longitude at a scale of 1:250,000.

The synoptic presentation of SLAR systems, in combination with an oblique angle of incident "illumination", has provided enhancement of geologic features such as faults, lineaments, joint systems, and different types of lithology which were never obvious on conventional aerial photography. Multiple imagery passes and preferred look-direction with SLAR enhances the detection of certain geologic features under a variety of terrain conditions.

Besides SLAR, microwave technology, both passive and active, is in use or planned for many diverse applications (Tomiyasu, 1974). The angular resolution of microwave sensors is determined by the antenna beamwidth; high resolution can be achieved only by antennas large relative to the wavelength. Hence the resolution of optical and infrared sensors will generally exceed that of microwave sensors.

Future of remote sensing

Most of the progress in remote sensing during the last decade appears to have been the result of improvement in three areas: sensors, sensor platforms (especially satellites), and automatic data handling and processing techniques. A review of the systems design for advanced scanners for earth resource applications is given by Mundie et al. (1975). Rapid advances in technology during the past decade lead to an optimistic outlook for the achievement of an improvement in scanner resolution from two to ten times, especially in the 8—12-μm region. The successor to ERTS-1 is LANDSAT-2, which was launched on January 22, 1975. NASA has recently announced that a third satellite in the LANDSAT series of earth resource satellites (LANDSAT-C) is scheduled to be launched in 1977. The recently approved LANDSAT-C will fill the gap between LANDSAT-2 and the EOS satellite series scheduled for launch starting in 1979 (see Mundie et al., 1975, table 1).

INTERPRETATION AND EXPLORATION STRATEGY

Increasing instrumental sensitivity for greater depth of exploration solves only part of the problem in mineral exploration; interpretation of the data presents the greatest challenge. While the natural field electrical methods — self potential, AFMAG, tellurics, magnetotellurics, and VLF — are employed in special situations, the bulk of electrical prospecting rests on electromagnetic and combined resistivity-induced polarization surveys. Recently there has been a considerable conflict about the validity of "classical" EM interpretation wherein the host rock surrounding the target conductor is assumed to be so resistive that it can be modelled as free space. Scaled physical and computer modelling has revealed that a free space assumption is invalid for almost any realistic host rock resistivity from 1 to 1,000 ohm metres (Ward et al., 1974). Conductive overburden rotates the phase and decreases the amplitude of anomalies due to buried inhomogeneities (Lowrie and West, 1965). Modelling also has revealed that conductive host rock can lead to an increase in amplitude and a rotation of the phase of an anomaly. In general, present methods of interpretation of electromagnetic data are invalidated by conductive overburden or conductive host rock.

Catalogues of curves for EM, resistivity, and IP data are commonly used for interpretation of geophysical field data (the forward method). However, recent advances in the application of the generalized linear inverse method of interpretation indicate the use of model curves may soon be redundant (Glenn et al., 1973; Inman et al., 1973; Ward et al., 1974). The inverse method avoids human bias in matching field curves to model curves and provides estimates of the reliability of the model or models selected to match the field data. Numerical techniques are also being developed to calculate electromagnetic scattering. Analytical solutions exist only for bodies of simple geometry — such as spheres and cylinders in a whole space — problems that are not usually encountered in the earth. Numerical techniques employed are finite difference, integral equation, and finite element (Haren, 1974); by these techniques models of irregular shape and inhomogeneity can be calculated, and the effects of surface topography, buried topography, overburden and faults can be determined.

When and how to use geophysics requires proper systems analysis. New developments in geophysical methodology must be made within the framework of cost-effectiveness of an exploration system. Exploration architecture provides the key to the application of exploration techniques which include geology, geochemistry, and geophysics (Ward, 1972). The philosophy of mineral exploration involves many factors that consist of decisions and activities which transform available resources (including human) into a new resource — mineral deposits (Bailly, 1972). Mineral exploration companies are turning to large-scale integrated programmes to improve the effectiveness of the search for ore, and statistical techniques (De Geoffroy and Wignall, 1970; Gaucher and Gagnon, 1973) are being applied to geological, geophysical, and geochemical data (as discussed in Chapter 11) to assist exploration management in the selection of prospecting areas and drilling targets in regional programmes. In considering new geophysical methodology in future exploration programmes, it is essential that geophysicists understand the overall strategy and objectives of the exploration system.

CONCLUSIONS

The increasing world demand for minerals ensures the steady growth of new geophysical instrumentation that will have the capability of greater depth of exploration for both airborne and ground techniques. However, the present methods that can detect down to a depth of 120 m will be used for some time to come in areas of the world that have never before been surveyed. As deeper exploration methods are developed and deeper drill holes are required, borehole exploration technology will have to be developed and employed to a greater extent than in the past. Geophysics will be used more universally for mapping geologic structures. As interest increases in ocean mining, the next 25 years will see a gradual increase in the development of marine geophysical instrumentation.

The great emphasis at the present time should be directed toward data interpretation and the use of the inverse method. Sophisticated interpretation techniques and computer modelling will allow for a better understanding of the effect of conductive host rock, conductive overburden, topography, buried structures, and faults. More research should be directed toward a better understanding of the induced polarization phenomena and sulphide differentiation. A pressing requirement for airborne surveys is a better electronic navigation system for flight-path recovery. Rapid developments in the field of solid state electronics, computers, and microprocessors will all aid in the development of portable and lighter-weight equipment.

The explorationist should maintain a close surveillance of the developments in remote-sensing technology. Remote-sensing and SLAR imagery will not find mineral deposits directly, but the study of lineaments and faults used in conjunction with geology could be useful indicators for directing future exploration programmes. Stress in vegetation due to mineral poisoning and the effect of hydrated iron oxides and clays in remote-sensing imagery afford interesting speculations of its future potential. One of the most exciting developments in infrared sensing in the range of 8—14 μm is the ability to distinguish rocks by their silicate content. As infrared-sensing technology develops, it may be possible to quantify rocks by direct measurement.

A review has been made here of some of the developments in the field of exploration geophysics. It is difficult to project into the future, but the most obvious advances will no doubt come from an integration of geophysical developments with those in geochemistry and geology.

REFERENCES CITED

Archer, A.A., 1974. Progress and prospects of marine mining. *Min. Mag.*, 130(3): 150—163.
Bailly, P.A., 1972. Mineral exploration philosophy. *Min. Congr. J.*, 58(4): 31—37.
Baker, R.N., 1974. ERTS updates geology. *Geotimes*, 19(8): 20—22.
Barringer, A.R. and McNeill, J.D., 1968. Radiophase — a new system of conductivity mapping. In: *Proceedings of the Fifth International Symposium on Remote Sensing*. University of Michigan, Ann Arbor, Mich., pp. 157—167.
Bernstein, S.L., Burrows, M.L., Evans, J.E., Griffiths, A.S., McNeill, D.A., Niessen, C.W., Richter, I., White, D.P. and Willim, D.K., 1974. Longe-range communications at extremely low frequencies. *Proc. IEEE*, 62: 292—312.
Bosschart, R.A. and Siegel, H.O., 1974. The Tridem three frequency airborne electromagnetic system. *Can. Min. J.*, 95(4): 68—69.
Brant, A.A., 1967. Some considerations relevant to geophysical exploration. *Can. Inst. Min. Metall. Bull.*, 60(657): 54—62.
Bylinsky, G., 1975. ERTS puts the whole earth under a microscope. *Fortune*, 91(2): 117—130.
Coggon, J.H., and Morrison, H.F., 1970. Electromagnetic exploration of the sea floor. *Geophysics*, 35: 476—489.
Collett, L.S. and Becker, A., 1968. Radiohm method for earth resistivity mapping. Canada Patent No. 759,919, issued October 1.
Collett, L.S., and Katsube, T.J., 1973. Electrical parameters of rocks in developing geophysical methods. *Geophysics*, 38: 76—91.
Collett, L.S. and Sawatzky, P., 1966. The Serson direct-reading proton-free precession magnetometer, 1. Shipborne use. *Geol. Surv. Can., Paper* 65-31, 32 pp.

Cook, D.R., 1969. The effective use of exploration techniques. In: F.E. Kottlowski and R.W. Foster (Editors), *Exploration for Mineral Resources. New Mexico Bur. Mines, Circ.*, 101: 25—31.

Darnley, A.G., 1970. Airborne gamma-ray spectrometry. *Can. Inst. Min. Metall. Bull.*, 63: 145—154.

Darnley, A.G., 1975. Geophysics in uranium exploration. Paper presented at *Canadian Prospectors and Developers Annual Meeting*, Toronto, Ont., March 10—12.

De Azevedo, L.H.A., 1971. Radar in the Amazon. In: *Proceedings of the Seventh International Symposium on Remote Sensing*. University of Michigan, Ann Arbor, Mich., pp. 2303—2306.

De Geoffroy, J. and Wignall, T.K., 1970. Application of statistical decision techniques to the selection of prospecting areas and drilling targets in regional exploration. *Can. Inst. Min. Metall. Bull.*, 63: 893—899.

Dey, A. and Morrison, H.F., 1973. Electromagnetic coupling in frequency and time-domain induced-polarization surveys over a multilayered earth. *Geophysics*, 38: 380—405.

Dolan, W.M., 1969. Geophysical detection of deeply buried sulfide bodies in weathered regions. In: L.W. Morley (Editor), *Mining and Groundwater Geophysics, 1967. Geol. Surv. Can., Econ. Geol. Rep.*, No. 26, pp. 336—344.

Edwards, R.N., 1974. The magnetometric resistivity method and its application to the mapping of a fault. *Can. J. Earth Sci.*, 11: 1136—1156.

Gaucher, E. and Gagnon, D.C., 1973. Compilation and quantification of exploration data for computer studies in exploration strategy. *Can. Inst. Min. Metall. Bull.*, 66(737): 113—117.

Ghosh, M.K. and Hallof, P.G., 1974. Geoprobe EMR-14 — a new multi-spectral EM induction system for delineating depths of permafrost. In: L.S. Collett and R.J. Brown (Editors), *Proceedings of a Symposium on Permafrost Geophysics. Natl. Res. Counc. Can., Tech. Mem.* No. 113, pp. 50—59.

Glenn, W.E., Ryu, J., Ward, S.H., Peeples, W.J. and Phillips, R.J., 1973. The inversion of vertical magnetic dipole sounding data. *Geophysics*, 38: 1109—1129.

Goldak, G.R., 1975. Underwater radiometry proving useful tool to locate uranium. *North. Miner*, 60(51): 53—54.

Grant, F.S. and West, G.F., 1965. *Interpretation Theory in Applied Geophysics*. McGraw-Hill, New York, N.Y., 583 pp.

Gregory, A.F., and Moore, H.D., 1974. Remote sensing and mineral exploration. *North. Miner*, 59(51): 43—44.

Hallof, P.G., 1957. On the interpretation of resistivity and induced polarization results. Ph.D. Thesis, Department of Geology and Geophysics, Massachusetts Institute of Technology, Cambridge, Mass.

Hallof, P.G., 1974a. The odds are against us. *Can. Min. J.*, 95(4): 43.

Hallof, P.G., 1974b. The IP phase measurement and inductive coupling. *Geophysics*, 39: 650—665.

Haren, R.J., 1974. The finite element method and other numerical techniques applied to the electromagnetic problem in geophysics (a review). *Bull. Aust. Soc. Explor. Geophys.*, 5: 1—8.

Henderson, R.J., 1970. Bidirectional-dipole resistivity profiling for detailed resolution of basement and sea floor structure. *Geoexploration*, 8: 97—104.

Hood, P.J., 1965—1975. Mineral exploration trends and developments. *Can. Min. J.*, 86—96: February issues.

Hood, P.J. and Ward, S.H., 1969. Airborne geophysical methods. In: H.E. Landsberg and J. van Mieghem (Editors), *Advances in Geophysics*. Academic Press, New York, N.Y., pp. 1—112.

Howard, J.A., 1973. Passive remote sensing of natural surfaces by reflective techniques. *Geoexploration*, 11: 133—139.

Howard, J.A., Watson, R.D. and Hessin, T.D., 1971. Spectral reflectance properties of *Pinus ponderosa* in relation to copper content of the soil, Malachite Mine, Jefferson County, Colorado. In: *Proceedings of the Seventh International Symposium on Remote Sensing*. University of Michigan, Ann Arbor, Mich., pp. 285—297.

Hutchinson, R.D., 1974. Exploration in the Canadian mineral industry to the year 1999. *Can. Inst. Min. Metall. Bull.*, 67(747): 57—60.

Inman, J.R., Jr., Ryu, J. and Ward, S.H., 1973. Resistivity inversion. *Geophysics*, 38: 1088 —1108.

Katsube, T.J. and Collett, L.S., 1973. Electrical characteristic differentiation of sulfide minerals by laboratory techniques (abstract). *Geophysics*, 38: 1207.

Keller, G.V. and Frischknecht, F.C., 1966. *Electrical Methods in Geophysical Prospecting*. Pergamon, Oxford, 519 pp.

Kermabon, A., Gehin, C. and Blavier, P., 1969. A deep-sea electrical resistivity probe for measuring porosity and density of unconsolidated sediments. *Geophysics*, 34: 554—571.

Koziar, A., Redman, D. and Strangway, D.W., 1973. Audio frequency magnetotellurics. In: *Symposium on Electromagnetic Exploration Methods*. University of Toronto, Toronto, Ont., Paper 20, 4 pp.

Lamontagne, Y. and West, G.F., 1973. A wide-band time domain, ground EM system. In: *Symposium on Electromagnetic Exploration Methods*. University of Toronto, Toronto, Ont., Paper 2, 5 pp.

Lathram, E.H. and Gryc, G., 1972. Metallogenic significance of Alaskan geostructures seen from space. In: *Proceedings of the Eighth International Symposium on Remote Sensing*. University of Michigan, Ann Arbor, Mich., pp. 1209—1211.

Lowrie, W. and West, G.F., 1965. The effect of a conducting overburden on electromagnetic prospecting measurements. *Geophysics*, 30: 624—632.

McNeill, J.D., 1971. Technical note for the interpretation of E-Phase system. *Barringer Research Ltd., TR 71-188*, 20 pp. plus figures.

Mero, J.L., 1972a. The future promise of mining in the ocean. *Can Inst. Min. Metall. Bull.*, 65(720): 21—27.

Mero, J.L., 1972b. Recent concepts in undersea mining. *Min. Congr. J.*, 58(5): 43—54.

Morat, P., 1974. Technical processes for offshore collection of magnetic and telluric data. *Phys. Earth Planet. Inter.*, 8: 202—206.

Morrison, H.F., 1974. Electromagnetic device for determining the conductance of a nearby body by a single supercooled inductor coil. U.S. Patent No. 3,836,841, issued September 17.

Morrison, H.F. and Dolan, W.M., 1973. Earth conductivity determinations employing a single superconducting coil (abstract). *Geophysics*, 38: 1214—1215.

Mundie, L.G., Hummer, R.F., Sendall, R.L. and Lowe, D.S., 1975. System design consideration for advanced scanners for earth resource applications. *Proc. IEEE*, 63: 95—103.

Parasnis, D.S., 1973. *Mining Geophysics*. Elsevier, Amsterdam, 2nd rev. ed., 395 pp.

Paterson, N.R., 1967. Underwater mining — new realms for exploration. *Can. Min. J.*, 88 (4): 109—117.

Paterson, N.R., 1971. Airborne electromagnetic methods as applied to the search for sulphide deposits. *Can. Inst. Min. Metall. Bull.*, 64(705): 29—38.

Paterson, N.R., 1973a. Extra low frequency (ELF) EM surveys with the EM-25 (abstract). *Geophysics*, 38: 188.

Paterson, N.R., 1973b. Some problems in the application of ELF EM in high conductivity areas. In: *Symposium on Electromagnetic Exploration Methods*. University of Toronto, Toronto, Ont., Paper 28, 9 pp.

Pemberton, R.H., 1962. Airborne electromagnetics in review. *Geophysics*, 27: 691—713.

Poley, J.Ph. and Van Steveninck, J., 1970. Delineation of shallow salt domes and surface faults by temperature measurements at a depth of approximately 2 metres. *Geophys. Prospect., Suppl.*, 18: 666—700.

Rumsey, I.A.P., 1972. Application of thermal infrared, colour infrared and side looking radar to mineral exploration. *Can. Min. J.*, 93(8): 56—60.

Sarkisov, G.A. and Andreyev, L.I., 1964. Results and potential of marine electrical prospecting in the Caspian Sea. *Int. Geol. Rev.*, 6: 1573—1584.

Seigel, H.O., 1974. The magnetic induced polarization (MIP) method. *Geophysics*, 39: 321—339.

Serson, P.H., 1961. Proton precession magnetometer. Canada Patent No. 618,762, April 18. Same patent issued under U.S. Patent No. 3,070,745, December 25, 1962.

Srivastava, S.P. and Folinsbee, R.A., 1975. Measurement of variations in the total geomagnetic field at sea off Nova Scotia. *Can. J. Earth. Sci.*, 12: 227—236.

Strangway, D.W. and Holmer, R.C., 1966. The search for ore deposits using thermal radiation. *Geophysics*, 31: 225—242.

Strangway, D.W., Swift, C.M., Jr. and Holmer, R.C., 1973. The application of audio-frequency magnetotellurics (AMT) to mineral exploration. *Geophysics*, 38: 1159—1175.

Sturrock, R.F., Shand, J.A. and Lokken, J.E., 1974. Canadian measurements of Project SANGUINE test transmissions. *Can. J. Earth Sci.*, 11: 755—767.

Terekhin, E.I., 1962. Theoretical bases of electrical probing with an apparatus immersed in water. In: N. Rast (Editor), *Applied Geophysics, U.S.S.R.* Pergamon Press, Oxford, pp. 169—195.

Tomiyasu, K., 1974. Remote sensing of the earth by microwaves. *Proc. IEEE*, 62: 86—92.

Vincent, R.K., 1975. The potential role of thermal infrared multispectral scanners in geological remote sensing. *Proc. IEEE*, 63: 137—147.

Wait, J.R. (Editor), 1959. *Overvoltage Research and Geophysical Applications.* Pergamon, New York, N.Y., 158 pp.

Ward, S.H., 1972. Mining geophysics: new techniques and concepts. *Min. Congr. J.*, 58(1): 58—68.

Ward, S.H. and Rogers, G.R., 1967. Introduction. In: *Mining Geophysics*. Society of Exploration Geophysicists, Tulsa, Okla., 2: 3—8.

Ward, S.H., Ryu, J., Glenn, W.E., Hohmann, G.W., Dey, A. and Smith, B.D., 1974. Electromagnetic methods in conductive terranes. *Geoexploration*, 12: 121—183.

Watson, K., 1975. Geologic applications of thermal infrared images. *Proc. IEEE*, 63: 128—137.

Whiteley, R.J., 1974. Design and preliminary testing of a continuous offshore resistivity method. *Bull. Aust. Soc. Explor. Geophys.*, 5: 9—13.

Yost, E. and Wenderoth, S., 1971. The reflectance spectra of mineralized trees. In: *Proceedings of the Seventh International Symposium on Remote Sensing*. University of Michigan, Ann Arbor, Mich., pp. 269—284.

Zonge, K.L. and Wynn, J.C., 1975. Recent advances and applications in complex resistivity measurements. *Geophysics*, 40: 851—864.

Chapter 13

DEVELOPMENT OF DEEP MINING TECHNIQUES

DONALD W. GENTRY

INTRODUCTION

At present most of the world's production of raw mineral resources is the result of surface mining; approximately 70% of the estimated world production of crude metallic and non-metallic ores and coal are mined by surface techniques. In the U.S. the surface mining industry represents approximately 90% of the total domestic mining capacity.

Bulk surface mining methods enable the economic exploitation of large, disseminated, low-grade, near-surface deposits. Mining rates on the order of 400,000 total tons per day and productivities of approximately 200 tons per man-shift have been attained in surface-metal mining operations. Productivities up to 500 tons per man-shift have been attained in large coal and industrial mineral operations. Direct mining costs for these production rates and productivities are in the range of $0.20 to $0.45 (U.S.) per ton. These large production rates, productivities and associated low mining costs are primarily the result of significant advances in equipment technology in recent years. The utilization of equipment such as 138-m³ stripping shovels, 169-m³ draglines, 19-m³ production shovels and 200-ton haulage trucks are having a tremendous impact on the overall economies of surface mining. The rapid technological improvements taking place in the surface mining industry is further evidenced by the fact productivities have been increasing over the past several years, even though the average grade or quality of the deposits being mined is declining. Although surface mining is considered to be more advantageous than underground mining from the standpoint of grade control, recovery, economy, flexibility of operation, safety and working environment, there are deposits which *must* be mined by underground techniques because of overriding surface usage, deposit size, surface waters, irregularity or depth below surface.

Unfortunately, growth in the underground segment of the mining industry has not been as dramatic as that of the surface component. With few exceptions the underground portion of the industry has witnessed little change in methodologies of extraction and only slight advancements in equipment technologies over the last 40 years. This is not to imply that the introduction of equipment such as raise borers, tunnel borers, rubber-tyred load-haul-dump units, continuous mining machines, etc., have not had a significant impact on underground mining. The point is that development in

TABLE 13-I

Average excavation cost (U.S. dollars) per cubic metre material excavated (crude ore and waste; from Allsman and Yopes, 1973)

Mineral	Surface mining (U.S. $/m^3)	Underground mining (U.S. $/m^3)
Metals	0.67	3.32
Non-metals	1.48	5.18
Coal	0.11	3.06

the total underground mining sequence has not progressed nearly as much in recent years as the surface mining segment of the mining industry. This is, in part, illustrated by the fact that underground productivity is only about one-tenth that of surface mining methods. Also, in comparison with surface mining costs, underground production costs are considerably higher, as shown in Table 13-I. These costs are averages and only represent relative comparisons between surface and underground cost magnitudes. Table 13-I is somewhat misleading in that the cut-off grades associated with surface and underground operations are significantly different; also the higher underground costs are offset, in part, by associated increased selectivity of extraction resulting from underground mining methods.

Table 13-II represents ranges in productivity and production costs for selected underground mining methods. The mining methods included in Table 13-II were selected for comparison because they represent the techniques utilized in the production of the bulk of the world's mineral resources extracted from underground mines. The wide ranges in production costs and productivities are the result of a number of factors. First, the

TABLE 13-II

Ranges in productivities and production costs (in U.S. dollars) for selected underground mining methods (data are result of personal communications and experience)

Mining method	Direct production costs (U.S. $/ton)	Productivity (tons/man-shift)
Room and pillar stoping	1.50—4.50	10—150
Cut and fill stoping	3.50—15.00	3—12
Sublevel stoping	1.25—6.00	10—30
Sublevel caving	4.00—8.00	12—28
Block caving	0.75—2.00	10—50

TABLE 13-III

Labour as a percentage of total production costs for selected underground mining methods
(from Boshkov and Wright, 1973; Dravo Corporation, 1974)

Mining method	Labour as a percentage of total production costs
Room and pillar stoping	44
Cut and fill stoping	57
Sublevel stoping	60
Sublevel caving	64
Block caving	55

data reflect an attempt to represent underground mining operations located
in various parts of the world. Mining costs and productivity rates also vary
with such factors as: (1) the nature of a given ore deposit, (2) width of
mineralization, (3) distribution of mineralization, (4) rock types, (5) method
of mining, and (6) degree of mechanization. For example, continuous mining
machines utilized in many underground room and pillar coal operations can
yield extremely high production rates. In contrast to the conventional
cyclical unit mining operations of drill, blast, muck and haul, the "continuous
miners" are capable of nearly continuous breakage and muck removal from
the mining face. Unfortunately the utilization rate of these machines is often
low because of the inability to continuously transport the broken material
from the continuous miner to a storage or loading facility and because of
the inability to provide continuous ground (roof) support (usually rock bolts)
within the required distance of the mining face. Another reason for the wide
range in values is the relative cost and productivity of labour in various
areas of the world.

Table 13-III shows the percentage of labour in the total production cost
for various underground mining methods. Obviously these values are averages
and may fluctuate considerably depending upon the degree of mechanization
and automation utilized at any given property. Nonetheless these figures show
that the underground mining industry is very labour intensive and thus the
cost and associated productivity of the labour force at any given mine will
have a significant effect on the overall total mining costs and productivities.

Mine operators, in general, are extremely reluctant to provide data on
actual production costs; however, it is somewhat easier to obtain percentage
cost distributions of any given mining production cost. Estimates of total
mining costs may then be calculated by assuming or obtaining data on the
following criteria: (1) labour rate per shift (including fringe benefits), (2)
productivity in tons of ore per man-shift, (3) labour cost as a percentage of
the direct operating cost, and (4) direct operating cost as a percentage of

the total mining cost (which includes overhead, depreciation, depletion, etc.).

Such information then allows estimates of total costs to be calculated as follows (Dravo Corporation, 1974):

$$\frac{\text{Estimated direct}}{\text{operating cost}} = \frac{\text{shift labour rate + fringes}}{\text{tons per man-shift}} \times \frac{1}{\text{percent labour/direct operating cost}}$$

$$\frac{\text{Estimated total}}{\text{mining cost}} = \frac{\text{estimated direct operating cost}}{\text{percent direct operating cost/total mining cost}}$$

The information presented clearly shows that underground mining is very labour intensive and has significantly lower productivity and higher production costs than surface mining operations. In view of this, how important will underground mining be in the production of the world's requirements for mineral resources in the future?

FUTURE OF UNDERGROUND MINING

Although the vast majority of the world's production of raw natural resources comes from surface mining operations, the production of certain minerals may be almost entirely the result of underground mining (e.g., coal, salt, potash, zinc). Many authorities agree that future conditions will force a reversal of the trend from underground to surface mining — with a gradual shift back to underground mining (Allsman and Yopes, 1973).

There are a number of reasons which support the hypothesis that the role and significance of underground mining will increase in the relatively near future. One such reason is the changing social attitudes toward environmental controls. The enactment and enforcement of environmental regulations regarding degradation of the land surface, and water and air pollution will affect surface operations and tend to increase production costs significantly. Many of the mining environmental problems are less apparent or severe, from society's point of view, in underground mining operations as compared to surface mining techniques.

Another factor is the relative shortage of easily mineable, near-surface deposits. By necessity the recent trend in exploration is toward the location of deeper ore deposits. This is evidenced by the fact that the depths associated with new mine projects are increasing annually. Underground mining around the world is still conducted at relatively shallow depths. The deepest mines in the world are only on the order of 3 km below the surface. Thus the potential for "deep" ore deposits is enormous. As a result of deeper exploration targets, improved geological, geochemical and geophysical techniques have been developed which greatly aid in the discovery of deeper mineral deposits located beyond the economic depths of current surface mining techniques.

Exploration and production costs increase with depth and consequently the mineralized targets must either be large, disseminated types of deposits which lend themselves to bulk mining techniques, or "high"-grade deposits which will support the extra costs of discovery and mining at these greater depths. Because of these increased depths, the state of existing technology and increasingly stringent environmental regulations, most of these new deposits will have to be mined by underground techniques in the future.

Based on recent world-wide discoveries, indications are that the future deeper deposits will generally be of a lower grade or quality. Under these conditions there will be a continuing trend toward fewer, but larger, mines producing large quantities of ore by bulk mining methods in order to reduce the total cost per unit of saleable product. Underground mines having daily production figures of 80,000—100,000 tons may easily become realities within the next five to eight years. Even in the case of deep "high"-grade types of deposits the increased mining costs associated with depth will require greater selectivity from highly mechanized underground mining methods yielding higher production rates and larger productivities.

As mentioned previously, deeper ore deposits will result in increased mining costs. It is not possible to discuss all the problems associated with increased mining depths in the space allocated to this chapter. Therefore, only some of the more significant problems representing major cost increases associated with deep underground development and extraction are presented.

Development of underground ore deposits

The point in time when the decision is made to mine a property until the time at which the property starts to produce ore on a continuous basis is referred to as the "development" period of the property. This is the time interval over which the orebody is prepared or developed for future exploitation. The development period for most underground mines is on the order of four to seven years. This, coupled with the fact that mining is a very capital-intensive industry, indicates that mining companies are committed to huge capital investments over long periods when bringing new properties into production (see Chapter 2). Thus it is extremely important to shorten the development period as much as possible.

The rate of underground development has increased in recent years due to the advancement and utilization of equipment such as raise borers, tunnel borers and load-haul-dump units. However, prior to the underground development of the orebody itself, access to the deposit must be achieved. As mining progresses to greater depths in the future, access through shafts will become increasingly important, and the time required to sink these shafts must be minimized.

Most new shafts currently being sunk for mining production purposes are large circular (6—10 m in diameter) shafts utilizing various types of ground support systems with concrete lining being most popular. These shafts are, by-and-large, conventionally sunk utilizing the drill, blast, muck unit operations.

There has been an increasing interest in recent years in bored shafts. These shafts are constructed by using mechanical boring machines somewhat similar to tunnel borers. These shaft borers have mounted cutters which break rock by either a shearing or abrasive action. Thrust is applied to the cutting head of the machine through jacks or by the weight of the drill string itself. The bulk of the shafts bored to date in the mining industry have been constructed primarily for ventilation or escapeway purposes since the bored shaft diameters are relatively small (1.5—5 m in diameter) and inadequate for production shafts through which large tonnages must be hoisted. There is also a problem in that bored shafts in the U.S. mining industry — utilizing equipment drilling blind (no pilot hole) and having reverse mud circulation — have only drilled to depths of approximately 825 m. Current shaft-boring technology limits bored shafts to soft or moderately hard rock formations.

Shaft-boring machines will no doubt receive much attention in the immediate future as a means of minimizing shaft-sinking construction time and therefore minimizing the time for development of ore deposits prior to production. The primary attractions of bored shafts are safety, reduced ground support requirements and the fact that the shaft construction is highly machine oriented and requires less than 40% of the manpower needed for conventional shaft sinking. However, before shaft borers become an economical and viable part of deep mine development programmes there must be technological advances in the equipment. Shaft-boring machines capable of boring large diameter circular, and possibly non-circular, shafts with cuttings removal from the collar of the shaft must be developed. These machines should also have the capability of providing continuous ground support and shaft-equipment installation as the boring advances. Only when these objectives are achieved will shaft boring provide for time and cost reductions in the overall mine development period.

Ground control

In order to effectively and efficiently utilize the mineral resources of the world it is desirable to maximize the recovery or extraction ratio for any given ore deposit. One of the aspects associated with increasing extraction ratios is the problem of ground control and support. Ground control becomes an increasingly important problem to operators as mining progresses to greater depths.

Until recently much of the published work in the field of rock mechanics has been related to analytical solutions. Unfortunately the theoretical solutions obtained often do not adequately reflect the real rock mass behaviour in operating mines. During the past several years many of those practicing in the field of rock engineering have been directing their efforts toward the application of the principles of rock mechanics to mine design and associated rock mass behaviour predictions. Results of work performed in these areas has been most encouraging and well received by the mining industry. Continued efforts must be made in the areas of predicting and controlling rock mass deformations and failures if the science of rock engineering is to accomplish its primary objectives — improve the safety and profitability of the mining operation.

There are several areas in the field of applied rock mechanics which require further investigation; one area is the determination of in-situ rock mass properties. An accurate, simple and efficient determination of in-situ rock mass strength and deformation properties is essential if meaningful mine design parameters are to be established. This information will help the design engineer produce safer, more economical mine design from the sophisticated numerical methods currently available.

The development of techniques or methods which adequately describe geologic environments is greatly needed. Geologic structure is of extreme importance in mine design and the prediction of rock mass behaviour. An effective method of quantifying geology must be established if computerized numerical models are to yield meaningful results for determining design parameters.

With the development of in-situ determinations of rock mass properties and the representation of geologic structure it will be possible to develop a better understanding of rock mass deformational characteristics. This in turn will allow the development of more effective instrumentation and warning systems for the detection of unstable areas within the mine. Although much sophisticated instrumentation is currently available, complete warning systems are largely undeveloped. Systems must be developed which are relatively inexpensive and can be installed and operated by mine personnel if future deep mining is to be conducted in a safe, efficient manner.

Ventilation

As future mining progresses to greater depths the means of providing adequate ventilation to underground workings will become more difficult, since surface rock temperatures increase with depth. The rate of temperature increase with depth is defined as the geothermal gradient which generally increases with depth at approximately a constant rate in a given locale. Natural influences on rock temperatures include such factors as: closeness of underground workings to igneous activity or hot water and gases, age and residual heat of original rock material, radioactive decay, climatic conditions, conductivity and thermal properties of the rock. Such heat sources may cause rock temperatures ranging from more than 50°C to below freezing (<0°C) in permafrost areas. Any or all of the above factors may cause temperature increases which in turn have an effect on the number and geometry of mine openings, as well as environmental design problems. These environmental problems are associated with the ventilation and air conditioning requirements for the comfort as well as the efficiency of mine workers.

We have previously mentioned the probability of an increasing number of underground high-tonnage, bulk mining operations in the future. If these large widespread tonnage operations use diesel-powered production equipment, ventilation systems will have to be developed and designed to handle enormous volumes of fresh air at low pressures.

The development of ventilation or air conditioning systems to cope with the problems of temperature increases with depth and the expanding use of diesel equipment and conveyor systems underground is not an insurmountable problem. Nevertheless, much additional work is required in this area, and the net result of providing these ventilation systems will no doubt be an increase in underground operating costs.

Underground working environment

The underground environment in which miners work is of extreme importance from a morale, safety, and productivity standpoint. The portion of the working environment which is of immediate concern is generally referred to as the process environment; the elements of the process environment which are of primary importance are gases, noise, dust, radiation and illumination (Dravo Corporation, 1974).

There are a number of controls currently available for dealing with these problems. For instance, the control of underground gases is normally dealt with through dilution and removal by utilizing fan-induced air currents or by utilizing chemical processes which affect gas formation. The problem of noise is much more difficult to handle. The elimination of the noise source or the enclosure of the noise source is extremely difficult without significantly interfering with or restricting the use of the equipment currently utilized in underground operations. The most widely used devices for noise control in the underground operating environment are ear-protectors. These protective devices generally take the form of ear-muffs or some type of ear-plug which is inserted into the outer ear. The overall effectiveness of these devices is limited. Many believe the most promising approach to noise control is to completely isolate the individual from the noise source. In most cases this would necessitate the use of remote-controlled equipment.

The use of remote control for continuous mining machines is not new. The basic electronic technology for designing safe remote controls is known throughout the mining world; however, large-scale commercial acceptance has been slow to develop. Perhaps the primary reasons for the lack of success have been inadequate operator training and failure to design easy-to-use controls. Short-distance remote control of mining machines using a cable link is the most common application. Many coal mining machines currently operating in Europe and Russia use cable pendant-type controls. Recently, effective radio remote control has been developed to supplement the cable system. The radio system consists simply of two components, the battery-powered transmitter unit and a radio receiver located on the machine. Some mine operators have found that remote-controlled mining machines have increased productivity because of higher utilization of the mining machine. Extraction ratios may also increase in room and pillar coal operations because continuous mining machine operators can stand in areas of relative safety while pillars are being removed. There is no doubt that much coal is currently being abandoned in the interest of safer operations. Thus remote-controlled continuous mining machines can aid in the conservation and recovery of underground coal resources.

Airborne dust is a problem encountered in virtually all underground operations utilizing mechanical or chemical force to drill, break, load or transport rock. Airborne dust may be extremely dangerous to the health of the underground worker. Minimizing the amount of dust in the underground working atmosphere is usually accomplished by suppression, ventilation and personal protective devices (respirators). Suppression measures include water sprays and foam to keep dust from becoming airborne. Removal of airborne dust is generally accomplished by main or auxilliary air currents. Often vacuum collectors and water sprays are used in conjunction with the primary ventilation system. Respirators must be worn by miners when dust concentrations exceed threshold limit values.

The problem of radiation is not unique to uranium mines, but may result from close proximity to uranium concentrations or from groundwater with unusually high radiation concentrations. Radiation is currently best controlled in the underground environment through effective ventilation systems utilizing successive filtration of the air stream.

One of the least-controlled aspects of the underground mine environment is illumination. Most agree that the level of mine illumination is directly related to the morale, safety and productivity of the mine worker. Yet, this aspect of the working environment has received only minimal attention from regulatory authorities and from industry.

There are many problems associated with the underground working environment which are not being efficiently or effectively treated at the present time. It appears that if the miner is to be protected from these environmental hazards in the future, it will probably have to be accomplished through either (1) isolation of the miner from the source of the environmental problems (i.e. remote control of continuous mining systems), or (2) self-contained personal environmental protection units such as those used by space travellers. In either case much additional research and development is required before effective solutions to these problems become available for implementation.

Material transport

One of the most important factors in the efficient operation of an underground mine is the haulage system. It has been through the efficient use and, generally, the combination of mine cars, track, belt conveyors and rubber-tyred haulage equipment that underground operations have been able to compete with the more popular strip or surface mining (Willson et al., 1973).

In recent years the conversion from underground rail transportation systems to conveyor systems and rubber-tyred equipment has progressed at a fairly rapid rate. The initiation of these systems, such as an extensible conveyor system in conjunction with an automated continuous mining machine, has resulted in significant increases in underground production rates. Belt conveyor systems offer several advantages over other haulage systems. For instance, belt conveyors have the ability to provide nearly continuous-flow transportation which enables the modern continuous mining machines to operate at maximum efficiency. Also, because the amount of material transported on a conveyor system depends on the belt speed, belt width, and weight of the material transported, belt conveyors can operate over a wide range of capacities. Another advantage of belt conveyor systems is the fact that they can be utilized in level or pitching terrain either underground or on the surface. Perhaps the biggest development in belt conveyor systems in recent years has been the advent of the wire-rope side-frame belt conveyor. The intermediate sections of this system can be suspended from the roof in underground openings and thus permit free passage of mine vehicles. The most unique advantage of the rope-side-frame belt conveyor is the extreme mobility offered by the system because of the relative ease with which it can be extended, retracted, dismantled, moved and reassembled.

Belt conveyor and rubber-tyred haulage systems may be inadequate for the vast tonnages contemplated from underground operations in the future. Much of the current research and development activity associated with underground mining is directed toward automated continuous mining machines which will provide a continuous flow of material from the working face. However, increased production rates and productivities can only be achieved if such continuous mining systems are fully integrated with continuous materials handling systems. Continuous materials handling systems are essential because any interruption in the steady flow of ore within the mine limits production capacities, requires increased manpower, and therefore adversely affects productivity.

The transport of solids through pipelines has been successfully accomplished with many different materials. Hydraulic and pneumatic transport systems are largely in the experimental stages and show promise for the future. Hydraulic transport of solids, utilizing water as the transporting medium has been successful in transporting hydraulically mined minerals in various parts of the world. Scott and Young (1968) state that there are certain natural advantages that favour pipeline transport of solids. These advantages are: (1) the hydraulic transport system economics are enhanced if large quantities of water are used in the mining system, especially if water is available at the starting point; (2) the pipeline transport system is least affected by weather for all but the most extreme cold weather conditions; (3) the compact pipeline offers distinct advantages if the physical size of the transport system is a limitation. To date, a general comprehensive theory of pipeline transportation of solids has not been universally accepted. With hydraulic transportation each slurry mixture is unique, and only actual pipeline pilot testing can determine its true flow characteristics.

Recently much attention has been directed toward the development of pneumatic transport systems, utilizing air as the transport medium. Such systems are currently in use for underground stowage of waste material and for muck removal from tunnel boring machines. These pneumatic systems are capable of handling over 300 tons per hour of minus 8-cm material (Clancey and Goode, 1973).

Hydraulic and pneumatic transport systems have never been fully perfected because of problems of size, weight, inability to handle coarse material and inherent inefficiencies. Recently, through increased research and development efforts, significant theoretical advances in hydraulic transportation have been achieved. These advances have led to the development of competitive vertical hydraulic transport systems. Also, horizontal hydraulic transport utilizing capsules has been proven experimentally to be attractive because of lower pressure drop than the conventional methods of solid transport in fluid media (Scott and Young, 1968).

Because of recent advances in pipeline technology and health and safety demands, fluidic transporation of solids appears very attractive for the future. Hydraulic and pneumatic transportation of solids offers the potential for satisfying all the requirements for an ideal materials handling system — that is, continuous removal of material, safety, reliability, economy and flexibility. The development of these systems could provide the key to a continuous mining, transport, and hoisting system that should greatly increase the capacities and efficiencies of future deep underground operations.

NEED FOR NEW CONCEPTS

Ultimately, if the mining industry is to produce the world's mineral and energy requirements in the future, great advances must be made in underground mine production rates and productivities. This is particularly important in view of increased costs that will result from the mining of deeper ore deposits. The economies of scale achieved by surface mining operations can only be achieved when the time intervals for individual unit operations can be stretched and unproductive time is reduced. This requires larger and larger equipment sizes with corresponding underground opening sizes. These larger openings must remain stable even under adverse geologic conditions. It is necessary to develop rapid stabilization techniques for these larger underground openings in order to utilize larger and less labour intensive mining equipment. Further developments are also essential in the area of equipment technology associated with mechanized or continuous mining systems if the industry is to achieve greater productivity, efficiency, and economy from deep underground mines.

Unfortunately, advances in these areas have been too slow in the underground segment of the mining industry. Efforts must be made by industrial and governmental agencies to increase research and development efforts in the area of underground mining technology. Researchers should be encouraged to break away from the concept of improving on old and sometimes antiquated technologies and develop totally new, revolutionary concepts of underground mining methodologies. If underground production rates and productivities are to increase, new, sophisticated and continuous mining systems must be developed. These concepts must be approached from a total systems design standpoint, perhaps utilizing some of the basic technologies mentioned previously. For instance, effort should be directed toward the development of a continuous, completely automated, mining machine for hard and soft rock (similar to those currently used in underground coal mines) which would provide for continuous breakage of material and be remotely controlled by a mine worker functioning in a controlled environment. Development of a continuous materials transport and hoisting system would also be required from the mining face to the surface. Examples of such continuous transport might be pneumatic or slurry systems. The advantages of these continuous total mining systems are the reduction in manpower requirements, which, in turn, increases worker productivity for a constant production rate, and also reduces the number of mine workers exposed to potential mine accidents as a result of the reduction in manpower requirements. Until such continuous, automated, safe mining systems are developed for underground mines, percentage production and productivity increases from the future underground mining industry will necessarily be limited to comparatively small quantities.

Underground mining in the future will witness increased use of hydraulic and solution mining techniques. The potential for hydraulic mining to date has been realized primarily in the coal industry. Hydraulic coal mining is presently conducted in Japan, New Zealand, Canada, U.S. and the U.S.S.R. Hydraulic mining technology is currently limited to soft, friable or poorly consolidated materials. Most of the research being conducted around the world in hydraulic mining is in association with *cutting* hard rock formations. The $1,406-7,030$-kg/cm^2 high-pressure jets needed to cut this hard rock require advancements in technology in association with large capital investments before high-pressure hydraulic mining becomes a practical mining tool. However, if the technology can be developed at a competitive cost, hydraulic mining offers the following potential advantages: (1) high productivity, (2) better grade control, (3) reduced ventilation requirements, and (4) reduced ground control costs.

Production of commodities by solution mining is receiving increasing attention because it offers a way to extract minerals from deposits which could not be mined and processed by conventional methods. To date, salt and potash represent the most general ores mined by this technique. The producing formations may be in the form of bedded evaporites wherein the salt member may be some 9—153 m thick or from salt domes in which great thicknesses of salt occur (Marsden and Lucas, 1973). Solution mining has been conducted at depths of nearly 3,050 m.

In solution mining the object is to completely dissolve the desired material (ore) underground and subsequently "hoist" this brine to the surface for further processing. The basic components of solution mining are: (1) access wells to reach the deposit, (2) adequate solvent supply for solution at the "mining face", and finally, (3) hoisting (with pumps) the resulting solution to the surface either through the access wells or an adjacent connecting borehole shaft. The process of solution mining, therefore, depends to a great extent on the availability of reasonably pure strata or formation of salt or potash to be mined, access well or wells, and adjacent control of the solutioning process so that enlargement of underground cavities can be reasonably well delineated (Marsden and Lucas, 1973). Solution mining may be conducted from a single access well or from multiple wells (gallery system) which are coalesced by using the hydraulic fracturing technique. The solvent most utilized is water because of its low cost and wide availability. Occasionally the water may be slightly altered by partially saturating it with rock salt or by adding small amounts of acid, but the solvent consists mostly of water.

Although rock salt has been mined by solution techniques for many years in many countries, only limited success has been achieved in solution mining of potash. There are significant differences in the solution mining of rock salt and potash. For instance, rock-salt deposits are relatively pure with respect to soluble minerals, whereas potash is never found in the pure state and contains substantial amounts of other soluble minerals. Also, much less potash is dissolved and brought to the surface for the same amount of water circulated through the ore zone. Rock-salt deposits are generally much thicker than most potash beds; this offers a decided advantage because bed thickness determines the yield from wells and consequently the economies of the process.

Experimental attempts at solution mining of potash have been made in England, Germany and other countries. However, it appears that only the operation of Kalium Chemicals Ltd. at Belle Plain, Saskatchewan, Canada, has been commercially successful.

The application of solution mining to other water-soluble evaporites and slightly soluble metals and minerals has been attempted, but only limited commercial success has resulted. The possibility of introducing, through boreholes, a slightly acid leach solution to a low-grade copper orebody and removing the copper-enriched solution from another borehole has been considered by several companies. The primary problems or unknowns associated with such a procedure are: (1) the path followed by the solution moving through a fractured rock mass; (2) the solubility of sulphide and oxide minerals in the reducing environment below the water table; (3) the effect of introducing an oxygen-rich acidic solvent into the environment; (4) the non-closed system represented by a fractured orebody may, and probably does, represent a potential loss of solutions to non-recoverable portions of the orebody or country rock; and (5) the precipitation of ferrous sulphate in flow channels causing an increasing back pressure in the system, and possible additional solution losses.

Solution mining techniques will undoubtedly improve in the future. Standard operational techniques of pumping solvents and extracting valuable minerals require much additional study. At present the future of solution mining of metallic mineral deposits through boreholes appears bleak. Significant advances in technology must evolve if current technical problems are to be solved. The development of in-situ leaching by more conventional techniques may result in the technical breakthroughs necessary to permit the application of solution mining to deep metallic mineral deposits. The potential advantages in the areas of safety, development time and costs, production rates, extraction rates, and environment for solution mining versus conventional mining techniques are obvious.

The field of in-situ leaching offers, perhaps, the greatest challenge and holds the greatest promise for future mineral production. In-situ leaching is a process by which the mineral is segregated from an insoluble host rock by selective leaching in place, without transport of the host material. In-situ leaching techniques are applicable to leaching oxide and sulphide ores of copper and other metals. Basically, the in-situ leaching process requires that: (1) the leach solutions come in contact with the ore minerals, (2) the pregnant solutions must be recovered and collected, and (3) provisions must be made to pump the pregnant liquors to the recovery plant. The two areas currently receiving the bulk of the research activity in in-situ leaching are: (1) development of an underground confined and enclosed leaching area or vessel, and (2) development of hydrometallurgical processes to dissolve the minerals, recover the solutions and extract the contained elements.

One of the most attractive advantages of in-situ leaching is the fact that smaller, lower-grade orebodies which could not economically be mined by conventional methods can be "mined" by in-situ processes. Since capital and operating costs are lower for in-situ processes, otherwise uneconomical orebodies can be considered for production. Other advantages of in-situ leaching are: (1) development time and expense are less than for conventional methods, (2) deep lower-grade deposits may be exploited due to reduced capital and operating costs, (3) recovery of additional values from previously "mined-out" areas of properties, and (4) a potential reduction in environmental and pollution problems.

The large low-grade deep disseminated ore deposits of the future may well be "mined" by the in-situ leaching process after the deposit has been fractured by nuclear detonations. The ability of nuclear explosives to fracture and break large volumes of rock efficiently and economically makes their use in underground mining systems potentially attractive — particularly with respect to in-situ leaching. For instance, much theoretical and economic consideration is being given to using a combination of nuclear explosives detonated in mineralized copper zones followed by leaching as a viable mining method. The nuclear explosion would produce an underground rubble-chimney of broken ore, probably below the water table. Presumably, after the chimney is filled with water and has reached hydrostatic equilibrium, oxygen, under pressure, is introduced near the bottom of the rubble-chimney. The increase in solubility of oxygen at high hydrostatic pressure is sufficient to initiate the oxidation of the primary sulphide minerals. The oxidation and dissolution of these sulphides produces enough heat to increase the temperature of the ore and water in the chimney as much as 100—150°C. The rate of dissolution of chalcopyrite becomes so rapid under these conditions that the rate of recovery is limited by the degree of exposure and other factors (Malouf, 1973). The pregnant liquors could then be collected in drifts located below the broken ore chimney and ultimately pumped to the surface where the copper is recovered by conventional metallurgical processes.

Another example of in-situ mining which holds much promise for the future is underground gasification of coal. The objective in underground gasification of coal is extraction of the thermal energy of coal in the form of fuel gases or the production of synthetic gases by the partial or complete combustion of coal and water-gas processing of the hot coke in place with air, oxygen, steam or their mixtures (Wang et al., 1973). The basic underground gasification operation can typically be divided into two parts: path preparation, and gasification. To date, the following underground gasification methods have been developed or proposed: (1) chamber method, (2) stream method, (3) borehole method, (4) directed boring method, (5) percolation method, and (6) nuclear fracturing.

A vast number of laboratory investigations and field experiments have been conducted in the U.S.S.R., the U.S., Belgium, Italy, Poland, France, Czechoslovakia, Japan and England on underground gasification of coal with only limited success. Only in the Soviet Union has the industrialization of underground gasification of coal materialized. Underground gasification techniques have not been fully developed, perhaps largely due to the many interdisciplinary scientific problems associated with physiochemistry, fluid mechanics, hydrology, geology, geophysics and rock mechanics. Much additional technologic and economic research and development are needed before underground gasification techniques will become commercially viable.

Due to the many scientific and engineering problems associated with in-situ coal gasification, the development of commercially successful techniques is a great challenge to mining engineers and scientists. The successful development of underground coal gasification techniques could minimize, if not eliminate, the problems of underground mining operations, increase "mining" depths and thus possibly reserves, improve ecological environments, and make possible the use of coal resources currently uneconomical to mine by conventional underground mining methods.

Another example of in-place underground mining for the future is in-situ retorting of oil shale. The "oil" in oil shale is really a solid insoluble organic substance called kerogen. When kerogen is heated to 315—480° C it "cracks" or decomposes to oil, gas and a carbonaceous residue of the spent shale. This process is called retorting.

The in-situ retorting of oil shale has been under investigation in Germany, Sweden, Estonia and the U.S. since the 1940s; however, only Sweden has developed it industrially. The major purpose in the underground retorting of oil shale is to extract the organic substance in the form of liquid fuels, with gaseous byproducts, by means of in-situ retorting through a borehole system (Wang et al., 1973). As with underground gasification of coal, in-situ retorting of oil shale involves complex problems in heat transfer, pyrolysiskinetics, and process control.

The three basic requirements for an ideal in-situ oil shale retorting operation are: (1) preparation of the underground retort chamber, (2) retorting of the oil shale, and (3) gasification of the carbonaceous residue. Since oil shale typically has little or no natural porosity or permeability, the first requirement for in-situ retorting is the development of sufficient fracturing in the proposed retort chamber to permit the passage of the heating media through the chamber and the flow of resulting liberated gases and oil vapours to the outlet boreholes. Fracturing techniques currently under development for this process are pneumatic and hydraulic fracturing, electrolinking, conventional explosive fracturing and nuclear explosive fracturing. The size, frequency, spacing, and other characteristics of the fracture patterns resulting from these various techniques will undoubtedly have an effect on the choice and design of specific retorting systems utilized.

Most oil shale retorting processes rely on conduction as the principal means of heat transfer within the retort chamber. This heat energy can be supplied internally by burning the oil shale, externally by injecting superheated steam or gas into the oil shale, or by electricity (Wang et al., 1973). Some of the underground retorting processes being investigated are: (1) electric heating, (2) combustion, (3) thermal injection, and (4) electro-carbonization. An example of one possibility which has been proposed is to utilize a combustion method in conjunction with a rubble-chimney produced by a nuclear deto- nation. In this situation the oil shale could be ignited at the top of the chimney with subsequent downward movement of the fire front, air, gas and oil. The gaseous products and shale oil could then be withdrawn from the bottom of the chimney through boreholes or collection drifts placed beneath the chimney.

Technology associated with the gasification of carbonaceous residue is extremely crude and only recently have engineers and scientists concentrated their efforts in this area. Much additional research and development is needed in all aspects of in-situ retorting of oil shale if an economically feasible technique is to become a reality. Successful development of such a technique would virtually eliminate most costly underground mine development and production methods currently envisioned; vastly reduce the materials han- dling problems; reduce spent shale disposal problems, and thus minimize environmental effects; and enable many countries to economically utilize existing low-grade oil shale deposits, and thus provide themselves with an increasing percentage of domestic energy requirements.

It would appear that some form of in-situ mining holds the greatest prom- ise as a means of obtaining the world's future mineral requirements from deep ore deposits. In view of current technology and other considerations it appears unlikely that current underground conventional mining techniques can be utilized at depths exceeding approximately 4,800—5,500 m — and then only under favourable conditions. On the other hand, it is conceivable that in-situ mining techniques could be utilized to depths of 6,100—9,200 m. Some additional potential advantages of in-situ mining techniques are: (1) elimination, or at least reduction, in costly underground development and operating techniques, (2) increased resource recovery and utilization, (3) increased safety, (4) increased production rates and productivities, and (5) reduced environmental complications. Unfortunately the technology and development necessary for commercial in-situ mining is in its infancy. Great advances in research and development are necessary before near-surface in-situ techniques are economically feasible — much less the utilization of these methods at 6,100—9,200 m of depth.

MANPOWER NEEDS

Thus far some of the environmental, technological and engineering factors requiring additional research and development, in conjunction with advances in future deep underground mining techniques, have been briefly outlined.

The necessary advances in engineering and technology can only be accomplished by people. This brings to light one of the most critical problems currently facing the mining industry — the procurement, allocation, and utilization of manpower.

The mining industry around the world is experiencing a severe shortage of qualified, university-educated mining engineering graduates. Very recently universities have experienced increases in enrollment in mining engineering and related curricula. There is little doubt that this is primarily the result of recent world-wide energy shortages and probable impending mineral shortages. Unfortunately these recent increases will not come close to meeting the industrial requirements for professional manpower in the foreseeable future. For instance, in the U.S. universities training mining engineering students graduate approximately 150 mining engineers each year for a domestic industry requiring ten to twenty times that number. It is imperative that industry and academia unite in establishing programmes which will lead to increased enrollments in mining engineering programmes throughout the world to meet the professional manpower needs of industry in the near future.

The manpower shortage in the mining industry is not limited to the professional labour force; there is a great need for mining technicians throughout the world. People trained in two- to three-year technical programmes concentrating on the more routine, applied aspects of mining are currently in great demand within the mining industry. Such programmes should provide manpower for positions as first-line supervisors in the unit and supporting operations of mining. Such a technical work force would help reduce the existing demands on the professional labour force and allow it to concentrate on the development of technologies and methodologies needed for increased production rates and productivities required for future deep mining operations.

Last, but certainly not least, is the unprecedented need within the mining industry for a skilled labour force. Technological advances resulting in more sophisticated production and control equipment within the mining industry currently requires, and will require in the future, a more highly trained, specialized labour force. Unfortunately, experienced miners are in extreme short supply. Many companies find that present prospective employees have had little or no previous mining experience. This, coupled with the fact that labour turnover rates are high, requires mining companies to establish elaborate, expensive, continuous labour training programmes. Efforts must be made to improve pay levels, working environment, and safety, in order to attract and retain people in the mining labour force.

CONCLUSIONS

There is much evidence to support the position that an increasing amount of the world's future production of metallic, non-metallic and energy resources must come from large, deep underground mines. If underground mining is to be a viable economic, effective and efficient means of producing these required resources, there must be significant advances made in underground mine capacities, production rates and productivity. These advances can only result from continued research and development in areas such as mine development, ground control and support, ventilation and air conditioning, working environments, safety, materials handling, and new mining methods. Emphasis must also be placed on continued development of underground equipment technology as well as the procurement, allocation and utilization of the professional, semi-professional and skilled labour force required in the minerals industry.

Over the short-term the underground mining industry is in dire need of a safe, economical, continuous, automated total mining system. Such a system should provide continuous breakage of material at the mining face as well as continuous materials transport and hoisting from underground to surface. This type of highly mechanized system would significantly increase underground production rates and productivities. Obviously, such systems must also be economically competitive.

It appears that some form of in-situ mining holds the greatest promise as a means of obtaining the world's future mineral requirements from deep ore deposits. It is conceivable that in-situ mining techniques could be utilized to depths of 6,100—9,200 m, which is well above expectations for current underground conventional mining methods.

The future of deep underground mining, under current economic, safety, and efficiency constraints, appears to rest on technologic and engineering advancements resulting from further research and development supported by large capital expenditures. Underground mining will be a satisfactory means of producing future mineral and energy resources *only* if the extraction techniques can do so safely and economically at high production rates and productivities.

REFERENCES CITED

Allsman, P.T. and Yopes, P.T., 1973. Open-pit and strip mining systems and equipment. In: *Mining Engineering Handbook*. SME/AIME, New York, N.Y., pp. 17-1—17-8.

Boshkov, S.H. and Wright, F.D., 1973. Underground mining systems and equipment. In: *Mining Engineering Handbook*. SME/AIME, New York, N.Y., pp. 12-1—12-12.

Clancey, J.T. and Goode, C.A., 1973. Underground haulage pipelines. In: *Mining Engineering Handbook*. SME/AIME, New York, N.Y., section 14.5, 7 pp.

Dravo Corporation, 1974. Analysis of large-scale non-coal underground mining methods. *U.S. Bur. Mines, Contract Rep.* No. SO122059, 605 pp.

Malouf, E.E., 1973. Leaching. In: *Mining Engineering Handbook*. SME/AIME, New York, N.Y., pp. 21-70—21-78.

Marsden, R.W. and Lucas, J.R., 1973. Specialized underground extraction systems. In: *Mining Engineering Handbook*. SME/AIME, New York, N.Y., section 21, 118 pp.

416

Scott, S.A. and Young, T.R., 1968. Hydraulic transportation. In: *Surface Mining.* SME/AIME, New York, N.Y., section 9.5, pp. 622—636.

Wang, C.S., Capp, J.P., Wane, M.T. and Boshkov, S.H., 1973. In-situ gasification and liquefaction mining systems. In: *Mining Engineering Handbook.* SME/AIME, New York, N.Y., pp. 21-78—21-96.

Willson, J.E., Lucas, J.R. and Adler, L., 1973. Underground haulage. In: *Mining Engineering Handbook.* SME/AIME, New York, N.Y., section 14, 53 pp.

SELECTED REFERENCES

Ackerman, D.H., 1970. Environmental planning in mineral development. *Min. Congr. J.* 56 (4): 40—46.

Brook, N., 1972. Water jet cutting likely to play a major role in ore mining. *Min. Mag.,* 126 (6): 450—452.

Brooks, F.H., 1973. Solution mining — a review. Paper presented at *Western Mining Congress,* Denver, Colo.

Budivari, S. and Potts, E.L.J., 1970. Rock deformation measurements for evaluating mine stability. *Trans. Inst. Min. Metall.,* 79: A37—A42.

Canadian Mining Journal. 1970. Accident prevention. 91 (9): 47—73.

Canadian Mining Journal, 1973. Mining operating and cost data — underground mines, mine surface plant data. Reference Manual and Buyers Guide. 153 pp.

Caverson, A.H. and Hanninen, C.C., 1973. Mine convergence measurements as an operating tool. *Min. Eng.,* 25 (5): 40—42.

Condolios, E., 1967. New trends in solids pipelines. *Chem. Eng.,* 45: 131.

Council on Environmental Quality, 1973. *Coal Surface Mining and Reclamation.* U.S. Government Printing Office, Washington, D.C., pp. 32, 101.

Engineering and Mining Journal, 1970. Mine designs for deep recovery. 171 (6): 156—163.

Fletcher, D.G. and Evans, V.A., 1973. Underground remote controlled loading: key to safety and efficiency. *Eng. Min. J.,* 174 (7): 78—79.

Gagnebin, A.P., 1970. Changing levels of mining technology. *Min. Congr. J.,* 56: 83—86 (November).

Gallimore, J.L. and Adler, L., 1972. The need for a new mining system. *Min. Congr. J.,* 58 (9): 24—29.

Geyer, H., 1973. How hydraulic excavators cut costs, raise efficiency in German salt mine. *World Min.,* 26 (5): 49—51.

Hougland, R.W., 1968. Improved safety and productivity in potash mining. *Min. Congr. J.,* 54 (7): 45—49.

Jaeger, J.C. and Cook, N.G., 1969. *Fundamentals of Rock Mechanics.* Methuen, London, 513 pp.

Janelid, I. and Kvapil, R., 1966. Sub-level caving. *Int. J. Rock Mech. Min. Sci.,* 3: 129—153.

Jewett, J.W., 1973. Economics of reclamation measures. *West. Miner,* 46 (6): 40—43.

Johnston, C.E., Surly, J.J. and Short, A.B., 1969. Transport considerations of underground operations over 100,000 TPD. *Min. Congr. J.,* 55 (4): 87—93.

Journal of Mine Ventilation Society of South Africa, 1972. Heat — a challenge in deep-level mining. 25 (11): 205—213.

Julin, D.C. and Tobie, R.L., 1973. Block caving. In: *Mining Engineering Handbook.* SME/AIME, New York, N.Y., pp. 12-162—12-222.

Karlsson, N.G., 1972. Exploitation of in-situ rock strength in order to minimize underground support requirements in Scandinavia. Paper presented at *8th Canadian Symposium on Rock Mechanics,* Toronto, Ont., 7 pp.

King, R.W.L., 1970. Rubber-tyred equipment in Australian metal mines. *Aust. Bur. Miner. Resour., Geol. Geophys., Rec.* No. 1970/5, 133 pp.

Murray, R.E., 1972. Rock removal methods in coal mines. *Min. Congr. J.*, 58 (10): 57—61.

Palowitch, E.R. and Malenka, W.T., 1964. Hydraulic mining research: progress report. *Min. Congr. J.*, 50 (9): 66—73.

Parris, T.D., 1969. LHD equipment ups production for INCO. *Min. Eng.*, 21 (6): 84—87.

Pillar, C.L., 1972. Block caving today and its potential in future mining operations. Paper presented at *Joint Meeting MMIJ/AIME*, Tokyo, Print No. T-II-d1, 7 pp.

Pillar, C.L., 1973. Block caving, a key to the economic extraction of deep seated ore. *West. Miner*, 46 (3): 30—31; 33—36.

Sandstrom, P.O., 1972. Application and optimization of sub-level caving techniques. Paper presented at *International Sub-Level Caving Symposium*, Stockholm, 15 pp.

Skimomura, Y., 1972. Recent trends in metal mining technology in Japan. Paper presented at *Joint Meeting MMIJ/AIME*, Tokyo, Print No. G-II-1, 12 pp.

Stewart, R.M., 1972. Mining equipment trends. Paper presented at *Joint Meeting MMIJ/AIME*, Tokyo, Print No. G-II-2, 9 pp.

Swift, R.L., 1971. Noise regulations and the individual. *Min. Congr. J.*, 57: 50—56 (December).

Walker, A.M., 1973. The transition to trackless mining. Paper presented at *75th Annual General Meeting of the Canadian Institute of Mining and Metallurgy*, 11 pp.

Ward, M.H., 1973. Engineering for in-situ leaching. *Min. Congr. J.*, 59 (1): 21—27.

Westlund, C., 1972. Speculations about cost structure and total economy in mechanized underground mining. Paper presented at *International Sub-Level Caving Symposium*, Stockholm, 14 pp.

Wimpfen, S.P., 1973. Mine costs and control. In: *Mining Engineering Handbook*. SME/AIME, New York, N.Y., pp. 31-2—31-50.

Zatek, J.E., 1963. Australians develop new hydraulic hoist. *Eng. Min. J.*, 164 (3): 102—103.

Zatek, J.E., 1968. Sub-level caving; how to use it; what are advantages, problems. *World Min.*, 21 (10): 76—78.

Zatek, J.E., 1970. Safety and economical aspects of raise boring. Paper presented at *1970 Southwest Safety Congress and Exposition*, Phoenix, Ariz., 8 pp.

Zatek, J.E., 1971. Underground noise control: the new challenge. *Coal Age*, Dec., pp. 11—102.

Chapter 14

MINERAL SUPPLIES FOR THE FUTURE — THE ROLE OF EXTRACTION AND PROCESSING TECHNOLOGY

HENRY E. COHEN

INTRODUCTION

Mineral processing is essential in preparing most metalliferous ores and industrial minerals to the standards of purity and physical consistency necessarily imposed by primary consumers. Slag-forming gangue must be eliminated from metalliferous ores to maximize furnace productivity, to minimize fuel consumption in smelting processes, and to reduce handling and freight costs; metallurgically harmful constituents of the ores must also be removed (e.g., arsenic in tin ores, copper in iron ores). Impurities often need to be removed from industrial minerals down to a few parts per million; for example, china clay for paper-making needs to be free of all traces of iron-bearing minerals for maximum whiteness, and it must not contain any mineral grains coarser than a few microns so as not to spoil the surface finish of the paper. Similar stringent specifications are often attached to ceramics, refractories, fillers, insulators and other products.

Thus mineral processing is technically necessary for transforming impure and low-grade ores into industrially acceptable raw materials. It is also necessary that the costs of mineral processing remain within the limits imposed by the industrial value of the final product. The cost balance must be achieved by combining various aspects of processing — controlled by the nature of the ore and by the capabilities of available technology.

The highest possible extraction of values and the simplest possible methods of treatment are two obvious concepts for mineral processing; these concepts imply incurring the least cost in extracting from the ore as much product as possible at the worst quality which the customer will accept. It would be naive to assume that any mining operation, privately or publicly owned, can afford to rob its assets by merely extracting the richest and easiest portions of an orebody in order to make a quick gain. In any but the most speculative enterprise of the "California gold rush" type, care is taken to plan a balanced extraction through a combination of mining and mineral processing which will permit full utilization of the asset (the ore in the ground) at an overall cost that will yield an adequate return on the investment.

GROWTH AND DEVELOPMENT OF MINERAL PROCESSING

The formula for achieving low processing costs in mineral production has undergone remarkable changes during the history of man's exploitation of mineral resources. Initially mineral processing could be described as labour intensive, but this would be without real meaning since during the first several thousand years of man's history capital, mechanical equipment, and power were unavailable. Increasing demands for minerals and metals over the centuries, supported by engineering developments, resulted in a gradual process of mechanization and expansion of production; in 1870 the opening of a mineral processing plant at Clausthal in the Harz Mountains, with the "staggering" capability of processing 500 tons of ore every 24 hours, was a milestone.

By the latter part of the 19th century mechanical processing was clearly taking over from hand dressing of ores; nevertheless, while processes required capital equipment, they remained very labour-intensive, especially for feeding machines and for transporting the ore from one process to another; manufacturers of shovels and wheelbarrows did well and continued to prosper for another fifty years. On balance, people were cheaper and more convenient than machines; they required less capital investment in the first place, and they were readily disposable during cut-backs in production. Moreover, most processes of mineral separation were designed empirically and required frequent manual corrections based on experience which remained an exclusive human function until the invention of computer memories.

The eventual need to choose between labour-intensive and capital-intensive processing methods arose largely from the increasing size of production units. While several thousand years of development preceded the 500-tons/day plant at Clausthal in 1870, only 80 years later, in the early 1950s, plants capable of treating about 100,000 tons of ore per day were in production in North America. The evolution was partly due to the exponential increase in production and partly due to the decrease in grade of the ore being treated. Not only were rich ores exhausted, they simply never existed on a scale large enough to meet mid-20th century demands for minerals.

Only low-grade ores occur in the earth's crust in sufficient tonnages for current rates of consumption. The mining of these lower-grade ores necessitates greater tonnages being treated for any given output, but with low value contents, the costs of treatment per ton must be smaller than for high-grade ores. One obvious method of lowering treatment costs is to increase the capacity of machines; capital charges and operating costs are scarcely increased by using larger sizes of equipment, and an operator can supervise a large machine as easily as a small one.

The growth of mechanization and the increased reliance on capital equipment were supported by the increased availability of energy and power based on natural fuels. From the late 1800s the development of coke, gas, and

electric power in geographically favoured areas allowed cheap power to supplement labour and eventually to replace it. This trend was most clearly
visible in extractive processes of pyrometallurgy which are heavily dependent
on fuel and energy, but in mineral processing generally, and especially in crushing and grinding, it permitted growth to unit sizes which were previously un-
attainable. The combination of improved engineering materials and design
with availability of cheap power was the essential pre-condition for various
permutations of reduced labour and increased capital equipment and power
which provided local optimization of production during the 20th century.

For an operation located near a source of cheap power, the preferred
processes would be energy-intensive. Cheap fuels and improved transport led
to the acceptance of the viability of mining projects at great distances from
consumers, depending on availability of markets, transport, fuels, and alternative ores in varying orders of importance.

At various times and for different commodities there were two opposite
trends which had similar effects on processing conditions. The price of a
commodity might decrease, or the grade of available ore might decline and,
to maintain a given level of revenue, a mine would need to increase the rate
of production and reduce the cost per ton treated. Alternatively, the price of
a commodity might rise significantly, causing the mine to lower its cut-off
grade so as to extend its assets (see Chapter 2); to meet contractual output
requirements and to "cash in" on the market, the mine would also have to
increase its throughput and, because of the lower feed grade, reduce the cost
per ton treated. Both trends would tend to encourage an increase in production
capacity, accompanied by a need to reduce the costs per ton treated. This
combined aim is most easily attainable by increasing the size of unit process
machines and by decreasing the number of processing units together with the
labour requirement per ton of feed.

The reduction in labour content began to acquire enhanced importance
after 1965 when wages began to escalate out of proportion to the rate of
inflation of other costs. The 1960s marked the approaching peak of capital-
and energy-intensive process development in replacing labour-intensive
methods. This development might well have continued into the 21st century,
with limitations on size imposed mainly by engineering materials and by problems of engineering design (in 1970 the size of grinding mills was limited only
by available capacities of motors and transmission gears to provide adequate
power).

THE THREATS OF SCARCITY

Fuel shortages caused a dramatic reversal in process requirements; the full
significance will become apparent only gradually. The developing fuel crisis
did not follow a simple clear line; it started in the late 1960s with growing

shortages of high-grade metallurgical coke which caused world-wide scrambles for coking coals, renewed interest in coal preparation and blending, use of alternative fuels (oil and oxygen injection in blast furnaces), substitution of electric smelting, and the relocation of smelters at ports or near fuel sources. Fuel oil was cheap and abundant, and was apparently destined to provide power for mines located in countries with access to supertankers or oil pipelines. By comparison, hydroelectic power remained a geographic eccentricity enjoyed by a few mines, notably in Norway, western Canada, and Zambia. The prospect of power from nuclear energy remained remote because of high costs. The impetus provided by the military potential of nuclear weapons merely served to emphasise the scarcity and low-grade of uranium ores, as well as the high costs of uranium extraction. Coupled with the high capital and operating costs of nuclear power stations, this seemed to confirm atomic power as a non-starter, or at best only as a very long-term prospect.

The real change came in 1973—1974 when the soaring price of oil forced the world to take a fresh look at the unacceptable face of nuclear power and to discover new life in the moribund coal industry. This was no longer a question of finding cheaper alternatives, but a matter of urgency to find any fuel with a reasonable certainty of supply and predictability of cost. Expensive power had come to stay and would eventually force a drastic reappraisal of the principles and design of processes for mineral recovery from ores. The word "power" had taken on new meaning with the realization by those controlling sources of energy that scarcity made almost any demand for increases in prices or in wages irresistible. Coal miners, power station workers, and oil sheiks had raised the price of energy to levels which threaten the precarious economic balance of mineral production. Thus, in the mid-1970s the position is one of expensive labour and expensive power having to be redeployed through judicious capital commitment for processing larger and larger tonnages of lower and lower grade ores and recycling more and more waste products to meet rising world consumption of metal and mineral raw materials. Redevelopment will have to contain elements of geographic relocation of mineral production relative to consumers and power supply. It will also require reassessment of the size of operations, possibly towards smaller units and minimum power, as well as transportation changes as compared with the earlier trend towards power- and transport-intensive production on a very large unit basis.

This new trend is bound to be reinforced by continuing strong political pressures for local development and self-sufficiency. The capricious inequalities in the geographic distribution of mineral deposits are at present interpreted as national assets to be exploited rather than as world resources to be shared; the increasing emphasis on fuller local exploitation of such assets results in decreasing willingness to export unprocessed ores and raw materials. Even if the irrational elements in this tendency are discounted, it cannot fail to contribute strongly to the redistribution of world production and the reassessment of optimal processing methods. Against this background mineral processing must face the strong probability that the world's population will continue to increase and that consumption of metals and minerals per world citizen will continue to rise towards the levels now attained in Europe (although it may never reach the levels current in the U.S.).

PROSPECTS FOR MINERAL PRODUCTION

That the earth's crust has a limited content of conventionally useful components is a strong probability. One can thus state with conviction that one or more of the following developments will have to occur:

(1) Dramatic new discoveries may be made of new resources deep in the earth, and engineering capabilities beyond present-day technology may make their extraction technically feasible. However, the economic viability of such supplies looks improbable.

(2) Very low-grade ores may be processed in vast tonnages. Such processing would have to occur at the mine to avoid transporting large tonnages of waste, but this type of development has been made very unattractive by high fuel prices. The costs of transporting fuel and supplies to remote mines and exporting concentrates to distant markets could become an increasing barrier to economic viability, even in the case of iron ores which are at present leading examples of giant operations in unit sizes of production and in ton-miles of annual transport.

(3) The necessary reduction in fuel and transport cost components may have to be achieved by changing to comparatively small mines and mills situated close to consumer areas. These mines would have to be geared to local consumption rather than to world markets, and the processing plants would have to be integrated so as to combine new production with recycling of locally generated waste materials. The feed for the concentrator, smelter, or industrial mineral producer would be mixtures of new local ores enriched by local recycled wastes. For example, in the case of iron and steel production, "mini" steel works with burdens enriched by local scrap metal might become preferable to the present situation which often combines a distant mine/concentrator with a semi-distant iron and steel works. This present situation comprises three significant transport cost components: concentrates from mine to smelter, steel from smelter to consumer, and scrap metal from consumer to smelter.

(4) Mineral processing may have to adopt revised principles of production — which aim, in order of priority, at lowest overall energy consumption, highest recovery of usable components, and cost economy — as against previous principles based largely on financial cost/profit balances. This change in emphasis may be further influenced by new demands for environmental protection and control. Society will have to define the overall "price" it is

prepared to pay for mineral resources. This "price" will have several complex and conflicting components to which priorities may be assigned on differing considerations of ethics, expediency, and need. The role of mineral processing in this context is to provide a choice of combinations of "price" components or terms of reference in any specific production situation. (Other aspects of "price" are discussed in Chapters 6 and 9.)

MEETING FUTURE REQUIREMENTS

From the point of view of ore characteristics which necessitate processing to make the material acceptable for consumer requirements, a number of basic variants can be distinguished. These are discussed below.

Ore quality

Ores of good "quality" occurring at locations remote from markets or user concentrations will have to be processed to very high specifications and small bulk. The remote location may impose less stringent (and thus less costly) environmental conditions, emission control, and tailings disposal. The largest possible production units would have to be chosen, compatible with the size of the orebody and with the capacity of world markets.

Alternatively, if ores of the above kind are of sufficient importance (for example, due to scarcity) and if the location can be made acceptable for human settlement, new centres of industry and habitation could be created near the deposit. This would be especially apposite if an underused source of energy were found within economic reach. Mineral processing could then be optimized only to the extent of making the operation competitive in world markets. An obvious example of this type of approach would be the transfer of most of the free world's ferrochrome production to the Bushveldt complex of South Africa which contains, in close proximity, a very large part of known chromite reserves and very large coal deposits. The continuing export of chromite concentrates from South Africa containing less than 50% of chromium (by weight) is an evident extravagance in energy terms.

When ores of unattractive qualities within normal concepts of production (low grade, difficult to separate, small size) are found close to existing centres of consumption or near adequate sources of energy, it would be possible to develop processing systems to permit their use on a basis of overall local advantage in competition with imports. The presence of a skilled labour force able to use sophisticated technology might aid the application of new processes and could help in replacing energy-intensive methods with capital-intensive ones.

Cut-off grade and extraction

Ores currently exploited under conventional commercial terms often have cut-off grades imposed which necessitate the deliberate selective abandonment of large tonnages of off-grade ore in the mine; low-grade chromite ores in Rhodesia are a typical example. Unmined off-grade ore is liable to be lost permanently since its extraction on its own is rarely likely to be economically viable. Thus, it is of importance to develop methods of processing which would permit the lowering of cut-off grades now, so that no off-grade ore needs to be abandoned in inaccessible locations. Three possible solutions to the problem include improved processing techniques; mining and stockpiling off-grade ore now for future treatment; and subsidies (on an international basis) for producing ores which are commercially sub-grade but which constitute vital future reserves of scarce commodities.

Many ores are currently exploited at concentrate grades and recoveries which are commercially satisfactory, but which are open to further improvement. If the value component is scarce, or is liable to become scarce with increasing consumption, steps should be taken now to raise the yield. The same argument applies to current methods which are excessively energy-intensive. Although development of new mineral processing techniques may be expected to contribute, it must be recognized that existing practices can be improved considerably; one of the most urgent needs is for the development of operational and control techniques which ensure greater constancy of performance at optimum levels.

By-products

Attention should be directed to ores which are now commercially exploited for certain main value components where available by-products are not recovered because of a lack of commercial incentives. Examples include the vanadium in titaniferous iron ores and tin in mixed base metal sulphide ores; later recovery of such minor constituents (for example, 1% vanadium or 0.6% tin) from the tailings of present production is much more difficult technically and economically. A system of incentives should be set up, either nationally or internationally, to persuade companies to extract the by-products from current production for sale to buffer stockpiles if immediate markets are lacking. It would be desirable to revise the mineral processing methods to optimize extraction of all main values and by-products; an early example of this is the change in mineral processing practice in certain South African gold mines to facilitate efficient extraction of by-product uranium.

The principle of first-hand extraction of all values and by-products may be open to question for some commodities which could be left safely in tailings until needed (e.g., quartz, mica, titanium), especially if currently known methods of recovery appear to be inadequate. However, if conservation of energy becomes an over-riding consideration, it should be noted that dewatering of tailings followed by deposition to be followed later by reclaiming and repulping entails considerable energy waste; handling of each tonne of solids necessitates handling some 5 or 6 tonnes of water. The energy waste is even more serious if a by-product could be extracted from the ore at a coarser size than the main values; not only is the by-product ground unnecessarily fine, but its recovery in the finer size may be more difficult and energy-consuming.

Environmental and energy constraints

It seems probable that most mineral processing operations situated in populated areas, and especially those in regions with high amenity values, will be increasingly subjected to stringent controls on emission of noxious substances and deposition of wastes (see Chapter 9). Fumes, acids, poisons, etc., will have to be totally contained or rendered harmless; solid tailings will not be permitted to discharge into streams, to be deposited in conspicuous piles, or allowed to fill large lagoons. Wherever such methods have been adopted in the past, they were justifiably based on minimizing costs; more environmentally acceptable methods can be expected to be more costly, especially if they are considered in isolation or only in the context of current mineral processing practice. As soon as effluent and tailings disposal methods become an extra charge on production, all processing steps must be re-examined to arrive at a new minimizing balance of overall costs. Process substitutions, or modifications which, in direct comparison, may have seemed unattractive in the past may become far more desirable if they entail environmental advantages. All possible alternatives need to be examined before the additional costs of environmental preservation or protection can be stated with confidence. This applies especially to the possible substitution of mineral processing and chemical extraction for more conventional pyro-metallurgical processes (e.g., in the production of copper). In calculating comparative "costs" of environmental constraints, it is necessary to include the differences in grade and recovery of the values, in capital and operating costs and in total energy requirements; this can become very complicated if the methods under consideration incorporate the recycling of scrap.

A significant part of total energy consumption in mining and mineral processing is made up of internal transport and materials handling, and there is a considerable potential for energy saving in integrating the two operations. While primary crushing is often carried out in the mine as an aid to more efficient transport and hoisting, there is no reason why this should not be extended to include all crushing, all or some grinding, and possibly some processes of concentration. Fine ore, dry or in slurry form, is hoisted more conveniently and at lower cost; if some form of pre-concentration (magnetic or gravitational scalping, bulk flotation, dry sorting) were feasible, the total tonnage for hoisting might be significantly reduced, and the tailings would be available for direct disposal in worked-out sections of the mine.

Further energy economies in mineral processing might be sought by developing processing routes which avoid any unnecessary fine grinding. Up to 70% of the total energy used in iron ore beneficiation (for example with magnetic taconites in the U.S.) is directly accounted for by processes of fine grinding the ores to particle sizes below 50 μm. The grinding processes themselves are, without exception, most inefficient; less than 1% of the input energy is utilized for actual mineral breakage, the remainder being mainly converted to heat. A search for more efficient comminution methods thus merits high priority. However, it is equally important to look at alternative methods of processing which might diminish the need for fine grinding. In the case of iron ores this might run counter to recent trends of producing the highest possible grades of concentrates. A coarser concentrate of lower iron content (higher silica) could be preferable if higher slag volumes were necessary for dealing with higher sulphur contents in the smelter fuel.

Coal utilization

With the probability of oil and natural gas remaining scarce and costly, mineral processing has an important contribution to make in facilitating the use of coal for energy supplies and for industrial processes such as roasting, reduction and smelting of ores, firing of refractories, etc. Many known coal deposits are commercially unacceptable due to high ash contents; others are metallurgically and environmentally unacceptable, with sulphur contents in excess of 1.5%, and often as high as 5 or 6%. Large tonnages of coal-bearing aluminous shales are known to exist in easily accessible deposits, but are not so far seriously regarded as potential fuel sources. High-sulphur coals remain undeveloped or are abandoned underground by selective mining. These various deposits could meet man's foreseeable energy requirements for hundreds of years in excess of anything that might have been hoped for from known sources of oil and gas (see Chapter 7).

There is a need for imaginative development of mineral processing, alone or in combination with chemical treatment, to bring these coals into effective use. An attractive additional possibility would be the exploitation of the "ash" content of low-grade coals as a source of alumina through processes which, although purely speculative, are at present conceivable. The coal concentrates would inevitably be fine-grained and would be eminently suitable for blending to yield formed coke for blast furnace use or other briquetted fuels. The fine concentrates would be directly usable for pulverized fuel burners and as reductants for various metallurgical processes.

Iron ores

Rather special considerations apply to future supplies of iron. There is no foreseeable shortage of iron ores on a global scale, and the wide geographic distribution of good ores minimizes any danger of a monopoly situation being forced on consumers in the future. In general, mining and mineral processing account for less than 10% of the total production costs and the total energy consumption for producing the metal, compared with about 70% for metallurgical extraction. Thus there would be a great potential saving in respect of both cost and energy consumption if a larger share of the transformation from ore to metal were transferred to the mining and beneficiation sectors. Attempts are needed to optimize the chemical grade and the physical condition of ores for the subsequent smelting process. Increased removal of alumina and of any other slag-forming constituents which affect slag fluidity would lower the temperature and energy requirements of smelting. In blast furnace practice, every 100-kg reduction in the slag volume corresponds to a saving of about 20 kg of coke, until a limiting slag volume of about 200 kg per tonne of metal is reached. Total removal of slag-forming constituents could be advantageous in cases where direct steel making can show a local cost/energy advantage. Purely in terms of slag removal, any process of physical concentration (magnetic, flotation, gravity, etc.) represents about 1% of the cost/energy requirements of comparable pyrometallurgical removal. Since there is no shortage of iron ores, it would be preferable to accept the somewhat lower metal recoveries associated with physical concentration methods in return for lower overall cost/energy balances.

Since the volume of consumption of iron is larger and its unit value is lower than those of most other metals, its basic cost of production bears a higher content of transport charges — long-distance bulk transport of ores or concentrates plus expensive transport of scrap metal from dispersed consumer areas back to the steel works. Even the richest iron ores (say 68% iron) contain some 32% oxygen and gangue plus at least 10% moisture and 5% fines as dust which tends to be lost in transit and handling. Thus, about 50% of the weight carried in bulk carriers represents wasted capital, operating costs, and energy consumption. The elimination of such waste seems a desirable objective. One possible solution would be to move the smelter to the mine if the latter is large enough to sustain metal production. Taking into account the resulting saving in transport of fuel, plus the subsequent export of metal, the theoretical saving might be about 20% in cost and energy. However, this would have to be supplemented by finding local sources of scrap metal or by using less scrap. Mineral processing could make a vital contribution here by providing alternatives to scrap in the form of supergrade concentrates or pre-reduced (metallized) feeds.

An interesting variant would comprise the establishment of iron and steel production at the iron ore mine (or at its shipping port) and simultaneously at the coal mine or its port. Ships would convey ore in one direction and coal in the other direction with maximum utilization. Australia would be a good example and could develop exports of finished or semi-finished products in place of current exports of ore and coal; mineral processing could be used to optimize the coal and ore for minimum transport costs and for maximum smelter efficiency. An alternative solution would be to develop new combinations of physical and chemical mineral processing to yield an enriched "ore" containing not less than some 90% iron for overseas shipment.

Other metals and minerals

World resources of many other mineral and metal raw materials are far more problematical than iron ore. Current supplies may be equal to or possibly larger than demand, but even if one accepts only conservative forecasts of the growth of demand through increasing world population or rising consumption per capita, very substantial new sources of many materials will have to be exploited. Since continued discovery of adequate new ores of commercial viability in present terms cannot be expected, it is probable that methods will have to be developed for extracting materials from much larger tonnages of very low-grade ores, especially for metals such as copper, zinc, lead, tin, nickel, cobalt, and manganese; the same applies to industrial minerals such as rutile, zircon, barytes, phosphates, and fluorspar. If low-grade ores of these commodities were to be the sole sources of supply and were to be treated essentially with present methods of processing, the price of the products would have to rise because of the greater tonnage processed per unit of value. The limits to such rises and thus the limits to increasing reserves by lowering cut-off grades would depend only on the price which the consumer, or society, is prepared to pay for the material in question compared with the prices of alternative materials which might be used for the same purpose.

It is possible that further progressive improvements can be applied to existing processes in order to lower their capital or operating costs, thus making current process technology economically acceptable for lower-grade ores. Such evolutionary improvements enabled gold mines to remain in production for three decades in spite of falling ore grades, fixed gold prices, and rising costs of labour, power, and materials. A combination of increased production efficiency and rising metal prices has made it possible for copper producers to lower their average ore grade from about 2.0% copper to about 0.5% copper over a span of less than ten years. However, a further lowering in average grade to, say 0.1% copper may be needed eventually to raise copper reserves to the levels of foreseeable demand; this would either require a remarkable increase in technical efficiency or it would have to be sustained by a five-fold rise in copper prices in real money terms (see Chapter 2). It seems probable that production from very low-grade copper ores will necessarily have to depend on cost supplementation through recovery of by-products; this is current practice for ores of about 0.5% copper, especially with by-products of precious metals such as gold and silver. It might be possible to consider humbler by-products such as feldspar or mica with a low-grade ore lying conveniently close to a consumer area. However, the scope for such supplementation is limited by the comparatively small volume of the market for such products; a single large-tonnage producer milling a low-grade (0.1% copper) porphyry copper ore at a rate of 100,000 tonnes a day could oversaturate the entire world market for feldspar. Similarly, although low-grade ores of the mineral kyanite, an important raw material for aluminium silicate refractories, could be supplemented by large tonnage by-product extraction of high-grade concentrates of mica and quartz, transport costs and limited market capacities would make it unacceptable unless entirely new tonnage-uses were developed for the by-products.

Recycling

An alternative to ensuring supplies of minerals through increased production is more complete recycling. Even now, for example, around 40% of copper and 70% of lead used in industrially advanced countries is recycled; there is no reason why these figures should not be increased, especially with more systematic waste disposal and collection systems. Mineral processing techniques of comminution, sorting, and separation are established components of recovery schemes now under development for waste products. A benefit in this respect is that increased consumption of a material leads to a greater tonnage and possibly higher concentration in waste products becoming available for recycling. This has advantages of scale in re-processing schemes in terms both of capital investment and operating costs. The main disadvantages lies in high dispersion, geographically and chemically, of consumer wastes compared to the relatively high concentration and purity of manufacturing wastes. At present it is far simpler and greatly more lucrative to recover copper from unused waste arising in the manufacture of cables than to recover the metal dispersed in motor cars in breakers' yards.

APPROPRIATE DEVELOPMENTS IN MINERAL PROCESSING

The various aspects of mineral extracting and processing discussed above form the basis for a review of the future requirements and possible developments. These are considered below in separate but related groups.

Improvements in existing process technology

Almost all processes of ore preparation and minerals separation are capable of significant improvement by being given feeds of more constant physical characteristics and chemical constitution. Potential improvements apply to all aspects of production, such as greater throughput, better recovery of values, higher grade of product, and lower unit costs. Since most ore deposits are inherently variable, uniformity of primary feeds (in both the short- and long-term) can be achieved only by processes of stockpiling and blending on a scale which is large enough to overcome ore variability and which is compatible with the rate of milling. Customary objections to this lie in high initial capital costs and inventory charges. It would be most desirable to investigate the technical potential for long-term production gains, especially in conjunction with moving comminution into the mine. Inventory accounting methods must be questioned, and investment concessions would be desirable to make the initial capital investment for ore blending more attractive.

Secondary feeds and in-process transfers are equally capable of contributing gains if they were subject to far greater constancy than is currently considered necessary. This is in striking contrast with accepted practice in chemical engineering processes where transfer constancy is almost axiomatic. Better control in mineral processing through in-line measurements is a special need for development, but the whole technology of materials handling and transfer between unit processes merits re-examination in an attempt to achieve greater simplicity and more constant balance with process rate requirements. In most mineral processing operations materials handling is second in cost (after fine grinding) and often accounts for more losses (of production) and maintenance costs than the combined unit processes of separation.

Existing processes in the separation of most ores are variously regarded as having attained grades and recoveries which meet present economic yardsticks of performance. For example, in the case of tin ores, recoveries of 70% are considered good, compared with around 90% for copper ores; recoveries near 100% could be attained in both cases, for tin through combining physical processing with chemical extraction, and for copper through even further improved flotation practice. The costs of production would be higher and would have to be justified by metal yield rather than by straight cost/profit balances; this indicates the need for governmental support in respect of any commodity which threatens to fall short of demand and should tie in with the arguments above in favour of lowering cut-off grade.

Cost and power consumption of existing processes per unit of product are almost without exception capable of improvement by operating the process more consistently at maximum throughput. This is mainly a problem of inadequate operational control; most unit processes await quantitative analysis of their variables and are mainly operated empirically. Neither the rate constants of their various functions nor the residence time characteristics for different components of the charge are known sufficiently to permit real optimization of operating conditions towards maximized production. Computer facilities exist now for analyzing and controlling complex interactions of multiple variables, but input information is inadequate due to the lack of on-stream sensors for feed parameters, mineralogical, and particle characteristics. This is a field of development deserving substantial effort and financial support.

Unit size increases of existing processes

The foreseeable demand for processing of ever lower-grade ores necessarily implies the treatment of very large tonnages of ore at lower costs per tonne treated. An important technical contribution will be the continued development of larger unit sizes of equipment. Individual machines with large throughput capacities can provide several important advantages in comparison with smaller multiple units; the capital cost is lower, building and civil engineering requirements are sharply reduced, process control is simpler, material handling is less complex and cheaper, power consumption per throughput tonne is less, and maintenance is simplified. For example, fine grinding is a necessity for many ores in order to achieve adequate liberation of the values; grinding mills are among the most costly items in capital, operating labour, and maintenance costs, as well as in power consumption. Up to the present single mill units with a throughput of 100 tonnes per hour are regarded as large, although 40 such mills would be needed working in parallel for a low-grade mine which might be treating 30 million tonnes per year. Single grinding units with a throughput capacity of up to 250 tonnes an hour are now being installed, and designs are proposed for a unit capable of throughputs of up to 5,000 tonnes per hour (equal to 100,000 tonnes a day, and thus equal to the 40 units of 100 tonnes per hour capacity). Such a mill might be 8 m in diameter, 15 m long, and absorb almost 30,000 horsepower. The engineering and power supply problems for such giants are formidable indeed by present standards; for example, the mill shell would have to be equipped with its own electric drive rotor, and its low speed of rotation would add problems of efficiency of electric power conversion.

Increase in size may also be anticipated for concentration processes. For example, the impressive multiplicity of small froth flotation cells which formed such a characteristic feature of all flotation plants built up to about 1950 has gradually given way to smaller numbers of much larger cells, culminating in the 1970s with single cells of 5 m diameter and height. The problems in this case lie in design for maintaining efficient slurry motion and air distribution throughout a contained volume of some 70 m$^3 \cdot$

In general terms it may be expected that wherever a very large, very low-grade operation is contemplated, its realization would be facilitated by avoiding multiplexing of equipment and associated process complications through scaling up of all unit processes to single machine sizes capable of handling the proposed throughput.

Unit size decreases of existing processes

Although mineral processing on a small scale is costly, it was suggested above that advantages of transport costs, energy consumption, and other cost reductions could accrue from comparatively small-scale mineral production closely coupled with local centres of consumption or local energy sources and smelters. A small scale of operation facilitates application of far greater precision and control to all functions of the processes used; this would seem to offer great scope for redesigning small-scale processes so as to minimize labour, materials handling, and power consumption, to maximize the extraction of values, and thus to compensate for inevitably high capital costs per throughput tonne. The highest possible grades of concentrates would seem to be generally desirable for such operations to minimize the costs of subsequent metallurgical or chemical processing. The throughput rate of the mineral processing stage would have to be chosen so as to form a cost-reducing link between size/grade optimization of the ore supply, and the optimized production rate of the metallurgical or other following processes. The costs of the latter are generally orders of magnitude greater than those of mineral processing.

New combinations of processes

If for no other reason, the high costs of fuel should cause a reappraisal of existing combinations of processes which were mostly chosen under conditions of cheap fuel as compared with labour and capital costs. In addition, the cost of new ecological constraints may further influence the economic balance of alternative processing routes. For example, if an important orebody is situated at a remote and arid inland site, dry pre-concentration may be desirable to reduce transport costs to the shipping port, coastal finishing plant, or consumer centre. The pre-concentration may cause dust emission which could be judged quite acceptable in the relatively uninhabited area, although it might be quite intolerable in a densely populated area. Dry processing may also result in a certain loss of values which would have to be taken into account; for an iron ore containing 40% iron, a 15% loss of values might be just acceptable, while such a loss would be quite unacceptable for a nickel ore with 0.8% nickel, even though the contained metal value of the two ores would be about the same.

Combinations of physical, chemical, and pyrometallurgical processes will have to be chosen to meet specific local conditions. The overall energy conversion factors for sequences of processes from ore to product can vary widely. A route chosen for minimum energy consumption can be in conflict with requirements for maximum extraction and/or minimum pollution. For example, pyrite in various ores could be regarded as a valuable fuel for pyrometallurgical steps such as roasting, sintering, pelletizing, calcining, etc., but pollution control may necessitate the physical removal (by selective flotation) of pyrite in its unreacted state. It could then be found that an alternative chemical route could replace the pyrometallurgical one to eliminate process complications caused by pyrite, but the overall cost and energy balances are bound to be different.

Development of new processes

Recent technological advances hold out hope of potential applications in mineral processing which might facilitate the future recovery of minerals and metals from a wider range of sources than traditionally viable ores. For example, superconducting magnets with very high field strengths and gradients should permit separation of feebly magnetic minerals from each other if they differ in susceptibility; more important, perhaps, they should permit separations of very small particles of paramagnetic minerals down to micron sizes which currently defy efficient processing. Superconductivity has almost reached the stage of industrial viability; in terms of processing costs and energy consumption it could well provide a more efficient substitute for established processes such as gravity separation or froth flotation through offering a better separation criterion. Obvious examples would be the separation of pyrite from coal, concentration of hematite ores, or even separation of sphalerite from galena; other possibilities would include the separation of cassiterite from wolframite, mica from feldspar, or iron-bearing gangue from asbestos.

The application of controlled low-temperature cooling to produce selective embrittlement of minerals is another example of new process technology; this could result in liberation through selective fracture and might be followed by concentration based on simple size separation. The development of low-temperature technology applications in the mineral field has remained almost unexplored; it would require examination of the relative efficiency of energy conversions. Such processes might have geographic advantages of application in regions that are very arid or very cold.

While the processing of very fine particles receives much attention, further development is needed. Many minerals are not efficiently recoverable in particle sizes smaller than about 10 μm; for example, in many tin ores the natural or accidental sizes of cassiterite particles range down well below that limit and as a result up to 40% of the mineral is not recovered from several ores. It is estimated that future world supplies of tin could be raised by 50%, partly by better extraction from producing mines and partly by bringing into production ores which are untreatable by current methods.

CONCLUSIONS

Inventive research effort is needed in the above examples and in many other directions to produce the necessary advances in mineral technology.

However, the increasing capital costs and decreasing incentives of new mining ventures progressively tend to make the industry more conservative in its desire to avoid risks additional to those inherent in the uncertain nature of orebodies and the fluctuating hazards of finance. This creates a serious hold-up in the flow of innovations into industrial practice. Inventors, universities, and other research organizations can usually show that their new processes work under favourable conditions, usually on a small and discontinuous scale. The initial "one-off-engineering" of the equipment is often of questionable quality or very expensive. Laboratory conditions are either of the small batch type or closed circuit recycling to avoid having to cope with large quantities of ore, so that they cannot fairly reflect the conditions of industrial through-put. Little realistic information is provided on throughput, grade of products, power consumption, and other factors of importance to the industrial viability of a new process. Equally there is no indication of long-term continuous running reliability. There is, therefore, at this stage a serious credibility gap which many inventions fail to bridge; even the successful ones may take up to twenty years to attain full industrial acceptance. Some five or six years of that time may be taken up in arranging support for the production of an industrial prototype for long-term applications testing on production lines at mines.

The present cost of properly designing and building such a prototype unit, with attendant requirements of patenting, pre-prototype tests, etc., is rarely less than £100,000. It can be considerably higher if associated novel-process technologies (high voltage, vacuum, nuclear radiation, etc.) need to be established and made to be sufficiently rugged for operation by unskilled personnel on a mine site.

This necessitates farsighted support by governmental sources to ensure that testing on a suitable scale can prove or disprove a new process with speed. If a process is of value it will not need further support; if it is useless it should be abandoned as quickly as possible so that inventive resources are not locked up in prolonged infertile projects. Neither manpower nor finance for research are so plentiful that they can be left to linger on failures. In the long run it is cheaper to prove or disprove a new process quickly with the help of development funds and to recover the costs subsequently from the users of successful innovations. The performance of applications testing should come under industrial management and the inventor organization should be relieved of the burdens of financing and development which are both beyond their competence. Implementation of these principles would result in a greater effective research effort and could reduce the overall costs of successful innovation.

It would also be desirable to offer industry special inducements (tax credits, capital write-off, etc.) for capital investment in process innovations, especially substitution of potentially more effective processes (in terms of mineral resource utilization) in place of conventional plant of lesser efficiency with unexpired capital life. Many mineral resources could be exploited more

436

effectively and their values extracted more completely if the operators were not tied to the capital life of processing designs made obsolete by changed circumstances. In the light of future resource requirements, this is a subject sufficiently important to merit international attention, especially in cases where potentially scarce commodities are being incompletely recovered and where commercial interfaces between producers and consumers may inhibit private improvement. The world as a whole can perhaps no longer afford incompetent treatment of mineral resources.

SELECTED REFERENCES

Brandi, H.T., 1973. Considerations regarding the siting of iron and steelworks plants in countries outside Europe. *Stahl Eisen*, 93 (12): 541—546 (in German).

Carlisle, D., 1953. Maximum total recovery through mining high-grade and low-grade ore together is economically sound. *Can. Inst. Min. Metall. Bull.*, 46 (489): 21—27.

Chapman, P.F., 1974. The energy costs of producing copper and aluminium from primary sources. *Met. Mater.*, 8 (2): 107—111.

Chopey, N.P., 1973. Cool it, then grind it. *Chem. Eng.*, 80 (20): 54—56.

Cottrell, A., 1974. The age of scarcity? *Trans. Inst. Min. Metall.*, 83: A25—A29.

Ehrlich, P.T. and Ehrlich, A.H., 1972. *Population, Resources, Environment*. W.H. Freeman, San Francisco, Calif., 500 pp.

Fahlstrom, P.H., 1974. Autogenous grinding of base metal ores at Boliden Akteibolag. *Can. Inst. Min. Metall. Bull.*, 67 (743): 128—141.

Fleming, M.G., 1973. Man and minerals — a viable contract. *Trans. Inst. Min. Metall.*, 82: A29—A39.

Fletcher, A.W., Jackson, D.V. and Valentine, A.G., 1967. Chloride route for the recovery of tin metal from low-grade concentrates. *Trans. Inst. Min. Metall.*, 76: C145—C153.

Halls, J.L., Bellum, D.P. and Lewis, C.K., 1969. Determination of optimum ore reserves and plant size by incremental financial analysis. *Trans. Inst. Min. Metall.*, 78: A20—A26.

Hart, L.H., 1973. Mineral science and the future of metals. *AIME Trans.*, 254: 105—110.

Heath, K.C.G., 1974. The right to mine. *Trans. Inst. Min. Metall.*, 83: A19—A24.

Herbert, I.C., 1973. Extractive metallurgy. *Min. Mag., Mining Annual Review, 1973*. London, pp. 227—253.

Hughes, J.E., 1972. The exploitation of metals. *Met. Mater.*, May 6, pp. 197—205.

Kruesi, P.R., Allen, E.S. and Lake, I.L., 1973. Cymet process-hydrometallurgical conversion of base-metal sulphides to pure metals. *Can. Inst. Min. Metall. Bull.*, 66 (734): 81—87.

Kuhn, M.C., Arbiter, N. and Kling, H., 1974. Anaconda's arbiter process for copper. *Can. Inst. Min. Metall. Bull.*, 67 (742): 62—73.

Lippert, K.K., Pietsch, H.B., Roeder, A. and Welden, H.W., 1969. Recovery of non-ferrous metal impurities from iron ore pellets by chlorination. *Trans. Inst. Min. Metall.*, 78: C98—C107.

Manners, G., 1971. *The Changing World Market for Iron Ore 1950—1980*. Johns Hopkins Press, Baltimore, Md., 384 pp.

McCullock, H.S. and Wilson, B.M., 1973. South African costs of equipment for the metallurgical industry. *Natl. Inst. Metall., Rep.* No. 1529, May.

Michell, F.B., 1973. Mineral processing. *Min. Mag., Mining Annual Review, 1973*. London, pp. 203—225.

Mining Magazine, 1973. The Kloof concentrator. 128: 76—87.

Palley, J.P. and Paige, P.M., 1972. Electrometallurgy: can electrowinning replace cement copper? *Eng. Min. J.*, July, pp. 94—96.

Staines, A., 1974. Digesting the raw materials threat. *New Sci.*, 61 (888): 609—611.

U.S. Department of Commerce, 1973. *World Iron Ore Pellet and Direct-iron Policy.* U.S. Government Printing Office, Washington, D.C., 12 pp.

U.S. National Commission on Materials Policy, 1973. *Towards a National Materials Policy. World Perspective.* U.S. Government Printing Office, Washington, D.C., 87 pp.

Wild, R., 1969. Iron ore reduction processes. *Chem. Process Eng. (London)*, 50 (2): 55—61.

Yannopoulos, J.C., 1971. Control of copper losses in reverberatory slags. *Can. Metall. Q.*, 10 (4): 291—307.

Appendix

UNDP-ASSISTED PROJECTS IN GEOLOGY AND MINERAL RESOURCES *

DANIEL A. HARKIN

(a) Africa

Country and project number	Project title	Project costs		Remarks
		UNDP ($000)	Govt. ($000 equiv)	
Burundi				
BDI/68/505	Mineral survey	1,097	212	
**BDI/71/517	Mineral survey (Phase II)	1,147	485	Large nickel laterite deposit discovered; now being investigated in detail
Cameroon				
**CMR/74/011	Mineral exploration in the Dja formation	330	200	
Central African Republic				
CAF/70/511	Investigation of limestone deposits at Fatima	261	48	
Chad				
**CHD/72/002	Mineral survey, training of personnel and strengthening of the Geological Survey	791	332	
Congo (Peoples' Republic of)				
PRC/66/506	Mineral exploration in the southwest	776	548	

* As of December 1974. Small-scale projects in which the UNDP contribution is less than U.S. $100,000 are not listed.
** Ongoing projects or projects about to begin.

Africa (continued)

Country and project number	Project title	Project costs		Remarks
		UNDP ($000)	Govt. ($000 equiv)	
Dahomey				
DAH/68/504	Strengthening of the geological and mining service	517	233	
**DAH/73/006	Mineral survey	324	343	
Egypt				
EGY/65/556	Assessment of the mineral potential of the Aswan region	1,680	1,951	
**EGY/72/008	Assessment of the mineral potential of the Aswan region (Phase II)	1,100	1,357	
Ethiopia				
ETH/67/517	Mineral survey in two selected areas	1,246	836	
**ETH/71/537	Strengthening of the Geological Survey	1,164	4,814	Government contribution represents the total budget allocation for the Geological Survey department
Gabon				
GAB/70/507	Mineral exploration in eastern Gabon	554	403	
Ghana				
**GHA/72/008	Assistance to the State Gold Mining Corporation at Tarkwa and Dunkwa	281	47	
**GHA/72/025	Assistance to the State Gold Mining Corporation at Prestea	1,986	826	
**GHA/74/011	Assistance to the State Gold Mining Corporation in exploration	600	320	
**GHA/74/012	Assistance to the State Gold Mining Corporation in exploration and management	439	296	
Guinea				
GUI/59/501	Resources development survey	114	—	Government services not costed
GUI/67/512	National mineral and geological centre, Conakry	679	345	

Africa (continued)

Country and project number	Project title	Project costs		Remarks
		UNDP ($000)	Govt. ($000 equiv)	
GUI/70/505	Preliminary investigation of Mount Nimba iron-ore deposits	806	380	Evaluation of known deposits
**GUI/72/008	Mining and geology laboratory at Conakry	700	1,100	
Ivory coast				
IVC/64/504	Mineral survey in the south-west	1,041	850	
IVC/69/519	Mineral survey in the south-west (Phase II)	702	899	
Kenya				
KEN/64/504	Mineral resources survey in western Kenya	605	387	
Lesotho				
LES/71/503	Exploration for diamonds	478	150	
**LES/73/021	Exploration for diamonds (Phase II)	1,172	588	
Liberia				
LIR/69/509	Mineral survey in the central and western regions	894	344	
Malagasy (Madagascar)				
MAG/63/503	Surveys of the mineral and groundwater resources of southern Madagascar	1,246	690	
Malawi				
MLW/70/013	Airborne geophysical survey	199	—	
**MLW/72/008	Ground follow-up to airborne geophysical survey in four selected areas	231	277	
Mali				
MLI/71/519	Mineral survey in the western region	174	53	
Mauritania				
MAU/69/504	Strengthening of the geological service and mineral exploration	375	125	

Africa (continued)

Country and project number	Project title	Project costs		Remarks
		UNDP ($000)	Govt. ($000 equiv)	
Morocco				
MOR/67/516	Potash exploration in the Khemisset Basin	1,100	2,537	Large deposits of high-purity rock salt
MOR/71/535	Mineral exploration in the Anti-Atlas	1,083	1,395	discovered in Berrechid Basin
**MOR/74/002	Assistance in exploration of mineral resources of the Anti-Atlas	100	194	
Niger				
NER/67/510	Mineral exploration in two areas	981	491	
NER/71/522	Mineral exploration in two areas (Phase II)	1,240	516	
Nigeria				
NIR/64/519	Aeromagnetic survey of minerals in the northwest	567	491	Project operations discontinued prematurely
Rwanda				
RWA/69/506	Mineral survey	1,388	371	
**RWA/74/001	Mineral survey (Phase II)	769	262	
Senegal				
SEN/63/504	Mineral survey	925	596	
SEN/66/507	Mineral survey (Phase II)	527	313	
SEN/71/517	Bridging operation for copper exploration in the Bakel-Gabou area	589	404	
Somalia				
SOM/62/504	Mineral and groundwater survey	594	280	Uranothorite deposits discovered at Bur attracted investment for further exploration
SOM/68/514	Mineral and groundwater survey (Phase II)	850	977	Uranium mineralization discovered in Mudugh area
SOM/71/523	Mineral and groundwater survey (Phase III)	1,297	929	Proving of Mudugh deposits

Africa (continued)

Country and project number	Project title	Project costs		Remarks
		UNDP ($000)	Govt. ($000 equiv)	
**SOM/74/015	Strengthening of the national geological service	345	2,729	Government contribution represents the total budget allocation for the national geological survey programme
Sudan				
SUD/67/528	Mineral survey in three selected areas	1,537	750	
SUD/71/007	Mineral survey	235	95	
Swaziland				
SWA/65/501	Mineral survey	478	559	
Tanzania				
URT/64/505	Mineral investigation of the Lake Victoria goldfield	626	385	
URT/69/019	Mineral exploration	137	—	Services of a geophysicist and equipment. Government services not costed
**URT/73/030	Mineral investigations in northwest Tanzania	216	197	Exploration over extensions of geological formations within which lateritic nickel was found in Burundi
Togo				
TOG/62/504	Survey of groundwater and mineral resources	1,327	520	Limestone and marble deposits located
**TOG/72/004	Strengthening of the National Bureau of Mineral Research	293	180	
Tunisia				
TUN/64/510	Mineral investigation of the Foussana Basin	923	500	Extensions of lead-zinc deposits found and investigated
Uganda				
UGA/60/503	Aerial geophysical survey	313	140	
Upper Volta				
UPV/63/504	Mineral and groundwater surveys	1,050	577	Evaluation of large manganese deposits at Tambao

Africa (continued)

Country and project number	Project title	UNDP ($000)	Govt. ($000 equiv)	Remarks
UPV/66/506	Feasibility surveys for mineral development in the northeast, and associated transport factors	1,051	354	
UPV/71/516	Mineral exploration in the north	1,346	641	
**UPV/74/004	Mineral exploration in the Boromo/Houndé Birrimian Trough	2,447	1,800	
Zaire ZAI/68/514	Mineral resources survey in the Bas-Congo	893	603	
Zambia ZAM/68/509	Detailed mineral exploration west of Broken Hill	858	746	
	Total costs (Africa)	49,794	39,971	

(b) Asia

Country and project number	Project title	UNDP ($000)	Govt. ($000 equiv)	Remarks
Bangladesh **BAN/72/012	Strengthening of the Geological Survey	421	47	
Burma BUR/61/501	Survey of lead and zinc mining and smelting	590	233	
BUR/71/010	Laboratory expansion of the Geological Survey and Exploration Corporation	205	—	Government services not costed
BUR/71/516	Post-graduate training in mineral exploration at the University of Arts and Science, Rangoon	959	973	Project executed in association with UNESCO

Asia (continued)

Country and project number	Project title	Project costs UNDP ($000)	Project costs Govt. ($000 equiv)	Remarks
BUR/71/519	Exploration and pilot development of alluvial tin/tungsten deposits in the Tenasserim coastal areas	1,068	868	
**BUR/72/001	Pre-investment drilling and training in selected areas	722	230	
**BUR/72/002	Geological survey and exploration project	1,411	1,064	
**BUR/73/013	Consolidation and expansion of post-graduate training in applied geology at Rangoon Arts and Science University	1,427	1,544	Project executed in association with UNESCO
**BUR/73/017	Offshore exploration for tin	1,399	888	
India				
IND/67/564	Mineral development in Madras State (Tamil Nadu)	945	345	Iron ore deposits discovered
IND/71/594	Mineral surveys — Uttar Pradesh	679	622	
**IND/71/615	Mineral exploration and development in Tamil Nadu	598	341	
**IND/74/037	Mineral surveys — Uttar Pradesh (Phase II)	404	654	
Indonesia				
INS/69/521	Offshore exploration for tin and tin ore dressing research	1,263	1,264	Additional offshore placer tin reserves located and tin recovery improved in central treatment plants
Iran				
IRA/60/501	Geological Survey Institute	1,566	2,000	Copper mineralization at Sar Chesmeh identified as of porphyry type
IRA/68/528	Geological Survey Institute (Phase II)	663	1,270	
**IRA/71/542	Technical support to Geological Survey Institute	111	346	
Israel				
**ISR/71/522	Offshore prospection for building and other industrial materials	485	347	

Asia (continued)

Country and project number	Project title	Project costs		Remarks
		UNDP ($000)	Govt. ($000 equiv)	
Jordan				
JOR/66/512	Establishment of a mineral exploration unit	425	364	
JOR/70/521	Phosphate exploration and beneficiation studies	894	651	
**JOR/73/010	Phosphate beneficiation pilot plant	713	573	
Korea				
**ROK/72/028	Airborne geophysical survey and follow-up	398	425	
Malaysia				
MAL/60/507	Surveys of the Labuk Valley (Sabah)	770	790	Multi-component natural resources survey, including mineral exploration. Mamut porphyry copper deposit discovered
Nepal				
NEP/68/003	Long-term planning for geological exploration and strengthening of Nepal Bureau of Mines	103	—	Government contribution not costed
**NEP/73/019	Mineral exploration	983	773	
Pakistan				
PAK/59/507	Mineral survey	1,682	290	New coal deposits discovered
PAK/70/553	Detailed exploration of uranium and other radioactive occurrences in the Siwalik sandstones in the Dera Ghazi Khan district	672	424	Projects executed by IAEA in association with the United Nations
**PAK/74/002	Detailed exploration of uranium and other radioactive occurrences in the Siwalik sandstones in the Dera Ghazi Khan district (Phase II)	547	1,095	
Philippines				
PHI/62/509	Institute of Applied Geology, Manila	704	699	Extensions of copper ore bodies located
**PHI/70/010	Marine geology	129	—	Government contribution not costed

Asia (continued)

Country and project number	Project title	Project costs		Remarks
		UNDP ($000)	Govt. ($000 equiv)	
Saudi Arabia				
SAU/69/523	Centre for Applied Geology, College of Petroleum and Minerals, Jeddah	1,016	1,776	Projects executed by UNESCO in association with the United Nations
**SAU/72/003	Centre for Applied Geology, College of Petroleum and Minerals, Jeddah (Phase II)	251	997	
Solomon Islands (U.K. protectorate)				
UK/64/536	Aerial geophysical surveys	985	586	Follow-up investigations also included. Rennell Island bauxite deposit discovered
South Yemen				
PDY/70/003	Mineral exploration	131	52	
Regional projects				
RAS/71/168	Technical support for regional offshore prospecting in Southeast Asia	696	115	These three projects executed in association with the Economic and Social Commission for Asia and the Pacific
**RAS/73/022	Technical support for regional offshore prospecting in Southeast Asia (Phase II)	1,056	1,469	
**RAS/72/079	Southeast Asia Tin Research Centre	768	7,089	Large government contribution accounted for by inclusion of all government equipment and premises of several co-operating major tin-producing countries
	Total costs (Asia)	27,839	31,204	

(c) Europe

Country and project number	Project title	Project costs		Remarks
		UNDP ($000)	Govt. ($000 equiv)	
Cyprus				
CYP/62/502	Survey of groundwater and mineral resources	1,340	1,595.	
Greece				
GRE/70/529	Exploration for uranium in central and eastern Macedonia and Thrace	364	226	Projects executed by the IAEA in association with the United Nations
**GRE/73/006	Exploration for uranium in central and eastern Macedonia and Thrace (Phase II)	230	961	
Poland				
POL/66/504	Subsurface exploration for potassium salts	995	1,406	
Turkey				
TUR/69/532	Mineral exploration in two areas	1,009	2,678	
TUR/72/004	Mineral exploration in two areas (Phase II)	571	1,605	
TUR/72/036	Exploration for uranium in southwest Anatolia	569	687	Project executed by the IAEA
Yugoslavia				
**YUG/73/010	Mineral exploration in the Socialist Republic of Montenegro	344	321	
	Total costs (Europe)	5,422	9,479	

(d) Latin America

Country and project number	Project title	Project costs		Remarks
		UNDP ($000)	Govt. ($000 equiv)	
Argentina				
ARG/63/512	Mineral survey in the Andean Cordillera	1,167	1,244	Discovery of new porphyry copper province

Latin America (continued)

Country and project number	Project title	Project costs		Remarks
		UNDP ($000)	Govt. ($000 equiv)	
ARG/66/523	Investigation of porphyry copper-type mineralization in the provinces of Mendoza and Neuquen	889	607	Several porphyry copper targets investigated to provide Government with technical data on which to invite private sector proposals
ARG/70/535	Mineral exploration in the northwest region	480	2,895	Operations largely carried by Argentinian personnel, with limited United Nations contribution
**ARG/72/008	Specialist group for mineral exploration in Patagonia and Comahue	209	564	
**ARG/72/032	Mineral exploration in the northwest region (Phase II)	736	2,291	
Bolivia				
BOL/61/506	Pilot mineral survey of the Cordillera and Altiplano	922	685	
BOL/63/508	Mining and Metallurgical Research Institute	808	780	
BOL/66/513	Development of the gold deposits of the Tipuani area	273	418	Project operations discontinued prematurely
**BOL/70/015	Mining and Metallurgical Research Institute (supplementary assistance)	170	—	Counterpart contribution not costed
BOL/70/527	Survey of the Mutún iron ore and manganese deposits	304	136	
**BOL/73/002	Earth resources technology satellite programme	150	139	
Chile				
CHI/59/502	Mineral survey	1,078	639	Iron ore deposits discovered
CHI/63/516	Mineral resources survey of the province of Coquimbo	507	298	
CHI/66/528	Detailed mineral investigation of selected zones in Atacama and Coquimbo provinces	1,105	610	Identification of Los Pelambres prospect as porphyry copper deposit and proving of reserves

Latin America (continued)

Country and project number	Project title	Project costs		Remarks
		UNDP ($000)	Govt. ($000 equiv)	
CHI/71/542	Mining and Metallurgical Research Institute	791	48	Projects executed by UNIDO in association with the United Nations
**CHI/73/037	Mining and Metallurgical Research Institute (Phase II)	1,762	6,737	Government contribution includes cost of premises and all equipment in new Institute, including Phase I bilateral contribution from the Belgian Government
Colombia				
**COL/72/001	Mineral exploration plan	1,270	491	
**COL/72/002	Base metals study in the central and western mountain ranges	769	858	
**COL/72/004	Emerald prospection in the provinces of Boyaca and Cundinamarca	375	240	
**COL/72/005	Evaluation of nickel laterites in the Departments of Cordoba and Antioquia	366	372	
**COL/72/006	Precious metals projects in the Atrato and San Juan river basins	410	750	
Costa Rica				
COS/72/004	Mineral survey	120	81	
Ecuador				
ECU/64/515	Survey of metallic and non-metallic minerals	819	572	Discovery of porphyry copper deposit at Chaucha; San Bartolomé silver deposit discovered and evaluated
ECU/69/526	Survey of metallic and non-metallic minerals (Phase II)	637	238	
**ECU/73/011	Detailed investigation of selected mineral prospects	214	717	
El Saldavor				
ELS/65/503	Assessment of mineral deposits in the north	640	342	

Latin America (continued)

Country and project number	Project title	Project costs		Remarks
		UNDP ($000)	Govt. ($000 equiv)	
Guatemala				
GUA/65/508	Mineral surveys in two selected areas	838	469	
Guyana				
GUY/62/505	Aerial geophysical survey	641	700	
GUY/66/511	Mineral survey (Phase II)	889	1,077	
Haiti				
**HAI/72/002	Mineral survey	233	118	
Honduras				
HON/69/501	Investigation of mineral resources in selected areas	1,057	624	
Mexico				
MEX/62/504	Survey of metallic mineral deposits	897	1,850	Aeromagnetic survey located new iron ore deposits; La Caridad porphyry copper deposit discovered
Nicaragua				
NIC/63/503	Mineral survey	739	355	
Panama				
PAN/65/504	Mineral survey of the Azuero area	830	546	Discovery of porphyry copper deposit at Cerro Petaquilla, followed by private sector discovery at Cerro Colorado
PAN/69/517	Mineral survey (Phase II)	570	525	Porphyry copper-type mineralization discovered in San Blas mineralized belt
	Total costs (Latin America)	23,665	29,016	
	Grand total	106,720	109,670	

REFERENCES INDEX*

*Only page references to text pages are made in this index. References to pages containing bibliographic details have been omitted. These details are given at the end of each chapter.

458

SUBJECT INDEX